Advances in Natural Fibre Composites

Contents

About the Editors

Prof. Raul Fangueiro is currently Professor and senior researcher in the School of Engineering at the University of Minho, Portugal. He is the Head of the Fibrous Materials Research Group of the same university with expertise in advanced materials (nano, smart, composites) and structures (3D, auxetic, multiscale) with 35 researchers. He is the mentor and coordinator of the FIBRENAMICS International Platform (www.fibrenamics.com), including 300 partners developing promotion, dissemination, technology transfer and research activities on fibre-based advanced materials. He has more than 135 published papers in international reputed scientific journals, 340 conference publications, 36 books and 14 patents. He is the scientific coordinator of several national and international research projects on advanced fibrous and composite materials, mainly for building, architectural and healthcare applications. He has supervised various Ph.D. and post-doc scientific works and is an expert in the European Technological Textile Platform and member of the editorial board of several leading international scientific journals on composite and fibrous materials.

Dr. Sohel Rana is currently a Senior Scientist and Leader of Composite Materials Group at Fibrous Materials Research Group (Fibrenamics), University of Minho, Portugal. He obtained his Bachelor's degree in Textile Technology from University of Calcutta, India, and Master's degree and Ph.D. in Fiber Science and Technology from Indian Institute of Technology (IIT, Delhi), India. His current research areas include advanced fibrous and composite materials, natural fibres, nanocomposites, electrospinning, multifunctional and biocomposites materials and so on. He is the author of one book, has edited six books and contributed to 14 book chapters, six keynote and invited papers, 35 papers in international peer-reviewed journals and more than 60 papers in international conferences. He also participates in the editorial board of several scientific journals and is a potential reviewer for numerous scientific journals including Green Chemistry, Composite Science and Technology, Composites Part A, Dyes and Pigments, RSC Advances, Composite Interfaces, Journal of Composite Materials, Journal of Reinforced Plastics and Composites, Powder Technology, Journal of Nanomaterials, Journal of Applied Polymer Science, Fibres and Polymers, and so on.

Potential of Hemp in Thermoplastic Biocomposites—The Effect of Fibre Structure

Kirsi Immonen, Panu Lahtinen, Panu Isokangas and Katariina Torvinen

Abstract Hemp is one of the annual crops whose use has increased in recent years through different applications. This work compares the properties of PLA-based biocomposites, where different hemp structures resulting from a side stream from hemp production were used, such as hemp fibre, hemp pulp, shives and hemp dust. We have also produced hemp-based nanofibrils and compared their effect on the viscosity with the aim of finding novel applications for hemp. A wood-based cellulose was used as reference materials in both the biocomposite and nanofibril studies.

Introduction

Hemp has been traditionally used in textiles and pulp and paper products, but its use in novel composites, non-wovens, and even medical applications has been increasing. The hemp fibre is located in the stem of the plant and its mechanical performance is comparable to glass fibre properties, making it a good choice for reinforcement in biocomposites [1]. It has especially been useful in thermosetting composites in woven mat form. The hemp in mat form can also be applied to thermoplastic PP-based composites and in making products that use compression moulding. Typical applications include automotive parts, furniture, and consumer products. There are already companies making non-woven type mats made with hemp blended with polypropylene or polyethylene fibre [2, 3]. Companies are also

K. Immonen (✉)
VTT Technical Research Centre of Finland Ltd., PO Box 1300, 33101 Tampere, Finland
e-mail: kirsi.immonen@vtt.fi

P. Lahtinen
VTT Technical Research Centre of Finland Ltd., PO Box 1000, 02044 Espoo, Finland

P. Isokangas
Co. Panu Isokangas, Gallen-Kallelankatu 7, 28100 Pori, Finland

K. Torvinen
VTT Technical Research Centre of Finland Ltd., PO Box 1603, 40101 Jyväskylä, Finland

R. Fangueiro and S. Rana (eds.), *Advances in Natural Fibre Composites*,
https://doi.org/10.1007/978-3-319-64641-1_1

making thermoplastic hemp compounds for injection moulding, using a polymer matrix of PLA, PP, ABS or starch-based polymers, depending on the end product [4, 5]. Research of totally bio-based hemp fibre polylactide (PLA) composites [6] has also been done with the aim, for example, of finding a proper surface treatment using alkali treatment to enable better compatibility of hemp with PLA [7]. Modifications to hemp shives has been made using NaOH, $Ca(OH)_2$ and EDTA, with the aim of producing lightweight composites [8]. The traditional use of hemp shives has been in chipboard, particle boards, and hempcrete (hemp shives in combination with lime) type applications [9]. The use of hemp shives with PLA in thermoplastic composites could open new possibilities for their use, as well as for using hemp dust or fines in combination with PLA [10].

In recent years, increasing interest has been shown in utilizing hemp fibre as a raw material for nanocellulose production [11–13]. Although a wide variety of studies exist, only a few include benchmarking with wood-based nanocellulose [14]. Yet, the fast rate of growth, chemical composition, and mechanical properties make it an attractive alternative to refining.

In this study, we used all hemp fractions (fibres, shives and dust) with and without modification in injection moulded PLA composites. Softwood cellulose pulp was selected as a reference material. Further, hemp pulp was prepared to observe the effect of the pulping process. Shives were also oxidized with the aim of improving the fibre polymer connection. Further, nanofibrillated hemp fibres were developed to explore the potential for hemp nanocellulose as alternative for wood nanocellulose.

Materials and Methods

Fibres and Polymer

In this study we prepared thermoplastic hemp-biopolymer composites using hemp bast fibres, hemp shives, and hemp dust as a side stream from fibre production (HempRefine). The hemp pulp was prepared at VTT using soda cooking. Cooking time was 90 min at 165 °C. Total alkali (NaOH) was 4.5 mol/kg. The liquid to solid ratio was 7.5.

The fibrillated hemp fibres were produced using a high-shear grinder Masuko Supermasscolloider MKZA10-15J. The samples were made of bleached hemp bast pulp, unbleached shives pulp, and TEMPO-oxidized hemp bast fibres.

The oxidized shives were prepared using ozonation. Medium consistency (12%) ozonation was carried out in a flow-through reactor at room temperature with ozone charge of 2%. After the ozone stage, the pulp was diluted to a 5% consistency with deionized water. After dewatering, the pulp was washed twice with cold deionized water with an amount equivalent to ten times the absolute dry pulp amount.

TEMPO mediated oxidation was carried out according to the method applied by Saito and colleagues [15]. The sample amount was 50 g and the pulp was suspended in 3 l of purified water. TEMPO (1 g) and NaBr (24 g) were used to catalyse the oxidation reaction with 10% NaClO (744 ml). The pH was kept at 10.3 by adding 1 M NaOH during the reaction. After pH level stabilized, the reaction was stopped by adding ethanol into the oxidized pulp suspension. Finally, the pH was adjusted to 7 by adding 1 M HCl. The oxidized hemp pulp was washed with deionized water and stored in a fridge at +6 °C before fibrillation.

PLA 3052D (NatureWorks) was used as a polymer matrix.

Hemp Nanomaterial Characterization

The fibrillated samples were characterized using the rheological and microscopic methods. The results were compared to the fibrillated hardwood-based cellulose nanofibrils.

The samples were dyed with 1% Congo red solution before imaging. The sample materials were mixed with dye in a ratio of 1:1, with further dilution of the dyed mixture on the microscope slide (ratio 2:1).

Apparent viscosity measurements were performed according to the method described by Kangas and colleagues [16]. The shear viscosities were measured by Brookfield rheometer model RVDV-III Ultra, using vane-type spindles. The samples were diluted to 1.5% fixed consistency with Milli-Q water and dispersed with an Ultra-Turrax disperser at 14,000 rpm for approximately 2 min. Viscosity measurements were performed in a 250 ml Pyrex beaker, and the temperature of the samples was adjusted to 20 ± 1 °C. The shear viscosity was measured at 300 measuring points at 0.5 rpm and 180 measuring points at 10 rpm.

Plastic Processing

Before compounding, the hemp fibres and pulp were compacted to a form that enables easy feeding of the material to the compounder. This compacting was made with modified pelletizer equipment developed at VTT for fibrous material treatment [17].

The materials were compounded with a Berstorff ZE25x33D twin screw extruder and injection moulded to a dog-bone-shaped test piece according to ISO 527, using the Engel ES 200/50HL.

3D Paste Printing

The TEMPO-oxidised hemp nanocellulose was used for 3D printing pastes. Pastes were formed with different proportions of talc, alginate and nanocellulose. Moreover, green leaf dye was used in the demonstration sample, which was designed by Susanna Kettunen (Lahti University of Applied Sciences). The total dry matter content of the pastes was 40–45%. The printing device was VTT's micro-dispensing environment that is based on nScrypt technology. The printer was equipped with a 0.5 mm diameter nozzle attached to a 10 ml syringe.

Composite Characterization

Composites were characterized visually and their mechanical properties, tensile and Charpy impact strengths were also tested. Tensile tests were performed according to ISO 527 using an Instron 4505 Universal Tensile Tester (Instron Corp., Canton, MA, USA) with a 10 kN load cell and a 5 mm/min cross-head speed. Charpy impact strength for unnotched specimens was tested according to ISO 179 using a Charpy Ceast Resil 5.5 Impact Strength Machine (CEAST S.p.a., Torino, Italy). The test specimens were kept in standard conditions (23 °C, 50% relative humidity) for at least five days before testing.

In order to see the fibre-polymer morphology in the composite, a SEM-analysis was made for the cross-cut surface using JEOL JSM T100.

Fibre length was analysed from injection moulded pieces by dissolving the PLA with Soxhlet-extraction. A small number of composite samples were placed into the Soxhlet thimble and fluxed with a hot solution of chloroform for 48 h. After PLA was removed from the fibrous material's surface, the particle size of fibres was analysed microscopically and using a Malvern Particle size analyser after dispersing the fibrous material in water.

Results

Visual and Morphological Results of Composite Materials

Pictures of injection moulded composite samples are presented in Fig. 1 and SEM-pictures and Fig. 2. The average fibre dimensions after thermoplastic processing are presented in Table 1 and Figs. 3 and 4.

Figure 1 presents the visual difference in PLA-hemp composites, where in mostly fibrous containing composites, the fibres come to the surface repeating the flow patterns of the material. The coarser material, shives, exhibits husks on the surface, giving a distinctive 'natural' effect to the product. A closer inspection of the

Fig. 1 Injection moulded PLA-fibre composites with hemp dust (*left*), shives (*middle*) and hemp fibre (*right*)

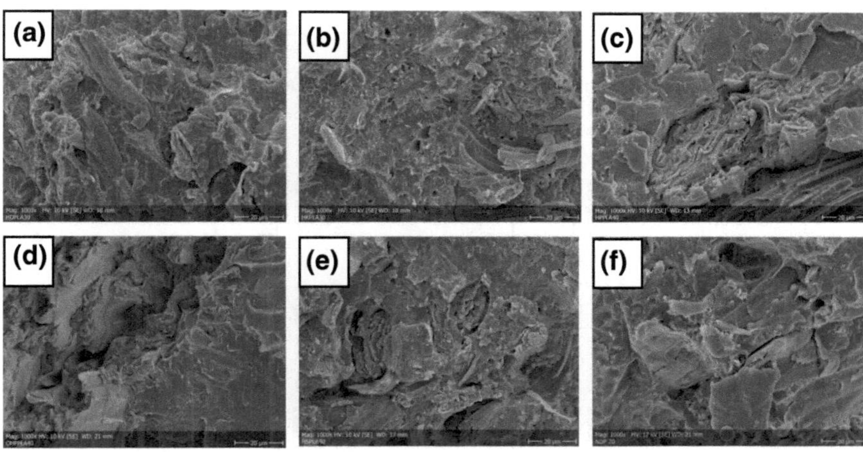

Fig. 2 SEM-pictures of injection moulded PLA-composites. **a** PLA-hemp dust. **b** PLA-hemp fibre. **c** PLA-shives. **d** PLA-oxidized shives. **e** PLA-hemp pulp. **f** PLA-cellulose pulp

Table 1 Average dimensions of the reinforcement material after injection moulding and PLA removal

	Mass median diameter (μm)	90% of the particles below (μm)
Hemp dust	40	143
Hemp fibre	404	1420
Shives	400	1290
Oxidized shives	253	917
Hemp pulp	113	605
Wood pulp	58	176

SEM-pictures in Fig. 2 shows how the fibrous materials are connected to the PLA matrix. The hemp dust in SEM-picture A shows fractures in between the fibrous

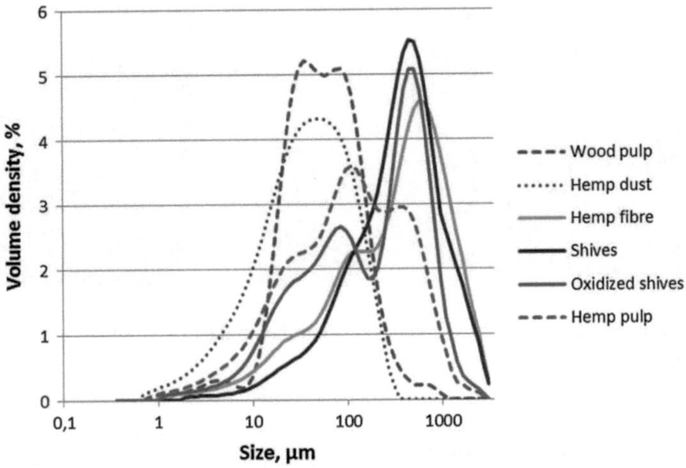

Fig. 3 Particle size distributions of reinforcement material after injection moulding and PLA removal

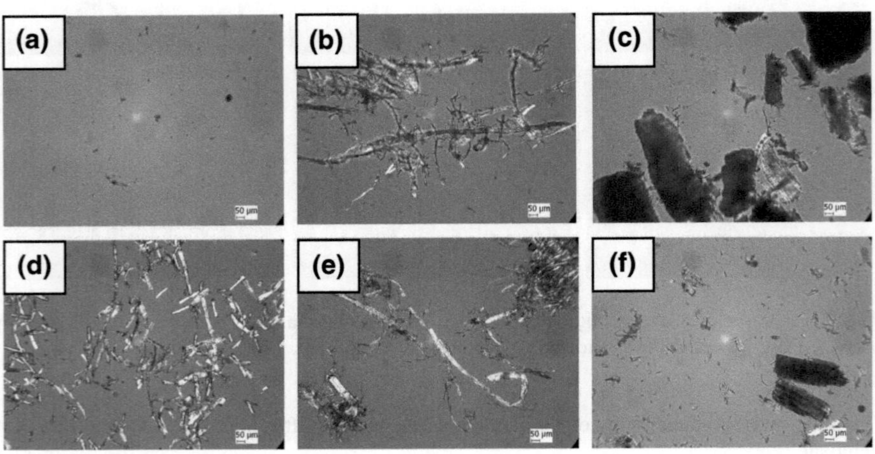

Fig. 4 Optical microscope pictures of reinforcements dissolved from injection moulded test bars. **a** Hemp dust, **b** hemp fibre, **c** shives, **d** wood pulp, **e** hemp pulp and **f** oxidized shives

material and the polymer matrix, but also some porosity in the polymer, which may be due to a smaller fibre fraction. Hemp fibres in SEM-picture B also some show porosity or unevenness in the polymer matrix. However, the fibres are quite closely bounded to the polymer. With pulp fibres (SEM-pictures E and F) there can be seen some small fractures between the fibres and polymer. Hemp pulp (SEM-picture E) is even more tightly bounded to the polymer than wood pulp (SEM-picture F). With shives, a larger coarse material can be seen inside the polymer matrix (SEM-pictures C and D) and oxidised shives are more closely bound to the polymer matrix (SEM-picture D) than the neat shives, which indicates better interaction of oxidized shives with PLA.

During compounding and injection moulding and due to high temperature and shear forces inside the process, the lingo-cellulosic fibre material breaks down [18]. For this reason, we removed the PLA matrix from the injection moulded samples and analysed the fibrous material or particles size distribution in addition to see the real reinforcing effect of fibres.

Figure 4 shows how the fibre material looks after dissolving PLA out. Even though Table 1 reveals the fibrous material size after injection moulding, Figs. 3 and 4 also show quite a lot of fine material. The particle size analyser used here better reveals the size distribution of the fine fraction, but due to the high aspect ratio in larger fibres, there is some uncertainty in the proportions of the larger size fraction, since the shape of the particle is not as well analysed using this method as with the FibreMaster. However it can be seen that the hemp pulp has the widest particle size distribution of all the samples and that it has longer fibres compared to wood pulp. After injection moulding, the longest fibres are the hemp fibres, followed by the shives. The oxidized shives are more broken than the neat shives, while the hemp dust shows the smallest fibre particles.

Results of Mechanical Tests

The tensile strength and Charpy impact strength (unnotched) results for hemp-biopolymer composites are shown in Table 2 and Fig. 5.

The results reveal the effects of fibre length on strength properties, as well as the comparison of hemp pulp vs. hemp fibre in regards to the effects of lignin, hemicellulose and other ingredients. Following the pulping process, it is mainly cellulose fibre left in the hemp, which seems to be stronger than lignin and hemicellulose containing fibre. When comparing the hemp pulp results with wood-based pine

Table 2 Tensile strength and Charpy impact strength (unnotched) results of hemp-PLA composites

	Tensile strength		Modulus (AutoYoung)		Strain at break		Impact strength	
	MPa	s.d.	MPa	s.d.	%	s.d.	kJ/m^2	s.d.
PLA ref.	63.0	0.6	3653	222	3.9	0.1	17.0	2.8
Pine pulp 40%	59.5	1.9	6950	356	2.4	0.1	9.9	2.2
Hemp pulp 40%	86.6	0.8	7440	596	1.9	0.1	19.9	3.2
Hemp fibre 40%	67.8	3.0	6607	184	1.2	0.1	9.8	2.5
Hemp shives 40%	51.1	1.4	6869	113	0.8	0.0	7.5	1.5
Oxidized shives 40%	59.9	0.3	6736	491	1.2	0.1	7.1	1.2
Hemp fibre 30%	67.4	1.5	5667	128	2.0	0.2	12.1	1.9
Hemp dust 30%	48.0	0.3	4486	93	1.6	0.1	9.8	1.1

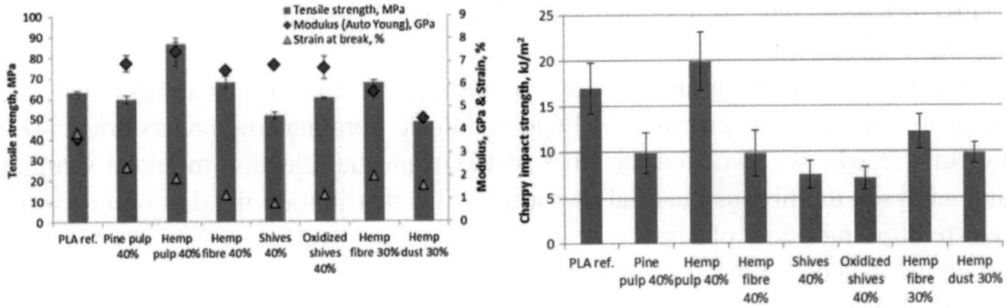

Fig. 5 Tensile strength (*left*) and Charpy impact strength (unnotched) results for PLA composites

pulp, we see up to a 45% improvement in tensile strength, a 7% improvement in modulus and a 100% increase in impact strength. These strength increases likely arise from the longer fibre length in the hemp pulp. Comparing the hemp fibres to hemp pulp the hemp fibres are longer than the hemp pulp, the higher lignin content and other ingredients of the hemp fibre appear to make the fibre weaker compared to hemp pulp. The tensile strength results for hemp fibre are 22% and impact strength results 50% lower compared to hemp pulp.

Even though the hemp shives have a fibre distribution close to that of hemp fibre, the shape and composition is different. Shives contain higher amounts of lignin (19–28% vs. 2–5%) and hemicellulose (31–39% vs. 7–19%) than hemp fibres [19], and these can have an effect on composite strength properties. Hemp shives had 24% lower tensile strength and 38% lower impact strength properties, but the tensile modulus was slightly improved (4%). The oxidation of shives improves the tensile strength by 16% still keeping other properties at the level of neat shives.

The hemp dust has the smallest particle sizes after injection moulding. Compared to other reinforcement materials, the mechanical strength values were clearly at a lower level, even with a 10% smaller addition level. Compared to neat PLA, only the tensile modulus was increased by 23%, but tensile strength (24%) and impact strength (42%) were lower in hemp dust at an addition level of 30%.

Results for Hemp Nanomaterials

Fibrillated samples made of TEMPO-oxidised (TCNF) native hemp pulp (CNF) and unbleached shives pulp (LCNF) are shown in Fig. 6. The hemp nanofibers formed viscous hydrogels, which are attractive as a reinforcing component, a rheology modifier, and a film forming material. One possibility is to use the hemp hydrogels as a component in 3D-printed objects, as we have demonstrated in this work in Fig. 8.

Optical imaging of the fibrillated materials in Fig. 6 showed that the TCNF sample contained more thin and long fibrils and fibril bundles, whereas the native

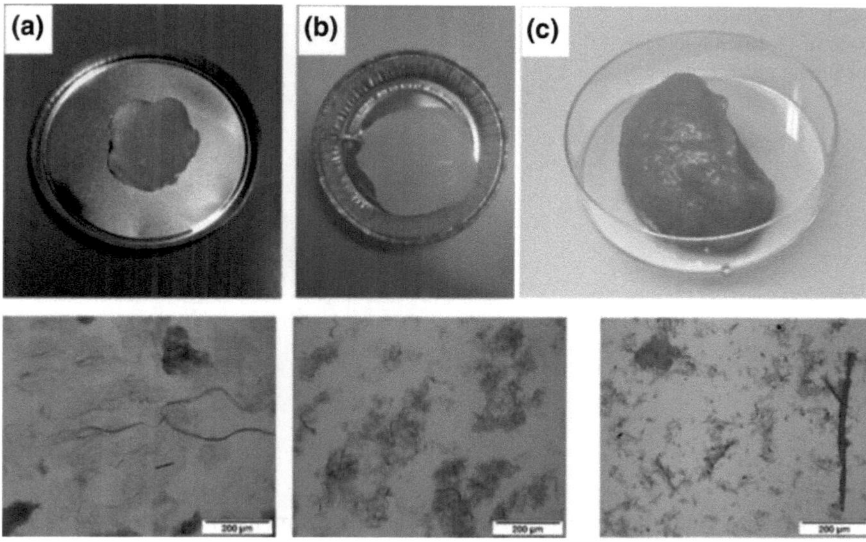

Fig. 6 Hemp nanofibers in water. a TEMPO-CNF, b native CNF and c unbleached shives CNF (LCNF)

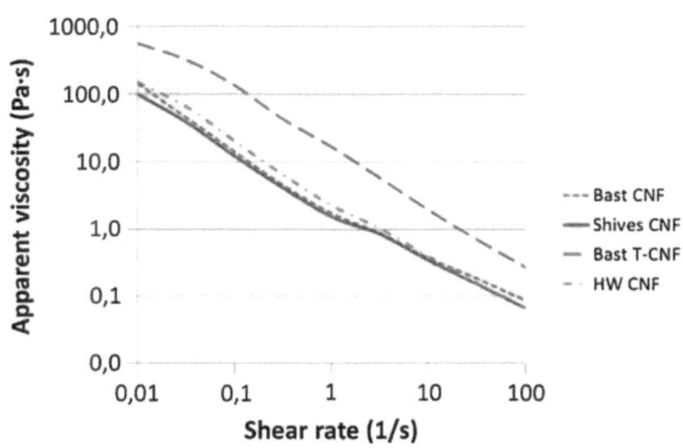

Fig. 7 Apparent viscosities of hemp and hardwood-based CNF samples

grade (CNF) and lignocellulose CNF contained shorter fibre fragments and residual fibres. The presence of long thin fibrils usually leads to higher gel strength and increased viscosity, as presented in Fig. 7. The TCNF sample had clearly a higher viscosity level compared to the unmodified samples. Samples made of hemp bast fibres and shives had comparable apparent viscosity to hardwood (HW)-based CNF.

The printing paste was formulated utilizing the cross-linking ability of alginate and the rheological as well as strength-enhancing properties of the TCNF. Dry matter content was further increased with talc filler. The successful combination of these materials is shown in Fig. 8. TCNF was used to improve the shape fidelity of the printing paste, which was enabled by the high viscosity. The visible lines on

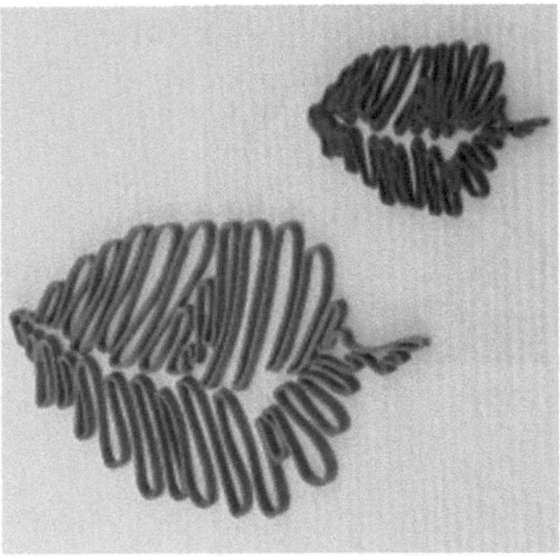

Fig. 8 3D-printed leaves made of biobased hydrogels containing green leaf dye

printed leaves could be printed accurately and their dimensional stability was also good after drying at room temperature.

Conclusions

The best biocomposite results were obtained for PLA-hemp pulp (40% fibre) with a tensile strength of 87 MPa, a Young's modulus of 7440 MPa, and an impact strength of 20 kJ/m^2. The results exceeded the values of PLA-pine pulp (40% fibre) due to the longer fibres. According to this study, the hemp-biopolymer composites have potential in injection moulded products due to the strength of the hemp bast fibres, the hemp pulp, and the easy processing of hemp shives in the composites. The materials also showed great potential in their visual aspect, in for example, furniture, decoration and construction applications. Hemp nanofibers formed viscous hydrogels, which has benefits as a reinforcing component, a rheology modifier, and a film forming material.

Acknowledgements The authors gratefully acknowledge funding from the NoMa-project, Novel structural materials with multi-scale fibre components, funded by the Finnish Funding Agency for Innovation (TEKES). Sari Asikainen and Jari Leino are acknowledged for performing pulping and ozonation trials.

References

1. Fan, M. (2010). Characterisation and performance of elementary hemp fibres: Factors influencing tensile strength. *BioResources, 5*(4), 2307–2322.
2. www.hempline.com.

3. www.realhemp.com.
4. www.hempplastic.com.
5. http://flaxcomposites.com/?page_id=451.
6. Sawpan, M. A., Pickering, K. L., & Fernyhough, A. (2007). Hemp fibre reinforced poly(lactic acid) composites. *Advanced Materials Research, 29–30,* 337–340.
7. Hu, R.,& Lim, J.-K. (2007). Fabrication and mechanical properties of completely biodegradable hemp fiber reinforced polylactic acid composites. *Journal of Composite Materials, 41*(13), 1655–1669.
8. Terpáková, E., Kidalová, L., Eštoková, A., Čigášová, J., & Števulová, N. (2012). Chemical modification of hemp shives and their Characterization. *Procedia Engineering, 42,* 931–941.
9. Building with flax and hemp—Brochure. www.grow2build.eu.
10. Spierling, S., Koplin, T., & Endres, H.-J. (2014). Hemp fines—An agricultural by-product for biocomposites? A holistic approach. In S. T. Smith (Ed.), *23rd Australasian Conference on the Mechanics of Structures and Materials (ACMSM23), Byron Bay, NSW* (Vol. II, pp. 875–880), 9–12 December. Southern Cross University, Lismore, NSW. ISBN: 9780994152008.
11. Alila, S., Besbesa, I., Vilar, M. R., Mutjéc, P., & Boufia, S. (2013). Non-woody plants as raw materials for production of microfibrillated cellulose (MFC): A comparative study. *Industrial Crops and Products, 41,* 250–259.
12. Peresin, M. S., Tammelin, T., Liukkonen, S., Hanninen, T., Pere, J., & Lahtinen, P. (2014). Exploring alternative raw materials sources for nanocellulose production. Congreso Internacional de Metalurgia y Materiales Samconamet/Iberomat/Materia 2014, Santa Fe, Argentina, Octubre 21–24, 2014.
13. Vartiainen, J., Lahtinen, P., Kaljunen, T., Kunnari, V., Peresin, M. S., & Tammelin, T. (2014). Comparison of properties between cellulose nanofibrils made from Banana, Sugar beet, Hemp, Softwood and Hardwood pulps. In *Proceedings of the ABTCP* 2014. The 47th ABTCP International Pulp and Paper Congress, Sao Paulo Brazil, October 7–9, 2014.
14. Fan, M., Dai, D., & Yang A. (2011). High strength natural fiber composite: Defibrillation and its mechanisms of nano cellulose hemp fibers. *International Journal of Polymeric Materials and Polymeric Biomaterials, 60*(13), 1026–1040. doi:10.1080/00914037.2010.551347.
15. Saito, T., Nishiyama, Y., Putaux, J.-L., Vignon, M., & Isogai, A. (2006). Homogeneous suspensions of individualized microfibrils from TEMPO-catalyzed oxidation of native cellulose. *Biomacromolecules, 7*(6), 1687–1691.
16. Kangas, H., Lahtinen, P., Sneck, A., Saariaho, A.-M., Laitinen, O., & Hellen, E. (2014). Characterization of fibrillated celluloses. A short review and evaluation of characteristics with a combination of methods. *Nordic Pulp & Paper Research Journal, 29*(1), 129–143.
17. Sivonen, E., & Valta, K. Patent application WO2012080574A1.
18. Peltola, H., Madsen, B., Joffe, R., & Nättinen, K. (2011). Experimental study of fiber length and orientation in injection molded natural fiber/starch acetate composites. *Advances in Materials Science and Engineering*, Article ID 891940, 7 pp. doi:10.1155/2011/891940.
19. Thomsen, A. B., Rasmussen, S., Bohn, V., Vad Nielsen, K., & Thygesen, A. (2005). Hemp raw materials: The effect of cultivar, growth conditions and pretreatment on the chemical composition of the fibres. Risø Report Risø-R-1507(EN).

Industrial Hemp Transformation for Composite Applications: Influence of Processing Parameters on the Fibre Properties

Vincent Placet, Camille François, Arnaud Day, Johnny Beaugrand and Pierre Ouagne

Abstract The main objective of this collaborative work is to characterize the influence of the processing stages of industrial hemp on the fibre properties. Transformation processes well suited for composite reinforcing textiles are considered. The work focuses on the different stages along the transformation chain of hemp, from the straw retting to the preform manufacturing. The main highlight is the predominant influence of retting on the tensile properties of individual fibres after their mechanical extraction from the stalks. Regarding the secondary processing, different technologies such as spinning, and use of natural binder systems are also proposed to produce yarns and woven fabrics. The effect on these secondary processing technologies and their parameters on the fibre properties are also characterised. The results show that the first steps of processing (retting and decortication) have the greatest impact on the tensile strength of hemp fibres.

Keywords Industrial hemp · Hemp fibres · Mechanical processing
Retting · Mechanical properties

V. Placet (✉) · C. François
FEMTO-ST, CNRS/UFC/ENSMM/UTBM, Université de Bourgogne Franche-Comté, Besançon, France
e-mail: vincent.placet@univ-fcomte.fr

A. Day
Fibres Recherche Développement, Troyes, France

J. Beaugrand
INRA, URCA UMR 614 FARE, Reims, France

J. Beaugrand
Research Unit BIA UR1268, INRA, Nantes, France

P. Ouagne
ENIT, EA 1905 LGP, Tarbes, France

R. Fangueiro and S. Rana (eds.), *Advances in Natural Fibre Composites*,
https://doi.org/10.1007/978-3-319-64641-1_2

Introduction

Today, in Europe, the well developed and marketed plant fibre continuous reinforcements (tapes, roving, fabrics and so on) for composites applications (PFCs) are mainly based on flax fibres. Alternatives to current commercial flax-based solutions are now considered and mobilize additional resources, like hemp. Considering their intrinsic properties, hemp fibres have a great opportunities for market capture in composite for secondary structural applications. So far, the main applications for hemp fibres are in the fields of pulp and paper (55%), insulation (26%), thermoplastic polymer composites (14%) and mulch (2.7%) [2]. The absence of hemp continuous reinforcement that would be available at an industrial scale could be connected to the lack of knowledge and engineering, in particular when compared to flax. This deficit for hemp is assigned to less intensive research activity, to a less important industrial production (in term of volume) and to technological barriers, such as fibre separation and the alignment of fibres throughout the transformation process. However, the knowledge on hemp and the readiness of the related transformation technologies have been progressing fast for these last few years thanks to the development of major research projects at national and European scales.

Currently, the processes used to separate the different vegetal fractions of hemp are mainly derivatives from the paper and technical textile industries. Technical fibres are generally extracted from the stalks using mechanical processes and lines made up of milling and cleaning systems. Among the different milling tools that are classically used, hammer mills are the most common [12]. They are efficient and they lead to clean technical fibres with a low content of shives. However, they are also very aggressive for the fibres themselves. They can lead to damages within the fibre wall, which is detrimental for the mechanical properties of individual fibres. Hernandez-Estrada et al. [6] showed recently that these first stages of processing (in particular decortication) have a great impact on the formation of dislocations in the fibres. Meanwhile, the preservation of the integrity of the fibres through the transformation stages and up to the composite scale is a real challenge and requires an optimization of both the primary and secondary processing. Scutching and hackling processes, which are commonly used for flax are less aggressive and also allows long and aligned technical fibres (line hemp) to be obtained [1]. This is, for sure, a processing route that has to be more used for hemp in view of their integration in composite materials. However, it is still required to adapt the harvesting and scutching procedures and machines to deal with the height and diameter of the stalks that are significantly greater than those of flax [1, 4]. In this work, we propose to characterize the influence of lab/pilot transformation processes that are currently developed for composite hemp-based reinforcements. Separation processes limiting the amount of damage caused to the fibres and secondary processes minimizing the twisting level at the scale of yarns while maximizing their tensile strength are targeted.

Over the last two decades, and since the renewal of interest in plant fibres, only few researchers have investigated the impact of processing on the mechanical properties of plant fibres. It still remains an open question, in particular when considering innovative transformation technologies. For flax, Van de Weyenberg et al. [13] showed that consecutive decortication stages of fibres (scutching, hackling…) change tremendously the mechanical properties of composites performance. They attributed this change to a modification of the mechanical and biochemical properties of the fibres. Thygesen et al. [15] studied the relationship between processing of cellulosic fibres (both hemp and flax) and fibre bundle strength. They studied processing methods that are traditionally applied for yarn production and also included retting, scutching, carding, and cottonization. They highlighted a monotonically decreasing relationship between the strength of fibre bundles and the number of processing steps, with an average reduction by 27% per processing step at the applied conditions. No large change in cellulose content and crystallinity were observed. So authors concluded that the reduction in strength must be explained by other changes in the fibres bundle ultrastructure. Indeed, in a bundle of fibre, the mechanical behaviour is controlled both by the performance of single fibres but also by the shear deformation at the interface between the fibres.

So, the overall objective of this work is to assess the influence of processing stages on the mechanical performance of single hemp fibres. The influence of processing parameters, including the retting level (an optional pre-processing parameter of the primary processing of stalks), extraction, combing, drawing and spinning on the mechanical properties of single hemp fibres are characterized. The gathering of results is the fruit of collaborative projects between several French labs.

Materials and Methods

Plant Material

Within this work, two types of materials were used to characterize the influence of processing on the mechanical properties of fibres, i.e. hemp stalks and combed fibres.

Hemp Stalks

The hemp stalks were supplied by La Chanvrière, a company in Bar sur Aube, France. Plants (Cannabis *sativa* L., cultivars ‘Fedora 17’) were cultivated in Laubressel in 2014 (GPS coordinates: 48° 18′ 00.1″N and 4° 14′ 12.0″E) on a chalky clay soil. Once mown, the straw was windrowed, turned over once, and laid in swatches on the ground for retting. Stalks were harvested after three different field retting times, 10, 39 and 75 days leading to three retting levels, noted R1, R2 and R3 respectively. For R3, stalks were turned over a second time, 60 days after moving, to homogenize retting. After reaching the targeted retting level, straws were balled and then stored under cover to prevent wetting. Hemp stalks used for experiments were collected at the same time, 136 days after mowing.

Hemp Sliver Laps

Long hemp fibres in the form of combed sliver laps were provided by an Italian company. Before being combed, the hemp fibres were carded and drawn. The slivers were then grouped together by a lap former to increase their size before being combed. The linear mass of the sliver laps was about 15,000 tex (14759 ± 613 tex). The length and diameter of single and technical fibres were measured on 30 samples randomly selected. The mean lengths of the single and technical fibres are 25 ± 14 and 83 ± 50 mm respectively. The mean diameters of the single and technical fibres are 20 ± 4 and 111 ± 33 μm.

Primary Processing of Hemp Stalks

Hemp technical fibres were extracted from the stalks by mechanical processing. This step was carried out by FRD® (www.f-r-d.fr) on a line made up of both milling and cleaning systems. This technology, developed at a pilot scale, is closed to flax industrial process. The same fibre extraction parameters were used for each sample. The setting of these parameters has been achieved through experiments done for several years and is considered as the best compromise between the fibre quality and the production yield. Fibre quality was evaluated with industrial standards developed by FRD, and by determining the residual content of shives in technical fibres and by measuring the fibre fineness.

The fibres extracted using this mechanical process are respectively named R1-MP, R2-MP and R3-MP depending on the initial retting level of the straws. The different stages and ways used to process the straws are schematically described on Fig. 1.

Secondary Processing of Hemp Fibres

An industrial drawing machine, adapted to long fibres, was used to reduce the linear mass of the sliver laps. Three drawing steps were used to obtain slivers with a linear mass of about 500 tex. After the first, second and third drawing steps, slivers with the following linear masses were respectively obtained: 3540 ± 139, 2204 ± 96 and 519 ± 49 tex. The slivers, which are assemblies of aligned fibres of finite length, do not possess generally enough cohesion to sustain further processing steps such as the textile architecturation (weaving, braiding…). So, they are generally

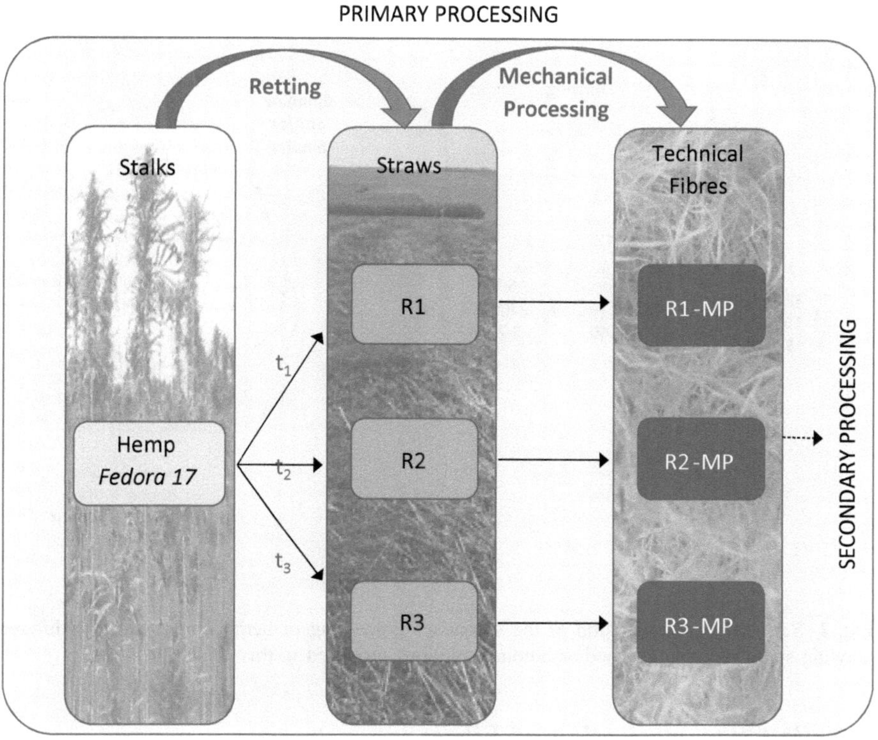

Fig. 1 Schematic representation of the main and different process stages used in this study for the fibre extraction

transformed into yarns. In this work, slivers were first transformed into non-twisted yarns (Y-NT, see Fig. 2). The fibres are slightly entangled during the drawing process. The failure load of the non-twisted yarns was 7.4 ± 1.9 N. This value is not high enough to manufacture a woven fabric or a braid. Thus, to increase the cohesion and the tensile performance of the non-twisted yarn, three other types of yarns were manufactured. Non-twisted yarns were bound with a natural adhesive constituent (Y-NTB). Low twisted unbound and bound with a natural adhesive constituent yarns were also manufactured (named respectively Y-LT and Y-LTB). Their load to failure were respectively: 19.2 ± 4.4, 19.7 ± 2.7 and 68.6 ± 18.6 N. A minimum of 15 N is generally required to manufacture woven fabrics, and the three types of yarns were therefore used for further processing steps.

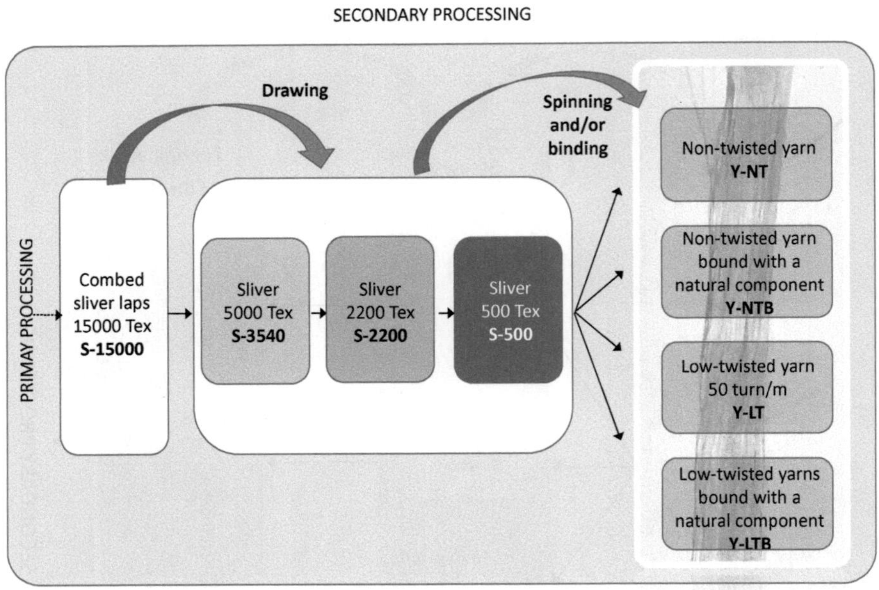

Fig. 2 Schematic representation of the secondary processing of hemp fibres, with the different drawing steps and spinning and/or binding solutions proposed in this study

Tensile Characterisation of Fibres

For each batch of fibres, thirty to fifty single fibres were manually extracted from the technical fibres using tweezers. They were glued onto thin paper support using adhesive (Loctite super glue) to facilitate their handling and examination prior to tensile testing. An electrodynamic machine (DMA Bose Electroforce 3230) was used to perform the monotonic tensile tests on fibres up to failure. The average width of each fibre was computed by obtaining ten measurements along its length using polarised light microscopy (Nikon Eclipse LV 150). The average width of each fibre batch was in the range of 20–25 μm. The paper frame supporting each fibre was clamped onto the testing machine and cut prior to the beginning of the tensile test. The clamping length was 10 mm. Fibres were tested at a constant crosshead displacement rate of 5 $\mu m\ s^{-1}$. The tensile tests were carried out at a temperature of 21 ± 2 °C and a relative humidity of 50% ± 5%. The applied force was measured using a 20 N load sensor, with a resolution of approximately 1 mN, and the displacement was measured using a LVDT with a resolution ranging from 0.1 μm. The sample elongation and the load were recorded continuously and the longitudinal mechanical properties (Young's modulus, ultimate strength and failure strain) of isolated hemp fibres determined. The strain was determined using the displacement measurements and the initial length of the fibre. The tensile stress was determined using the cross-section of each fibre and the apparent tangent modulus (E) was computed from the linear section of the first part of the stress–strain curve.

The effective cross-section area of the fibre was determined using the mean external diameter, assuming that the fibre is perfectly cylindrical and the lumen neglected. The mean values and standard deviation of these tensile properties were computed for each batch of fibres. Considering the highly skewed distribution that is generally observed for the tensile properties of plant fibres, a statistical analysis was also performed. The best distribution function for each property was identified.

Results and Discussion

Influence of Retting on the Tensile Properties of Fibres After Their Mechanical Extraction

Table 1 presents the values of bast fibre content and residual content of shives obtained after fibre extraction using the proposed mechanical process. The overall total bast fibre content in the straw is respectively 40.5, 50.7 and 42.5% for R1, R2 and R3. These high fibre contents contrast with data of literature that report a mean bast fibre content in hemp straw of approximately 30% (w/w) [7]. The field retting of straws could explain these high values. Indeed, ground retting lead to decohesion of stem tissues and decomposition of shives. The residue of shives can remain on the land after turning over and balling straw, inducing an increase of the relative content of bast fibres in straws.

The percentage of residual shives in bast fibres after mechanical extraction decreases slightly (from 7.75 to 5.35%) with the increasing level of retting. This result underlines once more the influence of retting on decohesion of tissues within the straw.

The results from the tensile tests on single fibres after their extraction from the straws (at the three different retting times) are shown in Fig. 3. The same settings of the mechanical processing were used after the retting step for each of the three batches. No significant difference can be noticed on the apparent elastic modulus (Table 2). The mean value is comprised for each retting level between 15.5 and 17.5 GPa, with a similar and high standard-deviation of approximately 11 GPa. A significant decrease of the strain at failure can be observed with the increasing level of retting, from 3.5% for R1-MP to 2.2% for R3-MP. This result was already observed by Placet et al. [10] in a previous study. As far as the tensile strength is

Table 1 Bast fibre content and residual content of shives after fibre extraction for each batch of straws

	Bast fibre content (%)	Shive content (%)	Residual content of shives in bast fibres after extraction (%)
R1	40.5	59.5	7.75
R2	50.8	49.2	5.45
R3	42.5	57.5	5.35

Fig. 3 Tensile properties of single hemp fibres after mechanical extraction and for the three retting levels. *Markers* experimental data, *Dash lines* identified cumulative distribution functions

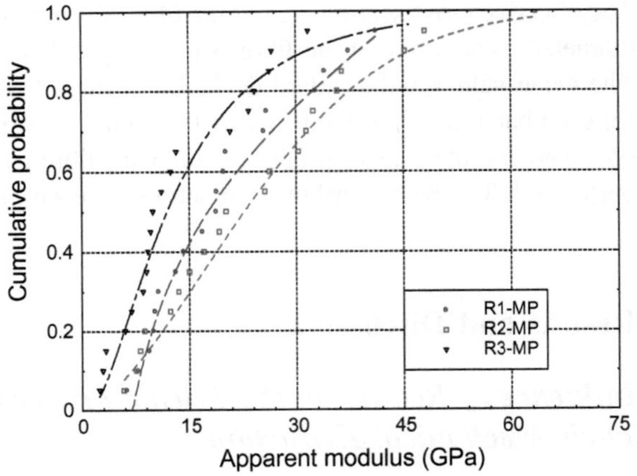

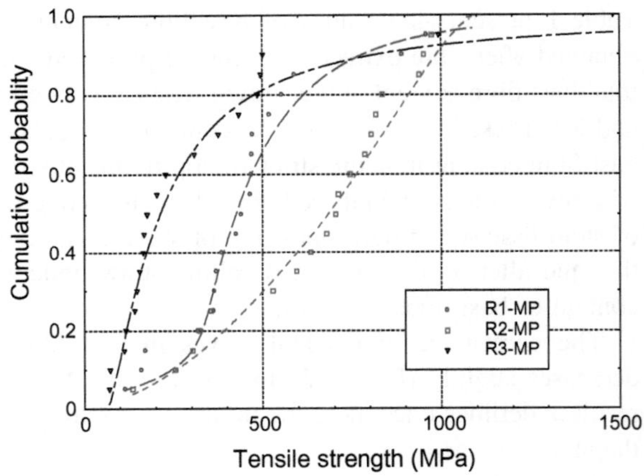

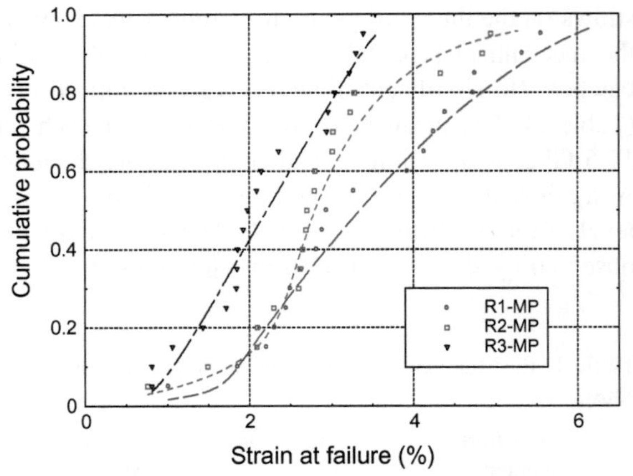

concerned, results show that the intermediate retting level (R2) leads to the highest value. Indeed, using these retting and primary processing settings (R2-MP), a tensile strength of 660 ± 265 MPa is measured at the scale of single fibres. This value is significantly higher to those measured when the stalks were lower or higher retted. Tensile strengths of 480 ± 245 and 340 ± 335 MPa were measured respectively for R1-MP and R3-MP. For R1-MP, the short time of exposure of the straws to the ground and thus to the microorganisms certainly does not permit the pectic content to be significantly attacked and removed by bacterial and fungal effectors. As a consequence, the bonding between the bast fibres both to each other and to the other constituents of the stalk remains strong. The further mechanical processing stage, aiming at separating the bast fibres from the ligneous shives, is aggressive and presumably highly damaging for the fibres. The damage formation can be directly related to the decrease in strength. Hernandez-Estrada et al. [6] showed recently that the first stage of processing (in particular decortication) has a greatest impact on the formation of damage. Hänninen et al. [5] also pointed out for flax that industrially processed flax fibres contain significantly more defects than green or retted ones.

Conversely, for R3-MP, it could be hypothesised that the fibre could have been processed from straws that had been over-retted. In addition to the removal of the pectic substances, the prolonged exposure time to the bacterial and fungal effectors could have presumably led to a degradation of the other constitutive polymers. This could induce a decrease in the mechanical performance of the fibres even before the mechanical processing. Placet et al. [10] showed that for over-retted straws, the decrease in the tensile performance of fibres can be attributed to a decrease in the crystalline cellulose index and to the degradation of hemicelluloses (a decrease in the relative content of xylose), which are thought to play an important role in the 3D organisation of wall macromolecules and the resulting mechanical properties. Liu et al. [8] also showed that a negative effect of field retting occurred after extended field retting (70 days in their retting conditions). This was attributed to the acceleration in the degradation of cellulose by the action of microorganisms.

The scattering in the tensile properties of R2-MP is high but expected and well-documented for plant fibres. The coefficients of variation (CoV) are on the same order of magnitude as generally measured for hemp fibres and other plant fibres such as jute, with a value of approximately 40–45% for the tensile rigidity and strength and 30–40% for the strain at failure [14]. This difference between the CoV in strain at failure on the one hand and rigidity and strength on the other hand is generally explained by the error when determining the effective cross-section of the fibres. Regarding R1-MP and R3-MP, the scattering in the strength values are very high with a respective CoV of 0.51 and 0.98. These values and the features of the distribution laws confirm that the mechanical processing on low-retted straws on the one hand, and over-retting on the other hand induce an increase in the number of flaws and flaw populations inside the fibres.

These results clearly indicate that field retting has to be perfectly tailored and controlled. If not, retting can be clearly detrimental for the quality of industrial hemp bast fibres and particularly for their mechanical properties. It is suggested that

Table 2 Tensile properties of single hemp fibres after mechanical extraction and for the three retting levels

	R1-MP	R2-MP	R3-MP
Apparent rigidity (GPa)	15.7 ± 11.6 0.9–41.5	17.5 ± 11.2 2.6–45.5	16.6 ± 12.7 2.9–43.3
Strain at failure (%)	3.5 ± 1.4 1.0–6.1	3.0 ± 1.1 0.8–5.3	2.2 ± 0.9 0.8–3.5
Tensile strength (MPa)	480 ± 245 115–1085	660 ± 265 135–1080	340 ± 335 70–1480
Diameter (µm)	25.4 ± 4.9 17.7–39.9	26.4 ± 5.3 17.0–36.0	28.9 ± 5.8 17.5–39.9

Mean ± SD, min…max values

an optimized period and level of field retting may be adopted to extract the fibres as gently and efficiently as possible.

Influence of Secondary Processing on the Tensile Properties of Fibres

Figure 4 and Table 3 synthesise the tensile properties measured on different batches of single fibres taken in yarns after their processing using the different technologies.

Fibres which were taken from the combed sliver lap (S-15000) have a mean rigidity of 18.2 ± 7.2 GPa, a strain at failure of 2.05 ± 0.8% and a strength of 325 ± 170 MPa. It can be observed that the three drawing steps applied to obtain slivers with a linear mass of about 500 tex induce a decrease of about 25% on the mean tensile rigidity and an increase in the average strain at failure of about 35%, without any significant change in strength. So, drawing appears to induce a softening of the longitudinal elastic properties of the single fibres. More surprisingly, twisting (at this level of approximately 50 turn/m) induces a slight increase in the tensile strength of single fibres, whether the yarn was treated with a natural adhesive or not, reaching an average value of 380 ± 180 and 395 ± 195 MPa for Y-LT and Y-LTB respectively. The origin of these phenomena (rigidity softening or hardening, increase or decrease on strength) observed under mechanical processing was not identified during this study and will be investigated in a forthcoming work by studying the microstructure of the fibres, in particular by measuring the cellulose crystallinity index and the cellulose microfibrils angle. Effectively, it is well known that the mechanical behaviour of wood and plant fibres are highly sensitive and dependent on their thermo-hygro-mechanical history [3, 9, 11]. So, it is not surprising that the transformation process influences their mechanical behaviour, since it involves various loading paths. One of the main difficulty, which researchers and industries are facing, is related both to the complexity of the transformation stages and the complexity of the thermo-hygro-mechanical behaviour of plant fibres. They

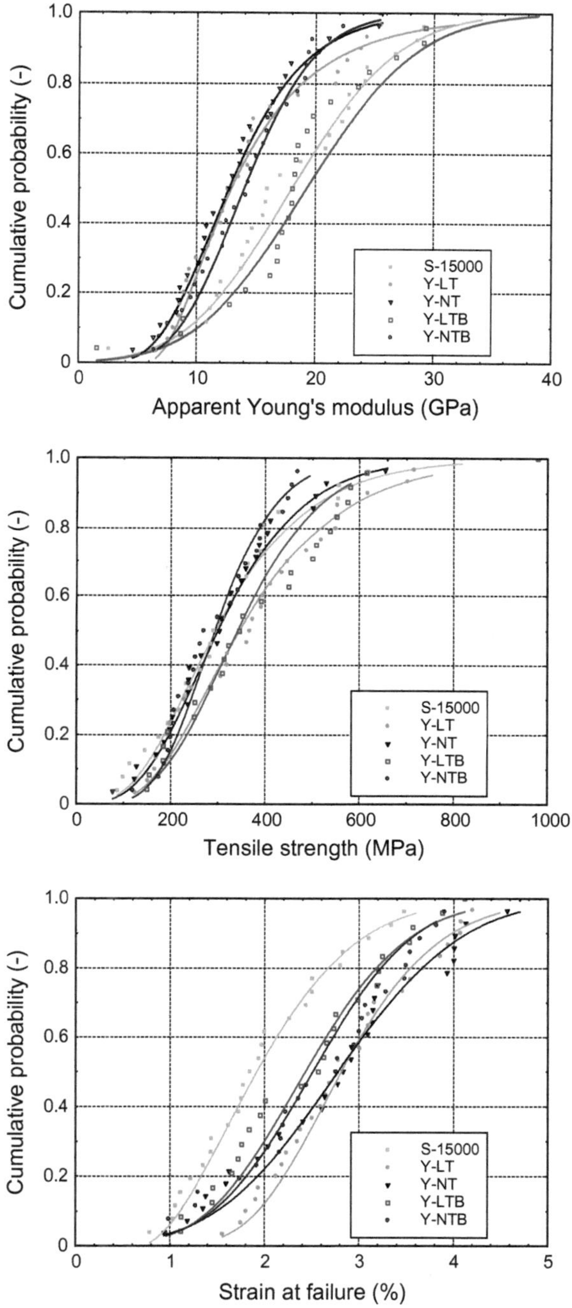

Fig. 4 Tensile properties of single hemp fibres after secondary processing. *Markers* experimental data, *lines* identified cumulative distribution functions

Table 3 Tensile properties of single hemp fibres after secondary processing

	S-15000	Y-NT	Y-NTB	Y-LT	Y-LTB
Apparent rigidity (GPa)	18.2 ± 7.2 2.6–34	13.5 ± 5.4 4.6–25.6	14.4 ± 4.8 6.4–25.5	14.7 ± 6.3 6.6–31.7	18.9 ± 7.6 1.6–38.7
Strain at failure (%)	2.05 ± 0.8 0.8–3.6	2.8 ± 1 0.9–4.7	2.7 ± 1 0.9–4.9	2.9 ± 0.8 1.6–4.5	2.5 ± 0.9 1.1–4.1
Tensile strength (MPa)	325 ± 170 90–820	320 ± 150 75–660	305 ± 105 120–495	380 ± 180 125–755	395 ± 195 150–980
Diameter (μm)	20.2 ± 4.5 13.2–35.2	21.5 ± 5 13.2–33	22.2 ± 4.5 14.6–33.2	24.7 ± 5 17.3–37.4	22.3 ± 4.4 14.8–32.9

Mean ± SD, min…max values

exhibit in particular couplings and stiffening phenomena under mechanical loading and/or environmental exposure that are still not fully understood [11]. It is also highly difficult to predict the loading paths and levels endured by the fibres during the processing stages, even if well mastering the process settings.

It can also be noticed that the treatment with a natural adhesive lead to an increase in the tensile rigidity. A mean value of 18.9 ± 7.6 GPa was measured for Y-LTB. An interaction between the natural adhesive and the constitutive polymers of the fibre wall can be assumed. Biochemical and surface analyses have to be done to confirm this hypothesis.

In general, it can be concluded that, even if secondary process stages can influence the tensile performance of single fibres, the chosen technologies have a lower impact than extraction step.

Conclusion

This work focused on the impact of the hemp transformation stages, from retting to yarn manufacturing, on the tensile behaviour of single fibres. Results emphasize the predominant role of retting on the further aggressive mechanical fibre separation steps and on the quality of the resulting fibres. It also points out the necessity to tailor the field retting to ensure a good quality and properties stability at the scale of industrially processed fibres. In light of these results, it can already be concluded that it should be essential to dispose of a tool indicating how to manage straw quality and retting before the mechanical processing of high-grade fibres. Further processing steps towards hemp reinforcement (such as drawing, spinning, binding with natural adhesive) can also influence the strength and rigidity of fibres but with a lower impact.

Acknowledgements Johnny Beaugrand is thankful for the support provided by the MATRICE state-to-country (Région Alsace Champagne-Ardenne Lorraine) program, France. Arnaud Day would like to thank, ADEME, Troyes Champagne Métropole and Région Alsace Champagne-Ardenne Lorraine for their financial support. The authors would also like to thank Groupe Depestèle (France) for their technical support and the use of their drawing and binding facilities.

References

1. Amaducci, S., & Gusovius, H. J. (2010). Hemp-cultivation, extraction and processing. In J. Müssig (Ed.), *Industrial applications of natural fibres—Structure, properties and technical applications*. Wiley series.
2. Carus, M. (2015). World-wide market data on hemp and other bast fibres—Status and outlook. In *12th International Conference of the European Industrial Hemp Association (EIHA), Wesseling, Germany*, May 20–21, 2015.
3. Guicheret-Retel, V., Cisse, O., Placet, V., Beaugrand, J., Pernes, M., & Boubakar, M. L. (2015). Creep behaviour of single hemp fibres. Part II: Influence of loading level, moisture content and moisture variation. *Journal Materials Science, 50*(5), 2061–2072.
4. Gusovius, H. J., Hoffman, T., Budde, J., & Lühr, C. (2016). Still special? Harvesting procedures for industrial hemp. *Landtechnik, 71*(1), 14–24.
5. Hänninen, T., Thygesen, A., Mehmod, S., Madsen, B., & Hughes, M. (2012). Mechanical processing of bast fibres: The occurrence of damage and its effect on fibre structure. *Industrial Crops and Products, 39,* 7–11.
6. Hernandez-Estrada, A., Gusovius, H.-J., Müssig, J., & Hughes, M. (2016). Assessing the susceptibility of hemp fibre to the formation of dislocations during processing. *Industrial Crops and Products, 85,* 382–388.
7. Legros, S., Picault, S., & Cerruti, N. (2013). Factors affecting the yield of industrial hemp—Experimental results from france. In P. Bouloc, S. Allegret, & L. Arnaud (Eds.), *Hemp: Industrial production and uses*. doi:10.1079/9781845937935.0000.
8. Liu, M., Fernando, D., Daniel, G., Madsen, B., Meyer, A. S., Ale, M. T., et al. (2015). Effect of harvest time and field retting duration on the chemical composition, morphology and mechanical properties of hemp fibers. *Industrial Crops and Products, 6,* 29–39.
9. Madsen, B., & Gamstedt, K. (2013). Wood versus plant fibers: Similarities and differences in composite applications. *Advances in Materials Science and Engineering*, ID564346, 14.
10. Placet, V., Day, A., & Beaugrand, J. (2017). The influence of unintended field retting on the physicochemical and mechanical properties of industrial hemp bast fibres. *Journal of Materials Science, 52*(10), 5759–5777.
11. Placet, V., Cisse, O., & Boubakar, L. (2014). Nonlinear tensile behaviour of elementary hemp fibres. Part I: Investigation of the possible origins using repeated progressive loading with in situ microscopic observations. *Composites Part A: Applied Science and Manufacturing, 56,* 319–327.
12. Sponner, J., Toth, L., Cziger, S., & Franck, R. R. (2005). Hemp. In R. R. Franck (Ed.), *Bast and other plant fibres*. Boca Raton, USA.
13. Van de Weyenberg, I., Ivens, J., De Coster, A., Kino, B., Baetens, E., & Verpoest, I. (2003). Influence of processing and chemical treatment of flax fibres on their composites. *Composites Science and Technology, 63*(9), 1241–1246.
14. Virk, A. (2010). *Numerical models for natural fibre composites with stochastic properties*. Ph.D. thesis, Plymouth University.
15. Thygesen, A., Madsen, B., Bjerre, A. B., & Lilholt, H. (2011). Cellulosic fibers: Effect of processing on fiber bundle strength. *Journal of Natural Fibers, 8*(3), 161–175.

Use of Sugar Cane Fibers for Composites—A Short Review

Nívea Taís Vila, Ana Luiza Musialak and Alexandre Ferreira

Abstract Natural fibers are widely used in a wide range of applications. Each fiber has unique characteristics and for this, studies are made to know the best application for each fiber. In the case of sugar cane fiber, in addition of being almost unlimited, it is commonly wasted in countries that cultivate it, which characterizes a fiber of very low cost. This article aims to portray the main properties of sugar cane, directing its properties to application in natural fiber composites. As part of this purpose, researches are presented in which bagasse fibers were applied in composites, with different purposes, such as thermal, physical and mechanical study, as well as practical applications, mainly as reinforcement in cements.

Keywords Sugar cane · Composites · Natural fibers

Introduction

The large amount of waste generated from industrial processes has generated great discussions in the social and environmental scope. Thus, development of alternative materials aimed at better utilization of these wastes has been in focus.

Sugarcane bagasse is the remaining substrate of the *Saccharum officinarum* processing in alcohol and sugar production mills. This is the largest residue of the Brazilian agroindustry, and has been the object of studies that aim to take advantage of the great potential of this material.

This paper aims to highlight the developments that have been achieved within the composite with the application of bagasse from sugarcane.

The limitation of raw materials derived from non-renewable sources calls for innovation and adaptation to the available resources and which the specific prop-

N.T. Vila (✉) · A.L. Musialak
Departamento de Engenharia Têxtil, Universidade Estadual de Maringá, Goioerê, Brazil
e-mail: nivea980@hotmail.com

A. Ferreira
Departamento de Engenharias, Universidade Federal de Santa Catarina, Blumenau, Brazil

R. Fangueiro and S. Rana (eds.), *Advances in Natural Fibre Composites*,
https://doi.org/10.1007/978-3-319-64641-1_3

erties are suitable for needed applications. In this context, the natural fibers, specifically the fibers of sugarcane reported here, play an important role since they have peculiarities and are extremely abundant in Brazil and are often wasted.

Sugar Cane Fibers for Composites

As regards to the properties of natural fibers, they vary by several factors beginning with the origin and composition of the fiber. The main properties that stand out in the natural fibers are the low specific weight, resistance to the abrasion and fatigue, stiffness and the low cost related to the abundance [1]. In general, natural fibers are composed of cellulose, hemi-cellulose, lignin, pectin, waxes and water-soluble substances [2]. Each of these components gives distinct properties to the fiber.

Even with varying amounts in each fiber, cellulose is the largest component of the vegetable fibers. The type of cellulose, including its geometry, molecular structure and degree of polymerization, significantly affect the chemical, physical and mechanical properties [2].

As a natural fiber, sugar cane waste fibers are compounds formed mainly by cellulose. There are different fibers derivate from sugar cane and they are mostly composed by 65% of cellulose [3].

For studying the use of different sugar cane waste fibers in composites, Stael performed NMR and stress-strain measurements in a EVA/fiber composite. Results showed that the mobility of the three tested fibers decreased with the addition of EVA and that indicates that there was a good interaction with the fiber. However, there was no significant change in mechanical analyses, which can be related to the way that the fibers are dispersed in the polymer matrix. The study indicates that the type of sugar cane fiber used does not affect its use in mixtures and composites [3] (Fig. 1).

Tita used lignin extracted from sugar cane bagasse to substitute 40% in mass of phenol in phenolic and lignophenolic pre-polymers to produce composites with thermoset matrices. After the extraction of the bagasse and preparation of the pre-polymers, the thermoset and composites were performed by compression of both. The fracture surface can be seen in Fig. 2. The fibers were modified with Succinic Anhydride and Lignin/Formaldehyde and treatment with NaOH solution. These modification did not show any improvement in the impact resistance properties, however, MEV image reveled a greater interpenetration of the matrix in the fibers beam [4].

It was found that lignin can substitute partially phenol in phenolic matrix, without prejudicing the impact resistance. However, the addition of lignin increase water uptake, which can be a negative point in this method but does not necessarily compromise its use [4].

Studies in polyurethane-composites with bagasse of sugar cane were presented by Mothé et al. [5], consisting of preparation by the melting mixing method. The analyses of this composite were performed with thermogravimetric measurements

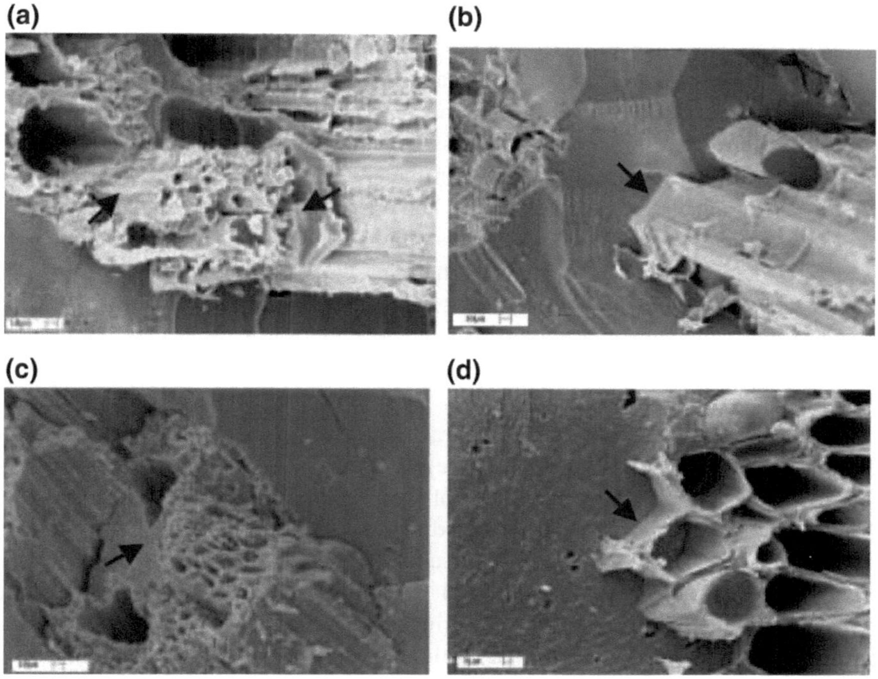

Fig. 1 Fracture surfaces of composites (1000×) **a** Fibers only extracted with cyclohexane/etanol; **b** fibers treated with 6% NaOH solution; **c** fibers only extracted with cyclohexane and **d** lignofenolic compounds reinforced with bagasse fibers subjected to ionized air treatment [4]

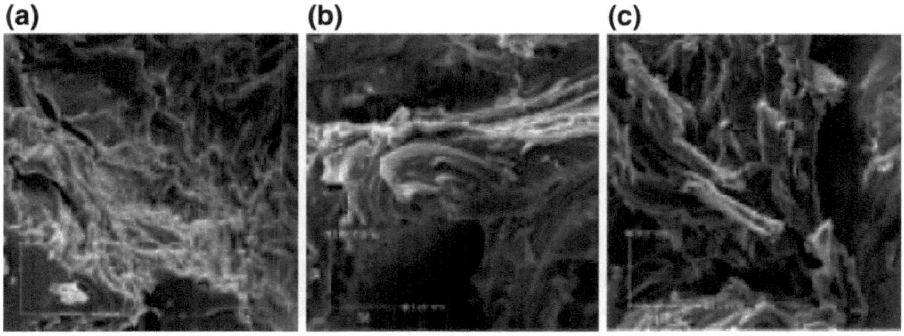

Fig. 2 SEM analysis of fracture region from tensile tests: **a** PU95% + BSC5%; **b** PU90% + BSC10% and **c** PU80% + BSC20%. *Magnification* 500× [5]

and microscopy. It was found that the thermal stability for different percentage of PU and BSC decreased when the percentage of BSC decreased. Therefore, the best performance regarding thermal stability was observed at a rate of 95%PU and 5% BSC [5].

Gilfillan, introduced starch and sugar cane fibre composites. Using potato and hydroxypropylated maize starch film produced by 'cast' and 'hot pressed' methods to evaluated physicochemical and mechanical properties for its use in composites with sugar cane. It was found that the hot pressed process was more adequate to prepare starch-fibre films, as this process changed the morphological aspects and improved the distribution of the fibers in the starch matrix as well as the thermal stability. Among the advantages in the hot pressed process, it is also stated less moisture uptake (5.5% less), more crystallinity (around 59%) and higher glass transition temperature and Young's modulus. Regarding the addition of the bagasse, enhancement in terms of lowering the moisture uptake and decreasing tensile strain were shown. Although it was not studied, it is suggested that grinded bagasse fibers would be a better option in terms of resistance improvement for starch films [6].

A composite created by lignin extracted from sugarcane bagasse and poly (3-hydroxybutyrate-co-3-hydroxyvalerate) (PHBV) was presented by Camargo. The lignin was obtained by extraction of steam-exploded sugarcane bagasse with NaOH, isolated at a low pH and separated in a press filter. The PHBV was supplied as it is. The composites were prepared in a mixer containing different proportions of lignin and PHBV. The results indicated that the glass transition temperature, the crystallization temperature and the melting temperature were not affected by the addition of lignin. The proportion of lignin over PHBV played an important role in the thermal decomposition profile, which is affected by the composite residual mass. Greater amounts of lignin entail an increase in residual mass. Furthermore, at a 50:50 (w/w) lignin/PHBV composites were stated to have no physical interaction leading to a decrease of mechanical properties. The fact that the lignin is obtained from renewable sources and at a low cost, generates a promising use of this composite where there is a call for biodegradability and no need of high mechanical properties [7].

Moubarik treated sugar cane in three different stages for its use as a reinforcement agent in low density polyethylene. The first stage consisted in using the air-dried sugar cane bagasse, previously cut into small pieces and treated with hot water for 2 h. This process was conducted aiming to extract the hemicellulose followed by an alkaline treatment to extract lignin. The third step was a bleaching with a sodium chlorite/glacial acetic acid mixture, removing the residual lignin and hemicellulose. This extraction process seemed to isolate cellulose fibers from sugar cane bagasse, with a great removal of lignin and hemicellulose indicated by FT-IR and CNMR spectrometry [8].

The composites were formed by adding Low Density Polyethylene (LDPE) in granules with sugarcane bagasse in an extruder screw. Different percentages of the bagasse and LDPE were tested. At a 25 wt% there was a notable increase of mechanical properties reaching 72% and flexural modulus increased by 85%, compared to LDPE alone [8] (Fig. 3).

Doost-hoseini tested sugar cane bagasse physical and mechanical properties in insulation boards. Three densities of homogeneous and multi-layered boards were produced. For the multi-layered boards, the fibers were separated in small and

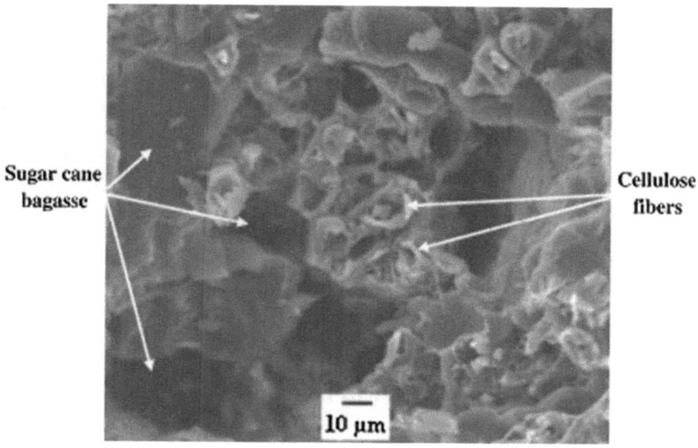

Fig. 3 SEM images of the tensile fractured surface of optimal cellulose fibers (20%)/LDPE composites [8]

large-sized. The boards were produced in hot press method, containing 12% of resin. The modulus of rupture (MOR) analysis indicated a better result for melamine-urea-formaldehyde (MUF) at a 0.5 g/cm^3 density, the higher tested. The lowest MOR was for the board with the lower density tested (0.3 g/cm^3) and made of urea-formaldehyde (UF). The sound absorption was observed to be very similar within the same board type, not varying much in terms of resin-type used. Yet, the resin-type influenced the acoustic of the board [9].

Several studies were performed utilizing sugar cane bagasse as cement reinforcement. Bilba performed a study in bagasse/cement composites targeting the performance due to the botanical components of the fibre, the thermal and chemical treatment applied, the fibre content and the water uptake. The composite samples were prepared mixing cement with bagasse fibres and water in different percentages of each material. In general, it was found that the addition of bagasse to the cement, retards the settling being less expansive in the mass. The thermal treatment used is expensive, therefore, making low cost material would need further investigation [10].

Onésippe investigated the thermal properties of cement composites reinforced with sugar cane bagasse fibres, ranging from 0.4 to 1 mm. The fibers went through a heat-treatment and a chemical treatment with alkaline solution. The matrix used was made of cement, sand, water and $CaCO_3$, bentonite, silica fume, acrylic styrene and cellulose pulp. Two composites were studied, one composed by the matrix and the rectified bagasse fibres (CBAGP) and one containing the matrix and the alkaline bagasse fibres (CBAGB). Results showed that CBAGP are weaker heat conductor materials, being prefered over CBAGB. Both CBAGP and CBAGB present a weaker specific heat than the matrix alone, exalting its application in construction materials. Figure 1 shows packed fragments of $CaCO_3$ onto the fibre, being the possible explanation for the loss of specific heat [11] (Fig. 4).

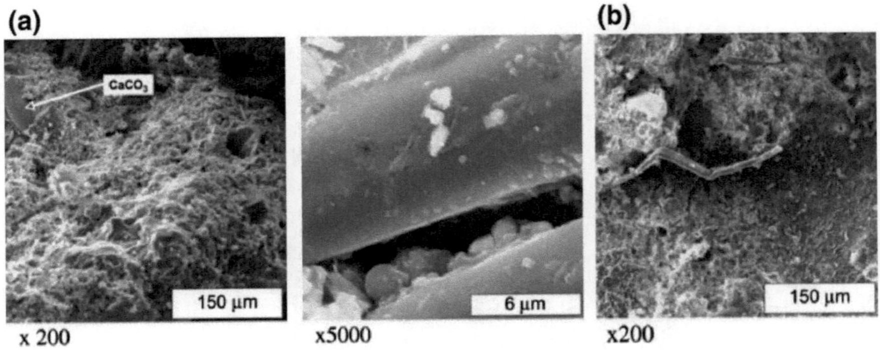

Fig. 4 SEM pictures of CBAGP **a** and CBAGB **b** with fibre content of 3% wrtc [11]

Frías correlated the activity of sugar cane bagasse ashes to pozzolans. This research was performed using three different sugar cane bagasse ashes (FBA, BBA and LBA) to investigate pozzolanic activity. The chemical evaluation of the three compounds shows the same oxides composition, being mainly silica and alumina. Per TG/DTA and FTIR analyses, quartz was the mineral most present in the ashes, and some minerals as cristobalite, kaolinite and gibbsite were not present in the three types of ashes. The calcining temperature evidenced that there is a difference in the morphology of the ashes. The results obtained for the pozzolanic activity, indicated excellent properties for the laboratory bagasse ashes, a low-medium activity for the FBA ashes and no activity for the boiler bagasse ashes. In prior to use the bagasse ashes, it is needed a reasonable management for the handling, remove of the containing materials and control of the calcining temperature of this waste [12].

De Soares studied the pozzolanic behavior of sugar cane bagasse by comparing it to amorphous and crystalline SiO_2. Grounded crystalline silica (quartz), amorphous silica (silica fume) and sugar cane bagasse ash (SCBA) used were consisted of SiO_2 by >94, >94 and 72.3% respectively. The reactivity of these three elements and other mixtures of silica fume and quarts was tested by a variation of electrical conductivity in calcium hydroxide solution. The conductivity tests indicated an increase in the reaction rate with an increase in silica fume content. The pozzolanic activity was investigated by immersing samples in lime solution, and the presence of C-S-H suggests pozzolanic activity in all the samples. Furthermore, cement pastes with partial replacement with SCBA, quarts and silica fume were tested, being noticed that samples prepared with SCBA showed reduced calcium hydroxide mass, close to quartz behaviour. The results suggest a better utilization of SCBA would be for inert constituents in cement composites [13].

Lamezon also investigated the Pozzolanic activity of SCBA in cementitious composites. This study was based on a comparison of SCBA from two different seasons to compare it. The first analysis of the ashes was performed using chemical composition analysis, grain size distribution, BET, FTIR, loss on ignition and X-ray

Table 1 Pozzolanicity tested in different methods [14]

Method	First harvest	Second harvest
Chemical composition and Luxan	X	–
Size distribution	X	X
DSC and modified Chapelle	–	X
PAI	–	–

diffraction. This analysis indicated different compositions and amorphicity on ashes from different harvests. However, that could be related to the incineration conditions. The pozzolanicity was evaluated by measuring the conductivity in a $Ca(OH)_2$ before and after adding the ashes. Other methods to evaluate pozzolanic effects was modified Chapelle and pozzolanic activity index (PAI). The Chapelle method consists in measuring the amount of CaO that reacts with the material after being boiled for 10 h. The PAI measures the pozzolanic degree in a cement a lime method. Table 1 shows the pozzolanicity of different harvest and different tests [14].

The author stated that this method to investigate the pozzolanic action was not consistent [14].

Sinab studied the mechanic behavior of reinforced concrete with sugar cane bagasse. Tests were performed in cement cylinders reinforced with sugar cane bagasse to determine its resistance. The bagasse particles used had 35% humidity and ranged between 15 and 25 mm. The fibers were treated with calcium hydroxide. Some samples are presented in Fig. 5. The concrete and fiber mixture was produced and arranged in cylindrical shape. The percentage of fiber in the mixture that presented a better result in terms of compressive strength was 0.5%. The addition of fibers in the cement compound has many advantages, such as lowering its density. However, increasing the fiber percentage on the cement lowers its resistance [15].

Fig. 5 Adhesion of fibers retained in sieves [15]

Conclusion

In this research, we studied the potential of fiber from sugar cane applied to composites. This short review was focused especially on composites targeted civil engineering. This study will serve as the basis for a research in this area of high interest.

It was verified by analysis of the bibliography given that many groups of researchers studied the possibility of these fibers in cement composites. Particularly the production of composites where the fibers are subject to chemical treatments, present promising results. Moreover composites were substantially studied according to their mechanical properties, but also according to other properties such as thermal properties. The use of sugarcane fibers presents itself as an excellent solution in the optimization of the weight/performance relation in fibers composite.

Acknowledgements The authors wish to thank the CENTRO DE TECNOLOGIA of UNIVERSIDADE ESTADUAL DE MARINGÁ for his support.

References

1. Jawaid, M., & Khalil, H. P. S. A. (2011). Cellulosic/synthetic fibre reinforced polymer hybrid composites: A review. *Carbohydrate Polymers, 86*(1), 1–18.
2. Bledzki, A., & Gassan, J. (1999). Composites reinforced with cellulose based fibres. *Progress in Polymer Science, 24*(2), 221–274.
3. Stael, G. C., D'Almeida, J. R. M., & Tavares, M. I. B. (2000). A solid state NMR carbon-13 high resolution study of natural fiber from sugar cane and their composites with EVA. *Polymer Testing, 19*(3), 251–259.
4. Tita, S., De Paiva, J. M. F., & Frollini, E. (2002). Resistência ao Impacto e Outras Propriedades de Compósitos Lignocelulósicos : Matrizes Termofixas Fenólicas Reforçadas com Fibras de Bagaço de Cana-de-açúcar (Vol. 12, pp. 228–239).
5. Mothé, C. G., De Araujo, C. R., De Oliveira, M. A., & Yoshida, M. I. (2002). *Thermal decomposition kinetics of polyurethane-composites with bagasse of sugar cane* (Vol. 67, pp. 305–312).
6. Gilfillan, W. N., Nguyen, D. M. T., Sopade, P. A., & Doherty, W. O. S. (2012). Preparation and characterisation of composites from starch and sugar cane fibre. *Ind. Crop. Prod., 40,* 45–54.
7. Camargo, F. A., Innocentini-Mei, L. H., Lemes, A. P., Moraes, S. G., & Duran, N. (2012). Processing and characterization of composites of poly (3-hydroxybutyrate-co-hydroxyvalerate) and lignin from sugar cane bagasse. *Journal of Composite Materials,* 46(4), 417–425.
8. Moubarik, A., Grimi, N., & Boussetta, N. (2013). Structural and thermal characterization of Moroccan sugar cane bagasse cellulose fibers and their applications as a reinforcing agent in low density polyethylene. *Composites Part B: Engineering, 52,* 233–238.
9. Doost-Hoseini, K., Taghiyari, H. R., & Elyasi, A. (2014). Correlation between sound absorption coefficients with physical and mechanical properties of insulation boards made from sugar cane bagasse. *Composites Part B: Engineering, 58,* 10–15.

10. Bilba, K., Arsene, M., & Ouensanga, A. (2003). *Sugar cane bagasse fibre reinforced cement composites. Part I. Influence of the botanical components of bagasse on the setting of bagasse/cement composite* (Vol. 25, pp. 91–96).
11. Onésippe, C., Passe-coutrin, N., Toro, F., Delvasto, S., Bilba, K., & Arsène, M. (2010). Composites: Part A Sugar cane bagasse fibres reinforced cement composites : Thermal considerations. *Composites Part A, 41*(4), 549–556.
12. Frías, M., Villar, E., & Savastano, H. (2011). Brazilian sugar cane bagasse ashes from the cogeneration industry as active pozzolans for cement manufacture. *Cement and Concrete Composites, 33*(4), 490–496.
13. De Soares, M. M. N. S., Garcia, D. C. S., Figueiredo, R. B., Teresa, M., Aguilar, P., & Cetlin, P. R. (2016). Comparing the pozzolanic behavior of sugar cane bagasse ash to amorphous and crystalline SiO_2. *Cement and Concrete Composites, 71,* 20–25.
14. Lamezon, P. G., Pádua, D., Panzera, T. H., Figueiredo, R. B., & Gomes, A. M. (2016). *Investigations on the pozzolanic effect of sugar cane bagasse ashes used in cementitious composites* (pp. 395–405).
15. Sinab, A. D. E. L., Alexander, J., & Saraz, O. (2016). http://www.revistas.unal.edu.co, pp. 1–8.

Evaluation of the Extraction Efficiency of Enzymatically Treated Flax Fibers

Jana De Prez, Aart Willem Van Vuure, Jan Ivens, Guido Aerts and Ilse Van de Voorde

Abstract This study evaluates the efficiency of the extraction of enzymatically treated flax fibers for application in composite materials. Three extraction parameters were introduced. Fiber Efficiency (FE) represents the yield of long fibers extracted after enzymatic treatment. Time Efficiency (TE) is introduced in order to evaluate the ease of extraction. Taking into account both Fiber Efficiency and Time Efficiency results in the introduction of the overall Extraction Efficiency (EE). The Extraction Efficiency is determined for flax fibers treated with pectate lyase, pectin lyase, polygalacturonase, endoxylanase and Viscozyme L as reference while comparing with green and dew retted fibers. Moreover, fiber fineness and transverse properties of the resultant composite materials were investigated. Polygalacturonase treatment appears to be the most promising. Results demonstrate the potential of enzymatic extraction of flax fibers with an increased transverse bending strength of the composite reinforced with polygalacturonase treated fibers (26.4 ± 3.8 MPa) compared to green flax fiber composite (12.1 ± 1.4 MPa) and dew retted fiber composite (16.9 ± 2.2 MPa).

Keywords Flax fibers · Enzymatic extraction · Fineness · Mechanical properties

---

J. De Prez (✉) · G. Aerts · I. Van de Voorde
Faculty of Engineering Technology, Department of Microbial and Molecular Systems (M²S), Cluster for Bioengineering Technology (CBeT), Laboratory of Enzyme and Fermentation Technology, KU Leuven, Ghent, Belgium
e-mail: jana.deprez2@kuleuven.be

A.W. Van Vuure · J. Ivens
Faculty of Engineering Technology, Department of Materials Engineering (MTM), Technology Cluster for Materials Technology (TC-MT), KU Leuven, Leuven, Belgium

R. Fangueiro and S. Rana (eds.), *Advances in Natural Fibre Composites*,
https://doi.org/10.1007/978-3-319-64641-1_4

Introduction

Flax fibers are one of the most promising fibers amongst natural fibers as reinforcement in composite materials, thanks to their high specific mechanical properties due to their high content of crystalline cellulose and small microfibrillar angle [8, 11]. Natural fibers are biorenewable, biodegradable and have a lower carbon footprint compared to synthetic fibers [12, 15]. Natural fibers show thus an interesting and eco-friendly alternative for glass fibers for use in composite materials. Natural fibers however need to be extracted from the plant. Traditionally, extraction occurs by dew retting followed by a mechanical extraction process. With dew retting, full-grown flax is harvested and laid down on the field during several weeks, in order to let microorganisms and moisture from dew and rain affect the flax stem. Through this process, fibers are loosened from the flax stem and separated from each other. After dew retting, flax plants go through a breaking, scutching and hackling process to completely separate fibers from shives, the woody particles coming from the flax stem. First of all, the dew retting process is extremely dependent on weather conditions and will result in an inconsistency in fiber quality. The mechanical extraction process will enhance the fineness and purity of the fibers but results however in fiber damage due to the extensive treatment [14].

In order to overcome drawbacks from the traditional dew retting method, enzymatic extraction has been suggested as an alternative. The use of enzymes will make it possible to selectively interact with the components responsible for holding the fibers together in the flax stem. A successful enzymatic treatment will separate fibers almost completely from the flax stem, which results in a less extensive mechanical treatment necessary and hence, less fiber damage. An enzymatic treatment can only be called successful if certain requirements are fulfilled. First of all, the long fiber yield resulting from the treatment and the time efficiency of the extraction are important aspects. Evaluation of the extraction efficiency of the enzymatic treatment is hence an important step to validate its potential. Other parameters that should be considered are the fineness of the fibers and the mechanical properties of the final composite material. This study evaluates the effect of flax treatment with pure enzymes, i.e. pectate lyase, pectin lyase, polygalacturonase and endoxylanase, compared to treatment with the commercial enzyme mixture Viscozyme L as reference, by determining the extraction efficiency, the fiber fineness and mechanical properties of the final composite material. The extraction efficiency is determined by introducing three new parameters. By taking all these parameters into account, in-depth knowledge is achieved, showing the potential of applying an enzymatic pretreatment process.

Materials and Methods

Enzymatic Treatment

Flax samples were provided by Verhalle from Belgium. Green flax of the Amina cultivar was harvested in Belgium in 2015. Enzymatic treatments were performed on whole flax stems dried during 24 h at 105 °C. 50 g of dried flax stem segments of 25–30 cm were incubated with an enzyme formulation containing 25 mM EDTA (pH 5.0). All treatments were performed during 24 h in a shaking incubator at 40 °C. The flax to volume ratio used during treatments was 1:20 (g:ml). Enzyme concentrations of 0.30 v/v % were used in case of pectate lyase, pectin lyase and endoxylanase treatment, with exception of 0.60 v/v % for Viscozyme L. Pectate lyase (Scourzyme L), pectin lyase (NS59049) and endoxylanase (Pulpzyme HC 2500) are pure enzyme solutions provided by Novozymes. Viscozyme L is a commercially available enzyme mixture from Novozymes containing hemicellulase, xylanase, arabanase, cellulase and β-glucanase activities and is included for comparison with the pure enzymes. Polygalacturonase is originating from *Aspergillus niger* and was purchased from Sigma-Aldrich. For polygalacturonase treatment, an enzyme solution was made according to solubility by dissolving 0.50 w/v % enzyme powder in 0.1 M acetate buffer (pH 5.0). Since most liquid enzymes are present in enzyme formulations at an average of 1 w/v %, polygalacturonase concentration of 0.60 v/v % was chosen instead of 0.30 v/v %. After enzymatic treatment, flax stems were submerged twice in cold tap water baths to remove enzymes and solubles. Flax stems were then dried at 105 °C for 24 h.

Extraction Method

Since mechanical extraction has a big impact on the fiber quality, strength and fineness, a careful manual extraction was applied to evaluate the effect of solely the enzymatic treatment. This way, other effects from mechanical extraction were eliminated. Fibers were manually extracted by pulling the fibers of the stem from top to bottom. The collected fibers were separated into long and short fibers prior to determining the weight of each fraction. The time needed to manually extract a certain amount of flax fibers was registered. Based on these data, the extraction efficiency parameters can be calculated.

Evaluation of Extraction Efficiency

In order to evaluate the extraction, three parameters are introduced. The first parameter, Fiber Efficiency (FE), describes the long fiber yield extracted from the plant in accordance to Eq. (1).

$$\mathrm{FE} = \frac{\text{Amount of long fibers extracted (g)}}{\text{Total amount of flax stems (g)}} \quad (1)$$

The amount of long technical fiber bundles available after extraction is important since only long fibers can be used for unidirectional composite materials. A low Fiber Efficiency would indicate a too extensive enzymatic treatment and thus degradation of the technical fiber and its strength.

The second parameter, Time Efficiency (TE), includes the time needed for the extraction and thus illustrates the ease of the extraction (Eq. 2). The formulation of this equation leads to a TE of 100% for a very easy extraction in the chosen lab conditions.

$$\mathrm{TE} = \frac{\text{Amount of long fibers extracted (g)}}{\text{Time needed for extraction (min)}} * 5 \quad (2)$$

A lower Time Efficiency reflects a more difficult extraction of fibers being not sufficiently separated from the flax stem. In contrast to the FE, a Time Efficiency of 100% is feasible thanks to the introduction of a factor of 5 into the equation. This factor illustrates a reference extraction, where 10 g of flax stems are manually extracted in 50 min.

Taking into account both Fiber Efficiency and Time Efficiency results in the introduction of the overall Extraction Efficiency, illustrated in Eq. (3).

$$\mathrm{EE} = \mathrm{FE} * \mathrm{TE} \quad (3)$$

The EE value enables us to compare different enzymatic treatments according to long fiber yield and extraction time. With a maximal EE value, the fibers are completely loosened from the flax stem so the woody core can easily be removed without pulling off the fibers from the stem.

Fineness Measurement

Fineness of fibers is determined with an automated image analysis method by measuring fiber width. Fibers (n = 35) are scanned with a resolution of 2400 dpi. The output of this method is a distribution curve of fiber fineness. The image from the scanner is processed using an algorithm which was developed by the

Department of Materials Engineering (MTM), KU Leuven in cooperation with prof. Toon Goedemé of the Department of Electrical Engineering, KU Leuven [10].

Production of Composite Materials

The unidirectional composite samples are produced with the Vacuum Assisted Resin Infusion method (VARI). Depending on the dimensions of the composite sample, a certain amount of dried flax fibers is weighed in order to produce a composite material with a fiber volume fraction of 40%. Fibers are carefully brought into the composite mold in order to produce a well aligned unidirectional composite. Spacers of 2 mm are used to ensure the composite sample will have the preferred thickness. Mold tools fitting in the mold cavities are laid upon the fibers. Sealant tape is applied around the whole mold in order to be able to apply vacuum with the vacuum bag. A resin distribution medium is placed at the resin inlet to guide the resin into the mold cavities containing fibers. At the outlet tube, breather fabric is added to retain the resin from going to the outlet immediately. The mold is covered with peel ply before closing it with the vacuum bag and placed on a heating plate.

The epoxy resin (Epikote 828LVEL) used for composite production is Bisphenol A diglycidyl ether and is mixed with 1,2-diaminocyclohexane (Dytek DCH-99) with a weight ratio of resin to hardener of 100:15.2. Before impregnation, the resin is degassed during 30 min to eliminate small gas bubbles. When 95% vacuum is reached, resin inlet is opened and the mold is filled until resin almost reaches the outlet tube. Resin infusion occurs at a temperature of 40 °C to ensure optimal flow. When impregnation is completed, curing is enhanced by raising the temperature to 70 °C which is held during 1 h. A post-curing step at 150 °C during another hour is required to complete composite production.

Determination of Transverse Bending Properties

Mechanical properties of composite samples are tested by performing three point bending tests using an Instron 5567 with a load cell of 1 kN. The crosshead displacement rate is 0.5 mm min^{-1}. The loading nose has a diameter of 6 mm while support rolls of 4 mm diameter are used. Samples have a thickness of 2 mm and width of 13 mm and are tested with a span length of 16 mm. These dimensions are different from the recommended span-to-thickness ratio of minimum 32 in ASTM D7264-07 [4]. The smaller ratio was chosen to ensure a high level alignment of the fibers in the test specimens. Tests were repeated six times for each composite type.

Results and Discussion

Evaluation of Enzymatic Treatments

In order to determine the potential of an enzymatic extraction procedure to substitute the dew retting process, assessment of the extraction efficiency is an important first step. For this purpose, the parameters Fiber Efficiency (FE), Time Efficiency (TE) and Extraction Efficiency (EE) were introduced as described in Materials and Methods. These parameters were determined for flax fibers treated with Viscozyme L, pectate lyase, pectin lyase, polygalacturonase and endoxylanase and are compared with the efficiency of fiber extraction for dew retted fibers as well as green fibers. Results of the efficiencies are shown in Table 1.

The Fiber Efficiency of the enzymatic treatments amounts to 33–34%. It should be noted that since the amount of long fibers present in flax is limited to circa 39% for Amina flax, the FE value cannot exceed this percentage. An FE value of 39% would mean only long fibers and no short fibers are collected after extraction. Depending on cultivar, the FE value is limited to the fiber percentage, which normally ranges between 20 and 35% [13]. To the best of our knowledge, no other reference in literature was found denoting the exact amount of fibers in that type of cultivar. Compared to the dew retted flax, which has a lower FE value of 24%, an improvement is definitely observed. A possible explanation for the low FE of dew retted flax could be because of a too high degradation of the long fibers during dew retting, resulting in a higher amount of short fibers. In contrast to enzymatic treatments, the dew retting process is uncontrollable leading to a possible degradation of the fiber itself.

Not only the long fiber yield but also the ease of extraction is evaluated by determining the time needed to extract the long fibers from the initial amount of flax stems. Dew retted flax is extracted with a Time Efficiency of only 35%. Treatment of flax stems with pectate lyase did not improve the Time Efficiency (35%) compared to dew retted flax. Endoxylanase treatment resulted in an increase in TE towards 44%, as well as Viscozyme L and pectin lyase treatment, which resulted both in a TE of 53%. Green fibers on the other hand were extracted with a Time Efficiency of 60%. It should be noted that the Time Efficiency illustrates the ease of extraction, but does not give an indication of the fineness of the extracted fibers. In

Table 1 Extraction parameters for enzymatically treated flax fibers compared to green and dew retted fibers

Treatment	FE (%)	TE (%)	EE (%)
Green fiber	35	60	21
Dew retted	24	35	8
Viscozyme L	34	53	18
Pectate lyase	33	35	11
Pectin lyase	34	53	18
Polygalacturonase	34	81	28
Endoxylanase	33	44	15

case of green flax fibers, extraction will result in technical fiber bundles that are not fibrillated as much as desired which explains the high TE for green fibers. One enzymatic treatment showed a higher TE compared to the time needed to extract the green fibers, namely treatment with polygalacturonase. Time Efficiency amounted to 81%, which indicates that the enzymatic treatment resulted in an almost complete separation of the fibers from the woody core.

Comparison of the Extraction Efficiencies confirms that dew retting is the least successful method (8%) while polygalacturonase treatment definitely resulted in a substantial improvement with the highest EE (28%) by producing the highest amount of long fibers in the shortest extraction timeframe.

Earlier research illustrated the importance of pectin degradation in enzymatic retting [1–3, 7, 9, 18]. Pectin degrading enzymes consist of e.g. pectin lyase, pectate lyase, polygalacturonase and pectinmethylesterase. Pectin lyase attacks methyl esterified pectin and results in eliminative cleavage of an 4,5-unsaturated galacturonide [17]. Pectate lyase results in a similar cleavage but prefers the unesterified pectate as substrate. Polygalacturonase is a pectinase that catalyses hydrolysis of α-1,4-polygalacturonic acid and thus degrades the pectin polymer in smaller compounds [6]. Endopolygalacturonase hydrolyses linkages randomly in the homogalacturonan polymer while exopolygalacturonase separates galacturonic acid from the non-reducing end of the polymer [16]. The enhanced TE of polygalacturonase treated fibers could indicate that homogalacturonan is the main component of the pectin fraction in flax responsible for holding the flax fibers together at the stem. However, endopolygalacturonase randomly hydrolyses the homogalacturonan polymer while the lyase enzymes gradually attack pectin, so polygalacturonase can be able to degrade the polymer more rapidly and thus have a bigger effect. Endoxylanase on the other hand is a hemicellulose degrading enzyme and hydrolyses β-1,4-glycosidic bonds in xylan from the interior [6]. The limiting effect on Extraction Efficiency of endoxylanase treatment could be due to the location of xylans. According to Bourmaud et al. [5], xylans are mostly present in the secondary cell wall, accompanied by glucomannans, while in the primary cell wall mainly xyloglucan appears. It should be investigated if degradation of xyloglucan by xyloglucanases has a more profound impact on extraction efficiency.

As mentioned before, enzymatic treatments cannot be evaluated based on only the Extraction Efficiency. Fineness measurements will further elucidate the effect of different enzymatic treatments.

Fineness of Extracted Fibers

In order to gain more insight into the effect of the different enzymatic treatments compared to dew retting, fineness of extracted fibers is also evaluated. Results of fiber fineness measurements with automated image analysis are shown in Table 2.

Fiber fineness measurements indicate an average fiber width of 218 μm for green fibers, while dew retted fibers have a fineness average of 140 μm.

Table 2 Fiber fineness measurement with automated image analysis

Treatment	Average fiber width (μm)	Minimum (μm)	Maximum (μm)
Green fiber	218	42	974
Dew retted	140	42	762
Viscozyme L	105	42	1143
Pectate lyase	160	42	1059
Pectin lyase	172	42	1080
Polygalacturonase	156	42	974
Endoxylanase	190	42	1292

Viscozyme L treatment resulted in fibers with a higher fineness of 105 μm, probably due to the multiple enzyme activities present in the enzyme mixture. Pure enzyme treatments yielded fibers with a lower fineness than dew retted fibers, but still higher than green fiber fineness. Pectate lyase and polygalacturonase treatment produced the finest fibers (160 and 156 μm), while pectin lyase and endoxylanase yielded fibers of 172 and 190 μm, respectively. Results of the fineness of extracted fibers confirm the potential of polygalacturonase treatment, which not only loosens the fibers from the flax stem but also separates the fibers from each other slightly better.

The minimal fiber width measured was 42 μm for each fiber. Maximum fiber width was lowest for dew retted fibers (762 μm), followed by green fibers and fibers treated with polygalacturonase (both 974 μm). Other enzymatic treatments resulted in a maximum fiber width from 1059 to 1292 μm indicating some fiber bundles were insufficiently treated. Distribution curves are included in Fig. 1.

Distribution curves show that in case of green fibers, pectin lyase and endoxylanase treated fibers the relative frequency gradually decreases with increasing fiber width. Dew retted fibers, Viscozyme L, pectate lyase and polygalacturonase treated fibers on the other hand show a higher amount of fine fibers measured with the automated image analysis. Especially treatment with Viscozyme L resulted in a high relative frequency of the finest fibers.

Transverse Bending Properties of Composites

Three point bending tests were performed on the composite materials reinforced with the enzymatically treated fibers. Figures 2 and 3 illustrate the transversal bending strength and bending modulus of the resultant composite materials compared to composite materials reinforced with green and dew retted fibers.

Results in Fig. 2 show an increase of transverse bending strength for all composites impregnated with enzymatically treated fibers compared to composites reinforced with green fibers and dew retted fibers, with exception of the composite based on endoxylanase treated fibers. Fibers treated with Viscozyme L resulted in a composite material with a bending strength of 27.7 ± 2.0 MPa. Polygalacturonase

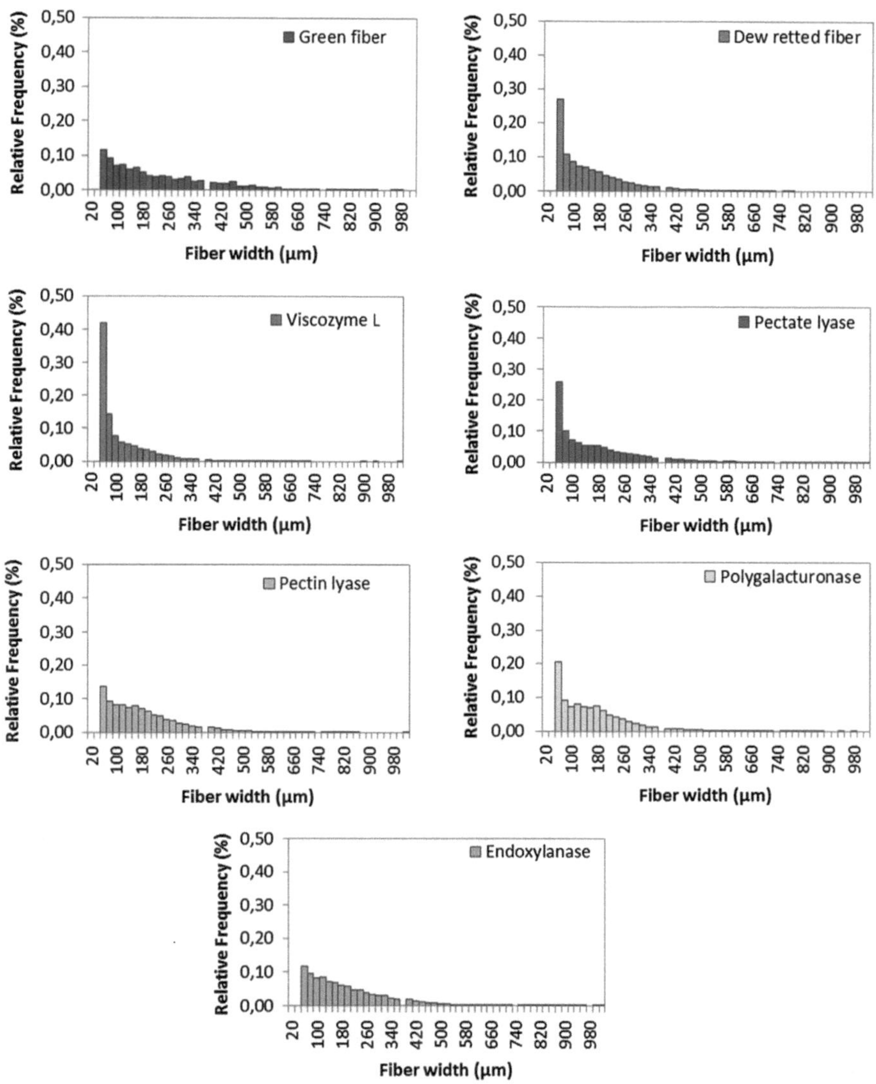

Fig. 1 Fiber width distributions of enzymatically treated flax fibers compared to *green* and *dew* retted fibers (Color figure online)

treatment succeeded to produce fibers resulting in a composite material with a strength of 26.4 ± 3.8 MPa, which is an improvement of 2.2 times the strength of the green fiber composite.

Pectate lyase and pectin lyase treatment also led to an enhancement of transverse bending strength compared to both green and dew retted fiber composites. Future experiments will indicate if the interfacial bonding between fiber and matrix is improved or if the internal strength of the technical fiber is enhanced due to the

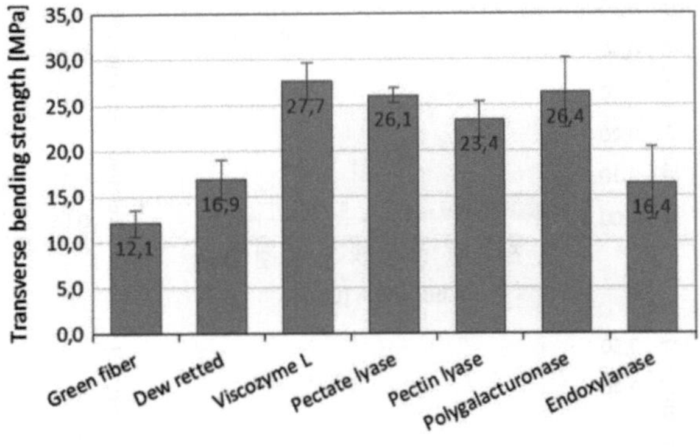

Fig. 2 Transverse bending strength of composites reinforced with enzymatically treated fibers

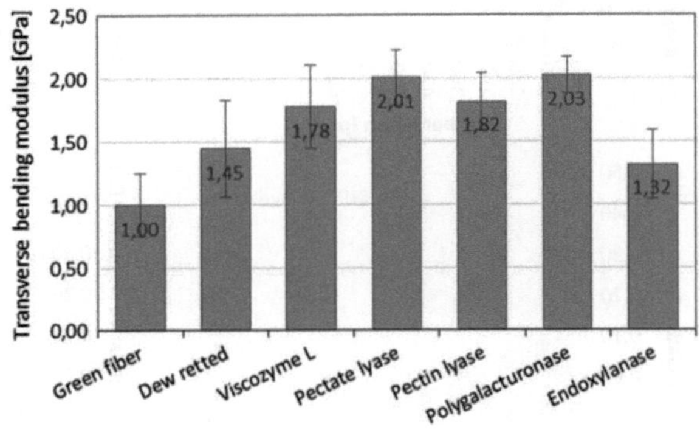

Fig. 3 Transverse bending modulus of composites reinforced with enzymatically treated fibers

enzymatic treatment. Most likely the fiber-matrix interface is improved, which should be verified by investigating the fracture surface and microscopy. Endoxylanase treatment on the other hand resulted in a composite material with a bending strength of 16.4 ± 4.0 MPa, which corresponds to an equivalent strength as the dew retted composite (no significant difference with $P = 0.78$). The limited effect of endoxylanase and pectin lyase treatment could be correlated to the amount of possibilities where these specific enzymes can attack and split. The fineness of the fibers treated with pectin lyase and endoxylanase was also lower than the other enzymatic treatments, which corroborates the lower transverse bending strength of the resultant composites. Increasing these enzyme concentrations probably would not improve the effect but combining them with for example polygalacturonase could lead to an enhancement and will be tested in further research.

Even though dew retted fibers have an average fiber width of only 140 μm, transverse bending strength of the composite only amounts to 16.9 ± 2.2 MPa. This could be due to impurities left on the dew retted fiber, which result in stress in the composite leading to an negative impact on the transverse properties.

Figure 3 shows an increase in bending modulus compared to the green fiber composite. Composites impregnated with pectate lyase treated fibers and polygalacturonase treated fibers show the highest bending modulus of 2.01 ± 0.22 GPa and 2.03 ± 0.14 GPa, respectively, along with composites impregnated with fibers after Viscozyme L and pectin lyase treatment. No significant difference was observed for the bending modulus of these composites ($P = 0.22$). Endoxylanase treatment did not result in a composite with a significant higher bending modulus compared to the green fiber composite ($P = 0.09$). Transverse properties are thus confirming the promising results for the enzymatic treatment of flax fibers with polygalacturonase and pectate lyase.

Taking into account Extraction Efficiencies, fineness of fibers and the transverse bending properties of the resultant composite materials, enzymatic treatment with Viscozyme L and polygalacturonase appear to be the most promising. The variety of enzyme activities present in the commercially available enzyme mixture Viscozyme L does not provide the knowledge of the contribution of each separate enzyme activity towards the separation of flax fibers from the stem or from each other. In order to understand the role and underlying mechanisms, pure enzymes are tested for their contribution towards the Extraction Efficiency, fineness of resulting fibers and transverse bending properties of resultant composites. Viscozyme L treatment resulted in the finest fibers (105 μm). However, polygalacturonase treatment resulted in fibers of 156 μm and in a composite material with a transverse bending strength and modulus as good as the composite material based on Viscozyme L fibers (no significant difference with $P = 0.47$). With the highest Extraction Efficiency of 28%, polygalacturonase treatment can be appointed as adequate alternative for the Viscozyme L treatment and illustrates that degradation of pectin by polygalacturonase is a major prerequisite for enzymatic extraction of flax fibers. The pectate lyase enzyme treatment also resulted in a similar fineness as polygalacturonase treatment and in no significant different composite properties ($P = 0.86$), but Extraction Efficiency only amounted to 11%. Likewise, composites reinforced with pectin lyase fibers did not significantly differ from polygalacturonase fiber composites ($P = 0.13$) but fineness was slightly lower (172 μm) with an Extraction Efficiency of 18%. Even though polygalacturonase treatment appears to be the most promising, pectate lyase and pectin lyase also have some potential for enzymatic extraction of flax fibers. Finally, endoxylanase treatment did not result in a composite performance as good as the other composites with enzymatically treated fibers. Degradation of xylans does not seem to be essential in the extraction of flax fibers but can play an important role when investigating moisture resistance of the fiber or the composite. The improved transverse properties of the other composites impregnated with enzymatically treated fibers is probably due to an improved interface between fiber and matrix. However, combining enzymes can result in synergy wherein endoxylanase can play an important role. Further research should investigate well-defined enzyme mixtures. To fully understand the role of enzymatic treatments, further determination of chemical components, morphological characterization and determination of mechanical properties like tensile strength is essential.

Conclusions

For introducing an alternative method for the traditional dew retting method, evaluating enzymatic extractions by determining extraction efficiency, fineness of fibers and composite performance is essential in order to validate its potential. Evaluation of the Extraction Efficiency showed that polygalacturonase treatment can enhance the ease of extraction by loosening fibers from the woody core. Fibers were obtained with an average fiber width of 156 μm. Transverse bending properties of composites were significantly improved when impregnated with pectate lyase and polygalacturonase treated fibers. On the other hand, Extraction Efficiency achieved with pectate lyase was markedly lower. Polygalacturonase treatment appears to be the most promising treatment taking into account Extraction Efficiency as well as fineness and transverse properties. Further characterisation of chemical properties, morphological properties and other mechanical properties like tensile strength is designated to completely understand the effect of enzymatic treatments in each aspect.

Acknowledgements The authors gratefully acknowledge sources of funding by VLAIO, the Flemish Agency for Innovation and Entrepreneurship (Vlaams Agentschap Innoveren & Ondernemen). They would also like to thank Elmar Janser from Novozymes for providing enzyme samples and flax company Verhalle, Belgium for supplying flax samples.

References

1. Akin, D. E., Condon, B., Sohn, M., Foulk, J. A., Dodd, R. B., & Rigsby, L. L. (2007). Optimization for enzyme-retting of flax with pectate lyase. *Industrial Crops and Products, 25,* 136–146.
2. Akin, D. E., Slomczynski, D., Rigsby, L. L., & Eriksson, K.-E. L. (2002). Retting flax with endopolygalacturonase from *Rhizopus oryzae*. *Textile Research Journal, 72*(1), 27–34.
3. Alix, S., Lebrun, L., Marais, S., Philippe, E., Bourmaud, A., Baley, C., et al. (2012). Pectinase treatments on technical fibres of flax: Effects on water sorption and mechanical properties. *Carbohydrate Polymers, 87*(1), 177–185.
4. ASTM. (2010). Standard test method for flexural properties of polymer matrix composite materials 1. *Annual Book of ASTM Standards, i,* 1–11.
5. Bourmaud, A., Morvan, C., Bouali, A., Placet, V., Perré, P., & Baley, C. (2013). Relationships between micro-fibrillar angle, mechanical properties and biochemical composition of flax fibers. *Industrial Crops and Products, 44,* 343–351.
6. Chen, H. (2014). *Biotechnology of lignocellulose: Theory and practice*. Beijing: Chemical Industry Press, Springer.
7. Evans, J. D., Akin, D. E., & Foulk, J. A. (2002). Flax-retting by polygalacturonase-containing enzyme mixtures and effects on fiber properties. *Journal of Biotechnology, 97*(3), 223–231.
8. Foulk, J., Akin, D. E., Dodd, R., & Ulven, C. (2011). Production of flax fibers for biocomposites. In S. Kalia, B. S. Kaith, & I. Kaur (Eds.), *Cellulose fibers: Bio- and nano-polymer composites. Green chemistry and technology* (pp. 61–95). Berlin, Heidelberg: Springer.
9. George, M., Mussone, P. G., & Bressler, D. C. (2014). Surface and thermal characterization of natural fibres treated with enzymes. *Industrial Crops and Products, 53,* 365–373.

10. Hendrickx, K., Romian, R. D., Goedemé, T., Van Vuure, A. W., & Ivens, J. (2015). Rapid and effective methods for the screening of flax fibres for composite applications. In *Proceedings of the 20th International Conference on Composite Materials* (pp. 1–9). Copenhagen, Denmark.
11. John, M. J., & Thomas, S. (2008). Biofibres and biocomposites. *Carbohydrate Polymers, 71,* 343–364.
12. Le Duigou, A., Davies, P., & Baley, C. (2011). Environmental impact analysis of the production of flax fibres to be used as composite material reinforcement. *Journal of Biobased Materials and Bioenergy, 5*(1), 153–165.
13. Stephens, G. R. (1997). *Connecticut Fiber Flax Trials 1994-95, Bulletin 946*. The Connecticut Agricultural Experiment Station, New Haven, October.
14. Van de Weyenberg, I. (2005). *Flax fibres as a reinforcement for epoxy*. Ph.D. thesis, KU Leuven, Belgium.
15. Van Vuure, A. W., Ivens, J., Verpoest, I., Fuentes, C., Osorio, L., Trujillo, E., … Depuydt, D. (2015). Natural fibre composites: Research highlights of the performance of bamboo, flax and hemp fibres. In R. Fangueiro (Ed.), *2nd International Conference on Natural Fibers* (pp. 1–5). Azores, Portugal.
16. Voragen, A. G. J., Coenen, G. J., Verhoef, R. P., & Schols, H. a. (2009). Pectin, a versatile polysaccharide present in plant cell walls. *Structural Chemistry, 20*(2), 263–275.
17. Yadav, S., Yadav, P. K., Yadav, D., & Yadav, K. D. S. (2009). Pectin lyase: A review. *Process Biochemistry, 44,* 1–10.
18. Zhang, J., Henriksson, G., & Johansson, G. (2000). Polygalacturonase is the key component in enzymatic retting of flax. *Journal of Biotechnology, 81,* 85–89.

Influence of Coupling Agent on the Properties of Polypropylene Composites Reinforced with Palm Fibers

Ingriddy R. Dantas, Noelle C. Zanini, Joyce P. Cipriano, Maria R. Capri and Daniella R. Mulinari

Abstract This chapter is focused on the development of palm fiber reinforced polypropylene (PP) composites by using thermokinetic mixer followed by injection molding. The effect of concentration (5 wt%) of coupling agent, Epolene (MA-g-PP) and pre-treatment with alkaline solution (10% w/v) was studied in details on palm fiber reinforced PP composites. Mechanical testing (tension and flexural) was carried out to determine the effect of fiber contents, treatment and compatibilizer. The chemical, physical and morphological properties of the fibers were examined by SEM, chemical composition, XRD and FTIR. The results showed that compatibilized composites have increased stiffness due to enhanced interfacial adhesion between the fibers and the matrix, as well as better homogeneity (better fiber dispersion) due to chemical bonding.

Keywords Palm fibers · Polypropylene · MA-g-PP · Alkaline treatment
Mechanical properties · Microscopy analysis

Introduction

Renewable materials are very interesting in many applications since they are environmentally friendly and naturally available [1]. Because of this, the use of natural fibers as reinforcement in polymers has experienced a tremendous growth in the study and development of composite materials used in structural and semi-structural applications [2].

I.R. Dantas · N.C. Zanini · J.P. Cipriano · D.R. Mulinari (✉)
Departamento de Mecânica e Energia, Universidade do Estado do Rio de Janeiro/UERJ, Resende-RJ, Brazil
e-mail: dmulinari@hotmail.com

M.R. Capri
Departamento de Engenharia Química, Universidade de São Paulo/EEL/USP, Lorena-SP, Brazil

R. Fangueiro and S. Rana (eds.), *Advances in Natural Fibre Composites*,
https://doi.org/10.1007/978-3-319-64641-1_5

Natural fibers have the advantages compared with traditional reinforcement material such as glass fiber, such as, low cost, low density, flexibility during processing, possibility of environmental protection and good mechanical properties [3–6]. The substitution of glass fibers by natural fibers has a lot of advantages that can be justified through ecological balances [7]. Besides, agro-industrial residues are generated in large scale and the proposal of re-using them is very interesting [8]. Among various natural fibers, palm fibers present advantages owing to its abundance and cost once it is a palmetto industry byproduct. Brazil is responsible for supplying 95% of palmetto to the external market, causing an increase in palmetto consumption which led to its disorderly extraction and to the extinction of palm native species.

In order to reduce predatory extraction and preserve native species, the growth of non-native species such as the Australian King Palm (*Archontophoenix alexandrae*) was started [9]. The palmetto can be extracted from several palm species, but the *Archontophoenix alexandrae* produces noble palmetto species [10]. For each extracted palm, approximately 400 g of commercial palmetto is obtained and the waste generated consists of 80–90% of the palm's total weight. However, natural fibers have poor compatibility with polymers, causing problems in the composite processing [11–13]. To solve this problem, various surface treatment methods as well as coupling agents and compatibilizers such as maleic anhydride grafted polymers, silane, etc. have been used to increase the compatibility between natural fibers and thermoplastic matrices, enhancing the composites performance [14–18]. These methods are usually based on the use of reagents, which contain functional groups that are capable of reacting and forming chemical bonds with the hydroxyl groups of the lignocellulosic material. A coupling agent acts as an interface between the reinforcement and matrix. Polypropylene grafted with maleic anhydride (MA-g-PP) has been used as a coupling agent in the composite preparation. The MA segment of the copolymer can react with the hydroxyl groups (OH) on the fiber surface. MA-g-PP has a good interfacial adhesion with PP phase, because PP is able to diffuse into the PP phase through micelle formation.

The objective of this work was to evaluate the potential of king palm fiber as the filler in composites using PP as matrix and also using MA-g-PP as the coupling agent, in order to improve interfacial fibers/matrix adhesion.

Experimental

Short cut raw Palm fibers (*Archontophoenix alexandrae*) were kindly donated by Biosolvit company. These fibers were dried at 80 °C for 24 h. They were then ground in a mill and sieved to obtain samples that passed through 40 mesh screens. Palm fibers were pre-treated with alkaline solution (10% w/v) for an hour under constant stirring at room temperature. After this treatment, the solution was filtered in vacuum filter and fibers were washed with distilled water. Then fibers were dried in an oven at 50 °C for 24 h. Polypropylene (PP) was used as the thermoplastic matrix. The grade used was H503 from Braskem with a density of 0.905 g/cm^3 and a melting temperature of 155 °C. To improve fiber/matrix adhesion, a coupling agent was used (maleic anhydride grafted polypropylene): Epolene C16 with an acid number of 2 and a viscosity of 2.85 cP at 190 °C.

Fiber's Analysis

The chemical composition of raw and pre-treated palm fibers was evaluated using the methodology applied for the fibers from sugarcane bagasse by Gouveia et al. [19]. The raw and pre-treated palm fibers were evaluated by scanning electron microscopy (SEM) using a HITACHI JSM5310 scanning electron microscope with a tungsten filament operating at 10 kV, employing the low vacuum technique and secondary electron detector. Samples were dispersed on a brass support and fixed with double-sided 3 m tape. Chemical structures of palm fibers were evaluated by FTIR. Spectra were obtained on an FTIR spectrophotometer (Perkin Elmer). Samples were prepared by mixing the materials and KBr in a proportion of 1:200 (w/w). For all spectra, 16 scans were recorded with a 4 cm^{-1} resolution.

Preparation of Composites

The raw and pre-treated palm fibers were mixed with the polymeric matrix (PP) at 5 and 10 wt% of the composition in a thermokinetic mixer model MH-50H, with the speed rate of 5250 rpm. After mixing, composites were dried and ground in a mill. Then, palm fibers/PP composites were placed in an injector chamber at 250 °C. The melted material was injected in required dimensions to a pre-warm mold in order to obtain the tensile specimens. The influence of the use of the coupling agent (MA-g-PP) in the preparation of composites was evaluated. The nomenclature and composition of samples tested are given in Table 1.

Mechanical Tests

Composites were analyzed in an "EMIC" universal-testing machine (model DL10000) equipped with pneumatic claws. For tensile tests, five specimens of composites were analyzed, with the dimensions in agreement with ASTM D 638 standard: 19 mm width, 165 mm length and 3 mm thickness at 2 mm min^{-1} cross-head speed.

In the flexural tests, the load was applied on the specimen at 1.3 mm min^{-1} crosshead speed. Five specimens were analyzed with dimensions in agreement with

Table 1 Composition of the composites and pure PP

Samples	Raw fiber content (wt%)	Pre-treated content fiber (wt%)	PP (wt %)	MA-g-PP (wt %)
PP	–	–	100	–
CP5%	5	–	95	–
CP10%	10	–	90	–
CPT5%	–	5	95	–
CPT10%	–	10	90	–
CPA5%	5	–	90	5
CPA10%	10	–	85	5

CP5% (5 wt% raw fiber reinforced with 95 wt% PP); CP10% (10 wt% raw fiber reinforced with 90 wt% PP); CPT5% (5 wt% treated fiber reinforced with 95 wt% PP); CPT10% (10 wt% treated fiber reinforced with 90 wt% PP); CPA5% (5 wt% raw fiber reinforced with 5 wt% MA-g-MA and 90 wt% PP); CPA10% (10 wt% raw fiber reinforced with 5 wt% MA-g-MA and 85 wt% PP)

ASTM D 790 standard: 25 mm width, 76 mm length and 3.2 mm thickness. The adopted flexural test was the 3-point bending method.

Results and Discussion

The raw and pre-treated palm fibers were characterized for the chemical composition, and the results are given in Table 2. It was possible to observe that pulping process reduced the hemicellulose and cellulose contents, resulting in the increase in strength of the composites, as cellulose has the highest mechanical properties among all components. These differences can be due to the removal of sugars and extractives during pulping process.

Figure 1 shows the FT-IR spectra of raw and pre-treated palm fibers. According to Mulinari et al. [20], the characteristics shown in the spectra of the palm fibers are due to its constituents. In the spectrum of raw palm fiber, the spectral profile is showing the characteristic peaks of lignin, hemicellulose, cellulose (strong wide band between 3610 and 3055 cm^{-1} assigned to O–H stretching vibrations of alcoholic and phenolic hydroxyl groups involved in hydrogen bonds and 2977–2830 cm^{-1} to C–H stretching vibration) and triglyceride (Ester carbonyl (C=O)

Table 2 Chemical composition of palm fibers

Chemical component (%)	Raw palm fiber	Pre-treated palm fiber
Cellulose	52.3 ± 0.6	61.7 ± 0.8
Hemicellulose	24.2 ± 0.5	20.2 ± 0.6
Lignin	21.5 ± 0.4	16.8 ± 0.5
Ashes	2.1 ± 0.1	1.6 ± 0.2
Process yield	100.1	100.3

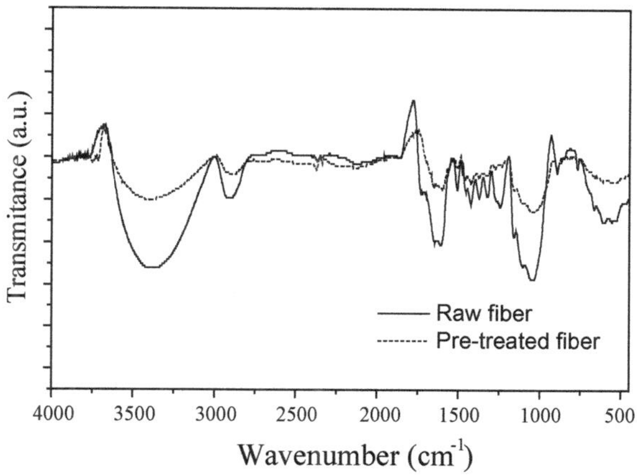

Fig. 1 FTIR spectra of raw and pre-treated palm fibers

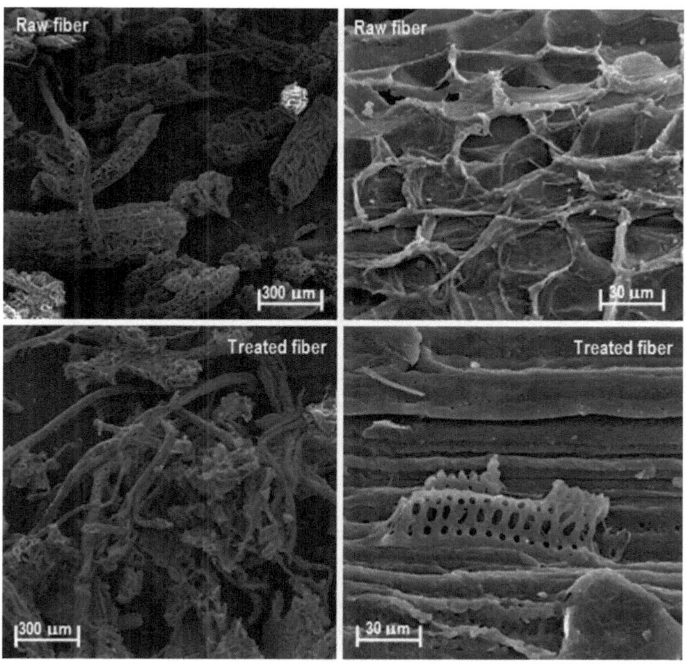

Fig. 2 SEM micrographs of raw and pre-treated palm fibers

functional group at 1730 cm^{-1} and stretching vibration of the C–O ester group at 1172 cm^{-1}). However a change in the FTIR spectra was observed for pre-treated palm fibers; the peak at 1730 cm^{-1} disappeared.

SEM micrographs of raw and pre-treated palm fibers are shown in Fig. 2.

The micrograph of fibers show cylindrical fragments with porous surface and others with smooth layer due to the presence of extracts, similar to some wax, which are present in the fibers in addition to amorphous contents such as lignin and hemicellulose. After pre-treatment, this layer is partially removed and this caused the reduction in the surface roughness of fibers. In addition, it was possible to observe a number of pits along the cell wall. These pits are responsible for transporting water and nutrients throughout various cells to the roots and leaves, and are hidden on the fibre surface. Composites were obtained through mixing for different time periods, due to the different composition of the fibers. This difference in the mix time can also cause differences in rupture and thermal degradation of fibers, and consequently affecting thermal and mechanical properties of composites differently [21].

Table 3 shows the influence of the MA-g-PP and the pre-treatment on the mechanical properties of PP matrix.

The tensile strength of PP composites was influenced both by the concentration of MA-g-PP and the pre-treatment. The composites compatibilized with MA-g-PP demonstrated better tensile strength as compared to the composites reinforced with raw and pre-treated fibers with an increase of 9.6 and 31%, respectively.

As expected, the fibre content influenced the tensile strength and modulus of the composites. This fact can be explained by the beneficial interaction between fibers and matrix. The presence of fibers and use of MA-g-PP resulted in an increase in Young's modulus by 22.5 and 32% as compared to the pure PP, respectively. This can be explained from the fact that the compatibilized composites showed better load transfer between the matrix and the reinforcing fibres at the interface owing to probable occurrence of mechanical anchorage or chemical interactions between anhydride and hydroxyl groups. Kakou et al. [22] observed similar behavior to understand the influence of coupling agent on the properties of high density polyethylene composites reinforced with oil palm fibers.

The interaction between fiber and matrix in the composites was analyzed by SEM after fracture. The study of the fracture surface through SEM revealed the areas in which possible failures have occurred. Figure 3 presents SEM images of

Table 3 Mechanical properties of the materials

Samples	Elongation at break (%)	Tensile strength (MPa)	Tensile modulus (MPa)	Flexural strength (MPa)	Flexural modulus (MPa)
PP	12.8 ± 1.6	25.1 ± 1.4	1021.3 ± 75.9	51.0 ± 0.4	1069 ± 35.1
CP5%	11.2 ± 1.0	28.8 ± 2.7	1181.5 ± 95.8	53.9 ± 0.6	1160 ± 56.8
CP10%	12.5 ± 0.4	27.5 ± 1.3	1254.4 ± 51.2	53.4 ± 0.8	1251 ± 48.5
CPT5%	10.9 ± 1.3	27.2 ± 1.1	1231.2 ± 81.2	55.1 ± 1.1	1115 ± 72.4
CPT10%	9.8 ± 1.1	33.1 ± 1.9	1299.1 ± 76.2	55.2 ± 1.6	1711 ± 86.3
CPA5%	10.6 ± 1.8	28.3 ± 1.6	1281.1 ± 78.1	55.8 ± 1.2	1199 ± 100.2
CPA10%	9.2 ± 1.6	37.2 ± 1.3	1348.2 ± 89.0	58.7 ± 1.2	1898 ± 147.2

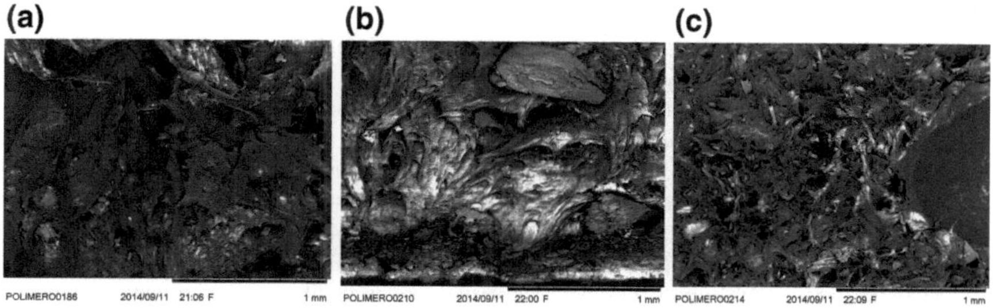

Fig. 3 SEM of the composites after tensile tests: **a** CP5%; **b** CPT5%; **c** CPA5%

composites. It is clear that the addition of MA-g-PP enhances the fiber/matrix adhesion, and limits fiber pull-out.

Flexural properties of the composites were also enhanced by fiber reinforcement, fibre treatment and MA-g-PP. This increase may be explained by a better fibers-matrix interaction under the compressive stresses during bending, developed in the transverse section of the flexural specimens. In this work, 10 wt% palm fibers and MA-g-PP compatibilized PP matrix was the optimum combination to achieve the best flexural modulus.

Conclusion

Palm fiber reinforced polypropylene (PP) composites were manufactured by a thermokinetic mixer followed by injection molding. This study showed that the *in natura* palm fibers are promising reinforcements for the fabrication of composite materials. Alkaline treatment and use of MA-g-PP can be effectively used to enhance the surface and mechanical properties of palm fibers for composites. The addition of raw fibers on the PP matrix presented a material with until 10% less polymer and with viable mechanic properties for determined applications of PP, where cost and tensile modulus are more important than ductility. The fibers treated and compatibilized with MA-g-PP influenced on the mechanic properties of the composites when compared composites reinforced with raw fibers.

Acknowledgements The authors thank FAPERJ (Processes E-26/201.481/2014 and E-26/010.002016/2014) for financial assistance granted.

References

1. Mulinari, D. R., Voorwald, H. J. C., Cioffi, M. O. H., & Da Silva, M. L. C. P. (2016). Cellulose fiber-reinforced high-density polyethylene composites—Mechanical and thermal properties. *Journal of Composite Materials, 1,* 1–9.

2. Pickering, K. L., Efendy, M. G. A., & Le, T. M. (2016). A review of recent developments in natural fibre composites and their mechanical performance. *Composites Part A, 83,* 98–112.
3. Arrakhiz, F. Z., El Achaby, M., Malha, M., Bensalah, M. O., Fassi-Fehri, O., Bouhfid, R., et al. (2013). Mechanical and thermal properties of natural fibers reinforced polymer composites: Doum/low density polyethylene. *Materials & Design, 43,* 200–205.
4. Hoto, R., Furundarena, G., Torres, J. P., Muñoz, E., Andrés, J., & García, J. A. (2014). Flexural behavior and water absorption of asymmetrical sandwich composites from natural fibers and cork agglomerate core. *Materials Letters, 127,* 48–52.
5. Jeencham, R., Suppakarn, N., & Jarukumjorn, K. (2014). Effect of flame retardants on flame retardant, mechanical, and thermal properties of sisal fiber/polypropylene composites. *Composites Part B, 56,* 249–253.
6. Nirmal, U., Hashim, J., & Megat Ahmad, M. M. H. (2015). A review on tribological performance of natural fibre polymeric composites. *Tribology International, 83,* 77–104.
7. Yan, L., Chouw, N., Huang, L., & Kasal, B. (2016). Effect of alkali treatment on microstructure and mechanical properties of coir fibres, coir fibre reinforced-polymer composites and reinforced cementitiuos. *Construction and Building Materials, 112,* 168–182.
8. Ribeiro, J. H. (1996). SOS palmito. *Revista Globo Rural, 3,* 24–26.
9. Simas, K. N., Vieira, L. N., Podestá, R., Vieira, M. A., Rockenbach, I. I., Petkowicz, L. O., et al. (2010). Microstructure, nutrient composition and antioxidant capacity of king palm flour: A new potential source of dietary fibre. *Bioresource Technology, 101,* 5701–5707.
10. Mulinari, D. R., Voorwald, H. J. C., Cioffi, M. O. H., Rocha, G. J. M., & Da Silva, M. L. P. (2010). Surface modification of sugarcane bagasse cellulose and its effect on mechanical and water absorption properties of sugarcane bagasse cellulose/HDPE composites. *BioResources, 5,* 661–671.
11. Pereira, P. H. F., Rosa, M. F., Cioffi, M. O. H., Benini, K. C. C. C., Milanese, A. C., Voorwald, H. J. C., et al. (2015). Vegetal fibers in polymeric composites: A review. *Polímeros, 25,* 9–22.
12. Paul, S. A., Joseph, K., Mathew, G. D. G., Pothen, L. A., & Thomas, S. (2010). Influence of polarity parameters on the mechanical properties of composites from polypropylene fiber and short banana fiber. *Composites Part A, 41,* 1380–1387.
13. Shalwan, A., & Yousif, B. F. (2014). Investigation on interfacial adhesion of date palm/epoxy using fragmentation technique. *Materials & Design, 53,* 928–937.
14. Pöllänen, M., Suvanto, M., & Pakkanen, T. T. (2013). Cellulose reinforced high density polyethylene composites—Morphology, mechanical and thermal expansion properties. *Composites Science and Technology, 76,* 21–28.
15. El-Sabbagh, A. (2014). Effect of coupling agent on natural fibre in natural fibre/polypropylene composites on mechanical and thermal behaviour. *Composites Part B, 57,* 126–135.
16. Liu, W., Xie, T., Qiu, R., & Fan, M. (2015). N-methylol acrylamide grafting bamboo fibers and their composites. *Composites Science and Technology, 117,* 100–106.
17. Essabir, H., Bensalah, M. O., Rodrigue, D., Bouhfid, R., & Qaiss, A. (2016). Structural, mechanical and thermal properties of bio-based hybrid composites from waste coir residues: Fibers and shell particles. *Mechanics of Materials, 93,* 134–144.
18. Mileo, P. C., Oliveira, M. F., Luz, S. M., Rocha, G. J. M., & Gonçalves, A. R. (2016). Thermal and chemical characterization of sugarcane bagasse cellulose/lignin-reinforced composites. *Polymer Bulletin, 21,* 1–12.
19. Gouveia, E. R., Nascimento, R. T., Souto-Maior, A. M., & Rocha, G. J. M. (2009). Validação de metodologia para a caracterização química de bagaço de cana-de-açúcar. *Quimica Nova, 32,* 1500–1503.
20. Mulinari, D. R., Cioffi, M. O. H., & Voorwald, H. J. C. (2010). Review on natural fibers/HDPE composites: Effect of chemical modification on the mechanical and thermal properties. In F. M. P. Willems (Ed.), Green composites: Properties, design and life cycle analysis (pp. 53–78). New York.

21. Vijayakumar, S., Nilavarasan, T., Usharani, R., & Karunamoorthy, L. (2014). Mechanical and microstructure characterization of coconut spathe fibers and Kenaf Bast Fibers Reinforced Epoxy Polymer Matrix Composites. *Procedia Materials Science, 5,* 2330–2337.
22. Kakou, C. A., Arrakhiz, F. Z., Trokourey, A., Bouhfid, R., Qaiss, A., & Rodrigue, D. (2014). Influence of coupling agent content on the properties of high density polyethylene composites reinforced with oil palm fibers. *Materials & Design, 63,* 641–649.

Effect of Time-Dependent Process Temperature Variation During Manufacture of Natural-Fibre Composites

Hossein Mohammad Khanlou, Peter Woodfield, Wayne Hall and John Summerscales

Abstract One of the key issues in compression molding of natural fibre reinforced polymer (NFRP) bio-composites is the thermochemical degradation of the fibre and matrix during manufacture. In our previous work, models of thermal penetration, melt infusion, thermal degradation and chemical degradation of flax/PLA bio-composite were used to propose the degradation boundaries for bio-composite manufacturing. This study proposes a thermal degradation model which accounts for effect of time-dependent process temperature variation during manufacture of green composites. Kinetic data are used to calculate degradation progress parameters, defining experiment process maps for identifying the effect of the temperature history on the degradation progress and effects on the tensile strength. The model also can express the tensile strength changes in comparison with other degradation parameters.

Keywords NFRP composites · Natural fibres · Thermal degradation
Thermo-chemical degradation · Heating and cooling rate
Mechanical properties

Introduction and Theoretical Background

Thermal and chemical degradation occur during thermal processing of natural fibre reinforced polymer (NFRP) bio-composites and are key processes influencing mechanical properties of the composite. To quantify the extent of degradation, in our previous study [1, 2], we defined degradation progress parameters (α) and consolidation progress parameters (κ) where $\kappa = 1 - \alpha_i = 0$ for the completely

H. Mohammad Khanlou (✉) · P. Woodfield · W. Hall
Griffith School of Engineering, Gold Coast Campus, Griffith University,
Gold Coast, QLD 4222, Australia
e-mail: hossein.mohammadkhanlou@griffithuni.edu.au

J. Summerscales
Advanced Composites Manufacturing Center, School of Engineering,
Reynolds Building, University of Plymouth, Plymouth, Devon PL4 8AA, UK

R. Fangueiro and S. Rana (eds.), *Advances in Natural Fibre Composites*,
https://doi.org/10.1007/978-3-319-64641-1_6

degraded state and $\kappa = 1 - \alpha_i = 1$ for the initial or desired state. To estimate the progress of PLA melting, heat up and PLA impregnation process, Eqs. (1–5) were proposed [1]:

$$\kappa_{heatup} \equiv \frac{h(T_{cen}) - h_s(T_0)}{h_f\left(T_{platen}\right) - h_s(T_0)} \tag{1}$$

$$\kappa_{heatup} \approx \kappa_{sensible} \approx \frac{T_{cen} - T_0}{T_{platen} - T_0} = 1 - \frac{4}{\pi}\sum_{n=0}^{\infty}\frac{(-1)^n}{2n+1}e^{-(2n+1)^2\pi^2 at/L^2} \tag{2}$$

$$\kappa_{latent} \approx \sqrt{\frac{8k_{cond}\left(T_{platen} - T_m\right)t}{\left(1 - V_f\right)\rho h_{fs} L^2}} \tag{3}$$

$$t_{heatup} \approx t_{sensible} + t_{latent} \tag{4}$$

$$\kappa_{impreg} = \frac{A_{impreg}}{A_{yarn}} \tag{5}$$

where T_{cen} is the temperature in the centre of the composite (i.e. half way between the two heated platens), T_0 is the initial temperature of the composite and T_{platen} is the temperature of the platen, $h_f(T)$ is the specific enthalpy of the liquid while $h_s(T)$ is the specific enthalpy of the solid, a is the effective thermal diffusivity of the composite, L is the distance between the platens (i.e. the thickness of the composite) and t is time, T_m is the melting point, k_{cond} is the effective thermal conductivity of the composite, ρ is the density of the matrix material, V_f is the volume fraction of fibre and h_{fs} the latent heat of fusion (melting), A_{yarn} is the cross-sectional area of a single yarn (i.e. fibre and inter-fibre porosity), A_{impreg} is the instantaneous area within the yarn wetted by the melt; and where $\kappa_{latent} = 0$ indicates all of the material is in the solid phase and $\kappa_{latent} = 1$ indicates that all of the matrix material is liquid.

The thermochemical degradation progress of the fibre and matrix are defined as Eq. (6–8) [1, 2]:

$$\kappa_{pyrol} = 1 - \frac{m_0 - m_t}{m_0 - m_f} \tag{6}$$

$$\kappa_{chem-fibre} = 1 - \frac{DP|_{t=0} - DP}{DP|_{t=0} - 1} \tag{7}$$

$$\kappa_{chem-matrix} = 1 - \frac{M_{n-avg}|_{t=0} - M_{n-avg}}{M_{n-avg}|_{t=0} - u_0} \tag{8}$$

where m_0, m_t and m_f are the initial, at time t, and final mass of the material, DP is degree of polymerisation, $M_{n\text{-}avg}$ is the number-average molar mass of polymers in the matrix and u_0 is the molar mass of a single monomer unit. DP and $M_{n\text{-}avg}$ are calculated as a function of the temperature history of the melt using first-order Arrhenius rate kinetics [3–5]. For the fibre, the degradation rate is expressed as:

$$\frac{dN}{dt} = K \times N \tag{9}$$

where

$$N = 1 - \frac{1}{DP_{fibre}} \tag{10}$$

and K is defined using the Arrhenius equation:

$$K = A\exp(\frac{-E_a}{RT}) \tag{11}$$

For the thermo-chemical degradation of the PLA matrix,

$$\frac{d\lambda_0}{dt} = k_d(\frac{\rho}{m_0} - \lambda_0) - \frac{k_c\lambda_0^2}{2} \tag{12}$$

where

$$\lambda_0 = \rho/M_n = \frac{\rho}{m_0 DP_{matrix}} \tag{13}$$

and k_d and k_c are degradation and recombination Arrhenius rate coefficients, respectively as given by

$$k_d = A_d\exp(\frac{-E_{ad}}{RT}) \tag{14}$$

$$k_c = A_c\exp(\frac{-E_{ac}}{RT}) \tag{15}$$

The activation energies and pre-exponentials for Eqs. (11), (14) and (15) are available [1, 2, 4].

Experiment

An experimental program was carried out using unidirectional flax fibre and PLA to fabricate 1 mm thick bio-composite plates in a compression moulding machine. The details of the experiment are described elsewhere [1]. The volume fraction of fibre was 50% and the plates were cut into tensile test specimens using a laser cutter. Tensile properties were measured for high temperature (set temperature:

170 °C) consolidation times of 15, 30, 45, 60 and 120 min. K-type sheathed thermocouples and a data logger were used to record the temperature history during compression moulding.

Results

Figure 1a shows measured temperatures of the upper and lower plates of the compression moulding machine, which are recorded during the manufacturing process for 120 min. The bottom moulding plate and the composite layup were not introduced to the moulding machine until the top mould surface temperature reached the set point value. It is clear that the temperature history deviates somewhat from the ideal model of instantly rising to the set temperature, staying at a constant temperature for 120 min and then instantly cooling to room temperature. The largest deviation from the ideal model occurs during the first 20 min of consolidation where the temperature overshoots to about 185 °C. There are also differences between the temperature of the upper and lower plate. This is particularly apparent during heat up where the temperature of the upper plate rises slowly before pressure is applied. During this stage (−27 min to 0 min on Fig. 1a) the composite is not in contact with the upper plate.

Figure 1b shows the findings from the models for degradation progress resulting from various factors and experimental data for the tensile strength of the flax/PLA bio-composite with different processing times. Comparing the dotted lines with the solid lines shows the effect of including the actual temperature history on the predicted degradation of flax and PLA. The thermochemical degradation for the dotted lines was found by numerically integrating Eqs. (9) and (12) using the measured temperature history in Eqs. (11), (14) and (15). It can be seen that the differences between constant temperature and non-ideal temperature history are within 10% error.

By non-dimensionalizing the tensile strength against the maximum measured strength we are able to obtain an additional experimental measure of degradation of the composite during processing. The scale of the experimental data shown in Fig. 1b for tensile strength has been selected so that vertical scale on the left hand side (κ) shows the degradation of the tensile strength (Eq. (16)). The trend line of experimental results also shows a similar trend to the changes of κ associated with thermal-chemical degradation of the fibre. The model shows that a processing time of less than about 20 min is desired to limit thermal degradation of the fibre and greater than about 10 min to allow for penetration of the matrix into the fibre.

$$\kappa_\sigma = \frac{\sigma_{\exp eriment}}{\sigma_{\max}} \tag{16}$$

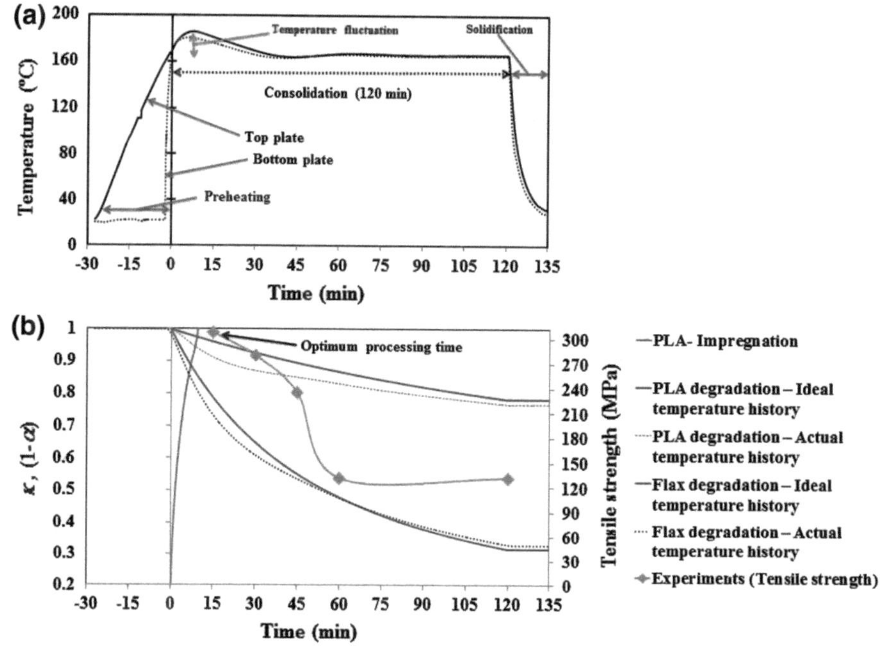

Fig. 1 **a** Measured processing temperature history for 120 min consolidation times, **b** Consolidation κ/time maps and experimental data for flax/PLA bio-composite. A value of $\kappa = 1$ or $\alpha = 0$ corresponds to the desired state of complete consolidation and no thermochemical degradation. The *solid lines* show ideal processing (isothermal) while *dashed lines* show calculations based on the actual temperature history of the melt for PLA and flax fibre

Simulation and Conclusion

To realize the effect of heating and cooling rate on the degradation of NFRP bio-composites, three various rates (at 1.5, 5, 20 K/min as slow, average and fast, respectively,) were selected (see Fig. 2a). In addition, the κ changes of PLA and flax fibre, which are resulted from the simulated heating and cooling rate to a processing temperature of 170°C, are shown in Fig. 2b. PLA shows insignificant reactions to the rate changes while flax fibre shows notable additional degradation during the heat up period for slow heating. In case of PLA, two processes (degradation and recombination) are responsible for the degradation progress at high temperature during the thermal processing [1, 2]. For PLA the final state of degradation (>150 min) is almost identical even though the slower heating/cooling case shows slightly lower values for κ during consolidation. The slight recovery (increase of κ for the slower cooling case) after 120 min for PLA can be due to monomer recombination factor (Eq. (12), 2nd term on the right) which is considered in the calculation of the degradation progress of PLA. However, there is not such a factor in the degradation progress model of the flax fibre (Eq. (9)), resulting

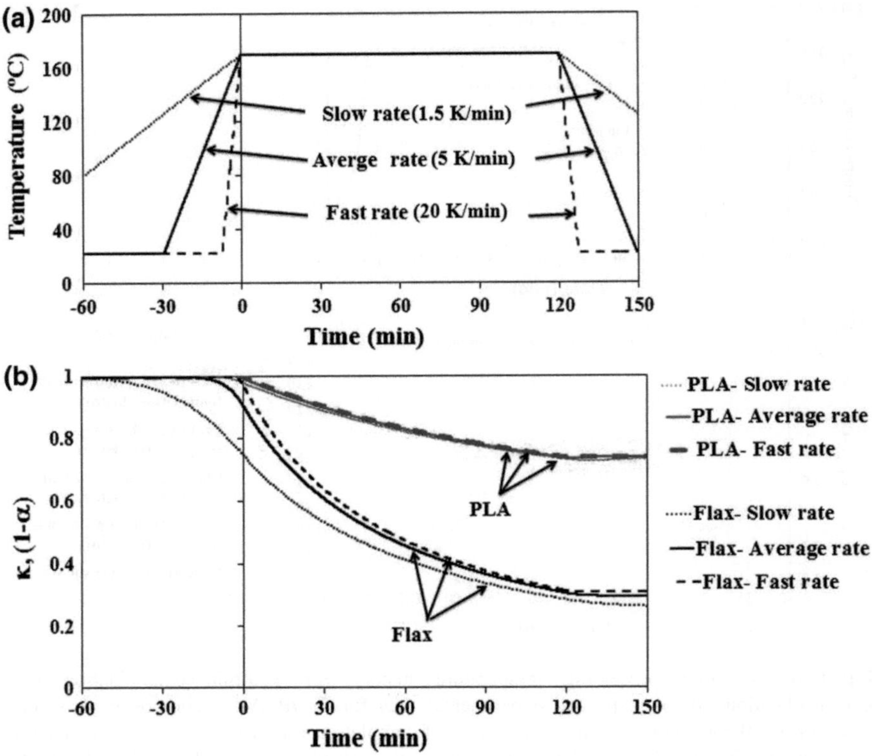

Fig. 2 **a** Three various rates of heating and cooling as slow, average and fast, **b** κ/time maps for PLA and flax at the various rates

in continuous degradation of the flax fibre. Moreover, the slow rate of heating and cooling represents a negative effect on the mechanical properties of flax fibre and consequently on the bio-composite; as fibre is mostly responsible for the mechanical properties of composites, according to the rule of mixtures [6]. Thus we have demonstrated that the present model is a useful tool for predicting the degradation effect of any temperature history to which the composite is subjected during manufacture.

References

1. Khanlou, H. M., Woodfield, P., Summerscales, J., & Hall, W. Consolidation process boundaries of the degradation of mechanical properties in compression moulding of natural-fibre bio-polymer composites. *Polymer Degradation Stability*. 10.1016/jpolymdegradstab201703004.

2. Khanlou, H. M., Hall, W., Heitzman, M. T., Summerscales, J., & Woodfield, P. (2016). Technical Note: On modelling thermo-chemical degradation of poly(lactic acid). *Polymer Degradation and Stability, 134,* 19–21.
3. Yu, H., Huang, N., Wang, C., & Tang, Z. (2003). Modeling of poly (L-lactide) thermal degradation: theoretical prediction of molecular weight and polydispersity index. *Journal of applied polymer science, 88*(11), 2557–2562.
4. Le Marec, P. E., Ferry, L., Quantin, J.-C., Bénézet, J.-C., Bonfils, F., Guilbert, S., et al. (2014). Influence of melt processing conditions on poly (lactic acid) degradation: Molar mass distribution and crystallization. *Polymer Degradation and Stability, 110,* 353–363.
5. Gassan, J., & Bledzki, A. K. (2001). Thermal degradation of flax and jute fibers. *Journal of Applied Polymer Science, 82*(6), 1417–1422.
6. Virk, A. S., Hall, W., & Summerscales, J. (2012). Modulus and strength prediction for natural fibre composites. *Materials Science and Technology, 28*(7), 864–871.

About Nonlinear Behavior of Unidirectional Plant Fibre Composite

Christophe Poilâne, Florian Gehring, Haomiao Yang and Fabrice Richard

Abstract At room condition and standard strain rate, unidirectional glass fiber reinforced organic polymers show linear behavior under longitudinal loading (the same with carbon fiber). Oppositely, plant-based reinforced organic polymers show often nonlinear behavior. We describe a viscoelastoplastic model based on eight independent parameters dedicated to simulation of plant fiber composite mechanical behavior. This model has been previously validated with flax twisted yarn/epoxy composite at room condition. We analyse now an unidirectional flax/epoxy composite at different strain rates to promote a mechanical behaviour with 'three apparent regions' visible in case of longitudinal loading. We show that adding of a strengthening phenomenon is a good solution to improve phenomenological model of plant fibre composite.

Keywords Plant fiber composite · Viscoelastoplasticity · Phenomenological modelling

Introduction

Imagine a baseball bat mainly made of long plant fibre composite. Imagine this baseball bat—initially straight—which become more and more curved as we use it, even in normal use. Because it will be more and more difficult to play with, we can

C. Poilâne (✉) · F. Gehring · H. Yang
Normandie Université, Esplanade de La Paix, 14032 Caen, France
e-mail: christophe.poilane@unicaen.fr

F. Gehring
e-mail: florian.gehring@unicaen.fr

H. Yang
e-mail: haomiao.yang@unicaen.fr

F. Richard
Université Bourgogne Franche-Comté, 25000 Besançon, France
e-mail: fabrice.richard@univ-fcomte.fr

R. Fangueiro and S. Rana (eds.), *Advances in Natural Fibre Composites*,
https://doi.org/10.1007/978-3-319-64641-1_7

say that this baseball bat has been badly designed! It could be the case when designers ignore the fact that the constitutive composite material is not only elastic.

Indeed, plant-based reinforced polymer presents often nonlinear mechanical behaviour at normal condition. This becomes particularly evident for tensile loading in fibre direction with the presence of a yield point which separates tensile curve in two regions. For convenience, we name the first region 'elastic', the second region being none-elastic [1]. This is the case for flax fibre reinforcement [1–4]. This is visible on experimental curves in articles that do not deal only with unidirectional reinforcement [2], but also with reinforcement by random mat of flax [5–7]. This is finally the case for other (than flax) plant fibre composite [5, 8]. The yield point occurs at a very low level of strain, between 0.1–0.3% according to experimental conditions and measurement methods [3, 9]. Some authors develop models to simulate this particular mechanical behaviour [8, 10–12]; this will give the possibility to engineers to improve the design of plant fibre composite parts.

In a previous work [1], we proposed a viscoelastoplastic model to study the nonlinear effects of plant fibre composite. The used material was made by epoxy resin and twisted yarn of flax, as reinforcement. We assume that this reinforcement is quasi-unidirectional (quasi-UD). We identified eight parameters to properly simulate the mechanical behaviour of flax/epoxy quasi-UD. The parameters were computed using inverse identification based on repetitive progressive loading (RPL) and creep test in the elastic region, at normal condition (room temperature, usual strain rate, normal humidity). The validation was done by creep test and relaxation test in the non elastic region. The 8-parameter model take into account viscoelastic and viscoplastic contributions. The model do not required of reorientation phenomenon, but we observed a contraction of the elastic region during loading. The first region of tensile curves is quasi-elastic and the second region is viscoelastoplastic.

Phenomenological Model

The aim of constitutive phenomenological model is to provide an accurate prediction of uniaxial mechanical response of flax fibre reinforced polymer (FFRP). We particularly aim at simulating the two-region mechanical behaviour of FFRP described in introduction. A viscoelastoplastic model about FFRP behaviour has been previously developed. The parameters of this model have been identified on 'UD' and 'quasi-UD' reinforcements based on twisted yarn [1]. UD are constituted by 'large' (more than 100 tex) non-woven yarns aligned in one direction only. Quasi-UD are constituted by 'small' (minor to 50 tex) yarns woven in two perpendicular directions: warp and weft. The weft yarns—which are ten times fewer in number than warp yarns—are needed for the ply handiness.

The total strain is partitioned in an elastic part (instantaneous reversible strain) and an inelastic part which is the sum of viscoelastic contribution (time-dependent reversible strain) and viscoplastic contribution (time-dependent irreversible strain):

$$\varepsilon = \varepsilon^{e} + \varepsilon^{in} = \varepsilon^{e} + \varepsilon^{ve} + \varepsilon^{vp}. \tag{1}$$

In the context of thermodynamics, physical phenomena can be described with a precision which depends on the choice of the nature and the number of state variables. The state variables are the observable variables and the internal variables. The standardized framework [13] assumes that mechanical behaviour is obtained when two potentials are defined: a free energy density ψ to define state laws and a dissipation potential Ω to determine the evolution of internal variables. Based on experimental results two potentials are proposed. The state laws can then be written as:

$$\sigma = \rho \frac{\partial \psi}{\partial \varepsilon^{e}} \tag{2}$$

$$X_i = \rho \frac{\partial \psi}{\partial \alpha_i} \tag{3}$$

where α_i and X_i variables represent inelastic phenomena, ρ is the mass density, and σ is the Cauchy's stress.

The evolution of internal variables is expressed as:

$$\dot{\varepsilon}^{in} = \frac{\partial \Omega}{\partial \sigma} = \dot{\varepsilon}^{ve} + \dot{\varepsilon}^{vp} \tag{4}$$

$$\dot{\alpha}_i = -\frac{\partial \Omega}{\partial X_i}. \tag{5}$$

The system of ordinary differential equations has been solved with an home-made simulation software, MIC2M [14, 15], using an algorithm based on the Runge-Kutta method. An inverse approach is used to extract the parameters from the experimental strain measurements. This approach consists of an optimization problem where the objective is to minimize the gap between the experimental strain and the numerical results. The minimization problem was solved using an algorithm based on the Levenberg-Marquardt method coupled with genetic approach implemented in MIC2M software [15]. To determine if the information is suitable for reliable parameter estimation, a practical identifiability analysis was performed on results [1]. This identifiability analysis is based on local sensitivity functions. Such functions quantify the relationship between the outputs and the parameters of the model. This approach led our team to model the mechanical behaviour of the FFRP by a phenomenological model with kinematic hardening taking viscosity into account. The viscoelastoplastic model was identified in the case of uniaxial test of unidirectional twisted yarn/epoxy composite [1]. The particularity of the reinforcement is the misorientation of constitutive fibres due to the use of twisted yarns. Based on experimental results, the free energy and dissipation potential are proposed in following equations:

$$\psi = \frac{1}{2\rho}E(\varepsilon^e)^2 + \frac{1}{2\rho}\sum_{i=1}^{3} C_i\alpha_i^2 \tag{6}$$

$$\Omega = \Omega^{ve} + \Omega^{vp} = \frac{1}{2\eta}(\sigma - X_1)^2 + \frac{1}{2K}\langle f\rangle^2 \tag{7}$$

with

$$f = |\sigma - X_2 - X_3| - \sigma_Y + \frac{\gamma_3}{2C_3}X_3^2 \tag{8}$$

where ρ is the mass density, E and σ_Y are the Young's modulus and the initial yield stress respectively, η and K are viscosity coefficients corresponding to elastic and plastic phenomena, respectively. C_1 is the viscoelastic stiffness. C_2, C_3 and γ_3 are hardening coefficients. C_2 characterizes linear kinematic hardening. C_3 and γ_3 refer to nonlinear kinematic hardening coupled to a contraction of elastic region to improve the unloading modelling in RPL tests. The state laws for mechanical behaviour becomes:

$$\sigma = E\varepsilon^e \tag{9}$$

$$X_i = C_i\alpha_i. \tag{10}$$

and the evolution of internal variables defined becomes:

$$\dot{\varepsilon}^{ve} + \dot{\varepsilon}^{vp} = \frac{1}{\eta}(\sigma - X_1) + \frac{\langle f\rangle}{K}\operatorname{sign}(\sigma - X_2 - X_3) \tag{11}$$

$$\dot{\alpha}_1 = \frac{1}{\eta}(\sigma - X_1) \tag{12}$$

$$\dot{\alpha}_2 = \frac{\langle f\rangle}{K}\operatorname{sign}(\sigma - X_2 - X_3) \tag{13}$$

$$\dot{\alpha}_3 = \frac{\langle f\rangle}{K}\left[\operatorname{sign}(\sigma - X_2 - X_3) - \frac{\gamma_3}{C_3}X_3\right]. \tag{14}$$

From a rheological point of view the model proposed in [1] is, for elastic contribution, a linear spring E, and for viscoelastic contribution, a classical Kelvin-Voigt model which comprises a linear viscous damper η and a linear spring C_1 connected in parallel. For viscoplastic contribution, a more complex model is required; it consists in adding two kinematic hardenings: a linear kinematic hardening and a nonlinear kinematic hardening. In addition, a coupling between translation and contraction of the elastic region during loading is added. Finally, seven inelastic parameters have to be identified: viscosity coefficient in elastic region η, viscoelastic stiffness C_1, initial yield stress σ_Y, viscosity coefficient in plastic region K, kinematic hardening coefficient C_2, nonlinear hardening C_3, and nonlinear hardening recall γ_3. The eighth parameter, namely the Young's modulus, was chosen from experimental measurement. The inverse method approach was used to extract constitutive inelastic parameters from the strain measurements from two tests: test A = repetitive progressive loading in tension, test B = creep in tension in 'elastic' region. The RPL is chosen to activate mainly viscoplastic phenomena and the creep test in 'elastic' region is chosen to activate mainly viscoelastic phenomena. Figure 1a coming from [1] shows RPL simulation (left) and creep simulation (right). The used parameters have been identified from unidirectional twisted yarn/epoxy composite tests at room temperature and standard strain rate ($10^{-6}s^{-1}$). The total strain (in red) is partitioned by three contributions (elastic in black, viscoelastic in dotted line, viscoplastic in blue). Elastic contribution is naturally activated all over the test. Viscoelastic contribution is very low at room condition and standard strain rate; creep test mainly shows elastic contributions, as expected.

At room temperature, the proposed model allows us to correctly simulate the behaviour of flax yarn reinforced epoxy composite in repetitive progressive loading, creep test (Fig. 2a) and relaxation test (Fig. 2b). The value of the eight identified parameters is given in Table 1.

In conclusion, for unidirectional twisted flax yarn/epoxy composite at room condition and standard strain rate, the first region of monotonic tensile curves is quasi-elastic and the second region is viscoelastoplastic.

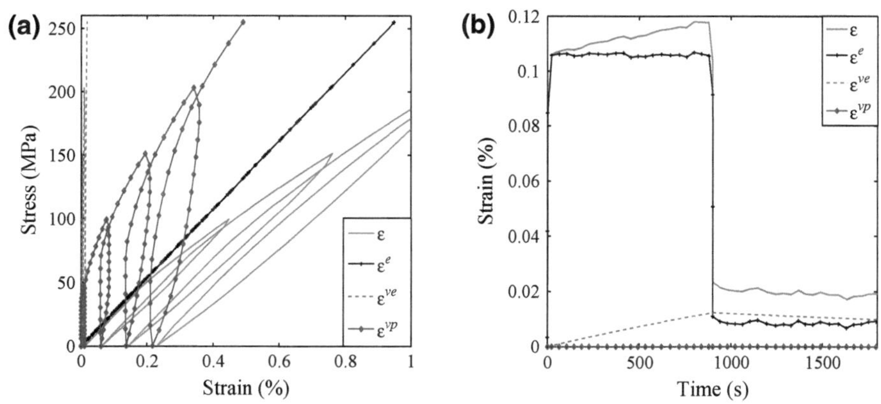

Fig. 1 Simulation response according to elastic contribution, viscoelastic contribution and viscoplastic contribution for **a** RPL test, and **b** creep test at 29; the tests were conducted on unidirectional twisted yarn/epoxy composite at room condition and standard strain rate

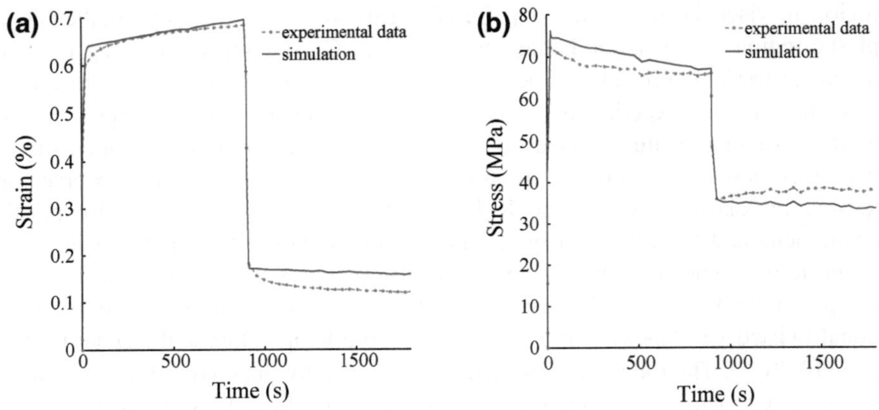

Fig. 2 Experimental data and simulation for **a** creep test at 126, and **b** relaxation test at 0.33; the tests were conducted on unidirectional twisted yarn/epoxy composite at room condition and standard strain rate

Table 1 Elastic and inelastic material parameters for UD twisted flax yarn/epoxy composite

Parameter	Definition	Identified value
E (MPa)	Young's modulus	2.69×10^4
η (MPa s)	viscosity coefficient in elastic region	1.78×10^8
C_1 (MPa)	viscoelastic stiffness	6.30×10^4
σ_Y (MPa)	initial yield stress	3.32×10^1
K (MPa s)	viscosity coefficient in plastic region	2.24×10^5
C_2 (MPa)	kinematic hardening coefficient	3.39×10^4
C_3 (MPa)	nonlinear hardening	6.85×10^4
γ_3	nonlinear hardening (recall)	9.64×10^2

Discussion

The phenomenological model presented in previous section did not need 'strengthening parameter' to correctly simulate standard flax fibre composite in normal condition. The idea of a strengthening phenomenon is due to one of the stronger assumption researchers make in case of bast fibre reinforced polymer: the possibility for microfibrils to reorientate themselves during longitudinal loading (microfibril of cellulose being the main component of bast fibres), even when fibres are trapped inside the matrix. The re-orientation of microfibrils has been demonstrated experimentally on elementary fibre and bundle of fibres under tensile test [16]. It has been correlated to experimental tensile curve by an inverse approach using finite element model [17]. Then the question of this reorientation when bast fibres are used as reinforcement in composite is very logical. Whatever the scale of the reinforcement—untwist at the scale of microfibril, untwist at the scale of yarn,

Fig. 3 Example of unidirectional ply of flax [19] (without matrix). Some of the fibres (elementary fibre or bundle of fibres) are not aligned in the longitudinal direction, but such misalignment is minor

in-plane reorientation at the scale of ply, unshrink for textile composite—if some reorientation of the reinforcement occurs with longitudinal loading we assume that the Young's modulus of the material has to be increase. But such Young's modulus increase was not required for simulating mechanic of viscoelastoplastic yarn reinforced epoxy composite [1].

To test our model for more complex cases than the previous one [1], the first need is to improve the direction of the fibres inside composite. Indeed, it is clear that twisted flax yarn as reinforcement is not the best candidate to activate reorientation of microfibrils inside a composite. It is known that in case of twisted yarn some of the fibres are oriented in the main direction [18]; but fibre orientation globally follows a statistical distribution with major part of the fibres aligned not in the longitudinal direction. A totally unidirectional flax reinforced polymer is now possible to make with industrial product. Figure 3 shows one industrial flax ply used to make such long fibre composite [19]. It is clearly visible that the fibres are mainly oriented in the longitudinal direction (the vertical one). In this product, there are no any weft yarn and no any sewing to link the fibres together. The process we use to make composite plates with flax and epoxy matrix is the hot platen press, as in [1]. The dry flax reinforcement was not treated before use, the objective of the analysis being not to obtain the highest properties but to analyse the mechanical behaviour of unidirectional flax composite. The reinforcement inside the final composite plate is constituted by a mix of elementary fibres and bundles of fibres (bundles as they are in flax stems). Consequently, this reinforcement is mainly oriented in the longitudinal direction of composite, which is the optimal organisation to analyse the mechanical behaviour.

Once the orientation of the reinforcement optimal, the increase of the specimen temperature is one possibility to make easiest the activation of viscous effects of flax composite (Fig. 4c) [1]. Firstly, the rigidity of epoxy matrix decreases with increasing temperature. Particularly, the mechanical properties of thermosetting

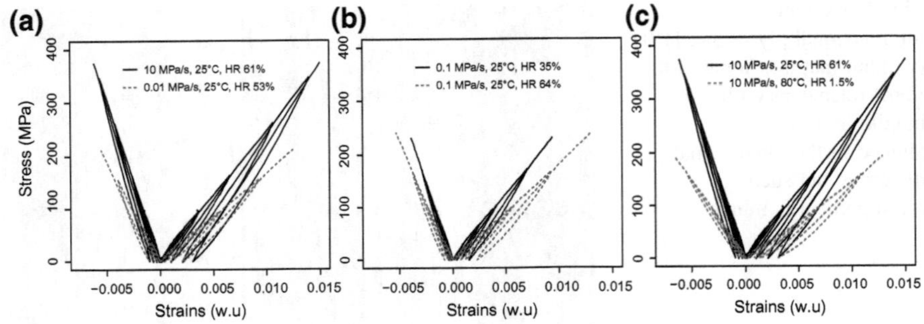

Fig. 4 Effect of **a** tensile speed **b** moisture only and **c** temperature/moisture, on RPL curves in longitudinal tensile of unidirectional flax composite

media drastically decrease when temperature nears to the glass transition temperature T_g of the material. Secondly, high temperature activates viscoelastic properties of bast fibres [20]. One macroscopic consequence of the temperature increase is to make possible a description of tensile curve by three apparent regions (see dotted line in Fig. 4c). Consequently, one simple idea for testing phenomenological model efficiency with unidirectional reinforcement is to increase the testing temperature. Note that when we do not set the relative humidity in oven during tensile test, the increase of temperature is correlated to a decrease of humidity, as shows in Fig. 4c. Note also that the increase of specimen moisture itself promotes viscous effects of flax composite—as shown in Fig. 4b—and makes also visible a tensile curve with three apparent regions [4]. This is a second way, but not simple, to test the efficiency of phenomenological model in one non trivial case. Another way, more simple, is to decrease the strain rate of the tensile test. Indeed when the strain rate is low, viscous effects exposed in [1] are more visible (see Fig. 4a). In other terms, the mechanisms which are responsible of viscous effects are more easy to activate with 'low strain rate', 'high testing temperature' or 'high specimen moisture'.

Eventually, to test our model with unidirectional flax composite, we chose repetitive progressive loading at four different stress rates (0.01 MPa $s^{-1}$to 10 MPa s^{-1}). For the lowest stress rate (Fig. 5b), experimental tensile curve seems to be constituted by three apparent regions instead of two, the first region being quasi-elastic, the third region showing increase of apparent tangent modulus. Let us add that the increase of apparent tangent modulus can be a priori described by the viscous parameters of previous constitutive model.

The previous phenomenological model was possible to fit on experimental data with the same set of inelastic parameters (but different values). Figure 5 shows the best fits we obtained for both extremal strain rates. The eight identified parameters are given in Table 2. It is clearly visible that the simulation of the test at the lowest stress rate do not correlate very well with the experimental data (Fig. 5b). In that case, the increase of apparent tangent modulus—visible on fourth and fifth load—was not possible to simulate. Moreover, indepth look on Fig. 5a shows that this behavior was neither possible to simulate on seventh load of the test at 10.

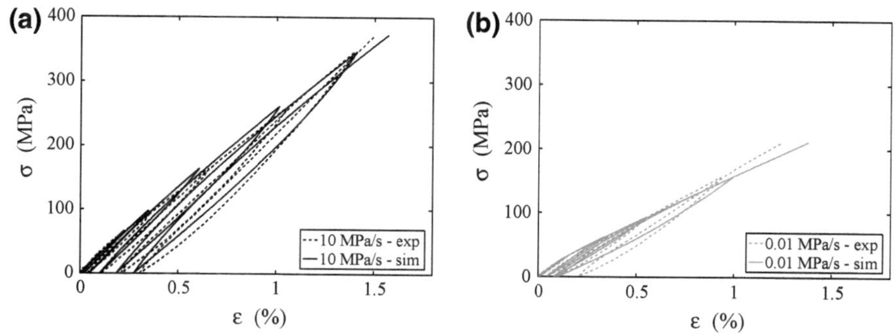

Fig. 5 RPL experiment and simulation with the phenomenological model presented in Sect. 2. The value of the eight parameters are given in Table 2

Table 2 Elastic and inelastic material parameters for unidirectional flax/epoxy composite

Parameter	Definition	Identified value
E (MPa)	Young's modulus	3.15×10^4
η (MPa s)	viscosity coefficient in elastic region	1.98×10^6
C_1 (MPa)	viscoelastic stiffness	7.72×10^4
σ_Y (MPa)	initial yield stress	2.38×10^1
K (MPa s)	viscosity coefficient in plastic region	1.92×10^7
C_2 (MPa)	kinematic hardening coefficient	3.92×10^4
C_3 (MPa)	nonlinear hardening	4.26×10^4
γ_3	nonlinear hardening (recall)	1.62×10^3

Something like a strengthening effect has not been taken into account with our model for a strain above 1 (depending on stress rate). Consequently to this underestimation of apparent rigidity, the permanent strains predicted by our model of are lower than the experimental ones, loop after loop.

Conclusion

We showed an apparent cyclic strengthening in tensile which is not possible to simulate with our previous viscoelastoplastic model. For an adapted material and specific tests, the difference between experimental data and simulation—once the viscoelastoplastic parameters identified—is noticeable. Low stress rates in tensile load were used here, but we make the assumption that high temperature tests or tensile tests on moisturized specimens should offer good possibilities also. Finally, we propose as strong assumption that the add of a strengthening phenomenon to the initial model [1] offers a good solution to improve simulation. By way of illustration, Fig. 6 shows the first result we obtain with the same data as used for Fig. 5

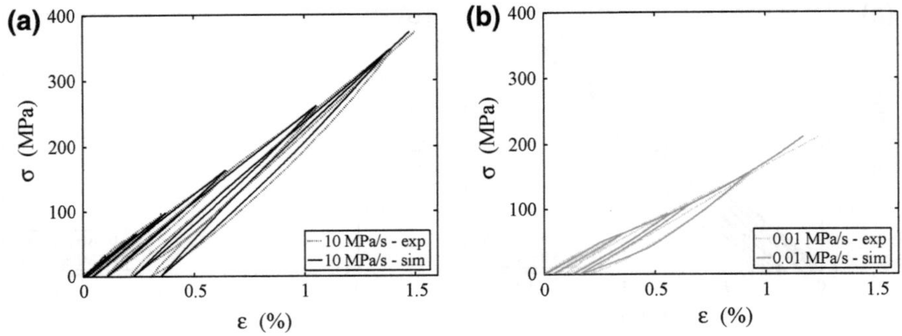

Fig. 6 RPL experiment and new simulation with strengthening parameters

when we replace the nonlinear hardening phenomenon by one strengthening phenomenon. The fitting of experimental data by simulation is clearly improved at the lowest stress rate (Fig. 6b). Although this new approach is not totally validated at this day, it offers an elegant solution because it seems easy to correlate the cyclic strengthening to one reorientation of the reinforcement by longitudinal loading for unidirectional flax composites. In a near future, analysis of such model will help at exploring the origin of the mechanical behavior of plant-based reinforced organic polymers.

Acknowledgements China Scholarship Council (CSC) is acknowledged for the financial support.

References

1. Poilâne, C., Cherif, Z. E., Richard, F., Vivet, A., Ben Doudou, B., & Chen, J. (2014). Polymer reinforced by flax fibres as a viscoelastoplastic material. *Composite Structures, 112,* 100.
2. Oksman, K. (2001). High quality flax fibre composites manufactured by the resin transfer moulding process. *Journal of Reinforced Plastics and Composites, 20,* 621.
3. Hughes, J. C., Carpenter, J., & Hill, C. (2007). Deformation and fracture behaviour of flax fibre reinforced thermosetting polymer matrix composites. *Journal of Materials Science, 42,* 2499.
4. Scida, D., Assarar, M., Poilâne, C., & Ayad, R. (2013). Influence of hygrothermal ageing on the damage mechanisms of flax-fibre reinforced epoxy composite. *Composites Part B-Engineering, 48,* 51.
5. Lebrun, G. (2013). Tensile and impregnation behavior of unidirectional hemp/paper/epoxy and flax/paper/epoxy composites. *Composite Structures, 103,* 151.
6. Cherif, Z. E., Poilâne, C., Momayez, L., & Chen, J. (2011). Optimisation d'un pré-imprègnè lin/èpoxy industrie. *Revue des Composites et des Matèriaux avancès', 21,* 119–128.
7. Cherif, Z. E., Poilâne, C, Momayez, L., & Chen, J. (2012). Pré-imprégnés lin/époxy: Influence des paramètres d'èlaboration sur les propriètès mècaniques. *Matèriaux & Techniques, 100,* 459–466.
8. Rubio-Lòpez, A., Hoang, T., & Santiuste, C. (2016). Constitutive model to predict the viscoplastic behaviour of natural fibres based composites. *Composite Structures, 155,* 8.

9. Shah, D. U. (2016). Damage in biocomposites: Stiffness evolution of aligned plant fibre composites during monotonic and cyclic fatigue loading. *Composite Part A-Applied Science, 83,* 160.
10. Andersons, J., Modniks, J., & Sparnins, E. (2015). Modeling the nonlinear deformation of flax-fiber-reinforced polymer matrix laminates in active loading. *Journal of Reinforced Plastics Composites, 34*(3), 248.
11. Chiali, A., Zouari, W., Assarar, M., Kebir, H., & Ayad, R. (2016). Analysis of the mechanical behaviour of flax and glass fabrics-reinforced thermoplastic and thermoset resins. *Journal of Reinforced Plastics Composites*, 1–16.
12. Sliseris, J., Yan, L., & Kasal, B. (2016). Numerical modelling of flax short fibre reinforced and flax fibre fabric reinforced polymer composites. *Composites Part B-Engineering, 89,* 143.
13. Halphen, B., & Nguyen, Q. S. (1975). Sur les matériaux standard généralisés. *Journal of de Mécanique, 14,* 39.
14. Richard, F. (1999). *Identification du comportement et évaluation de la fiabilité des composites stratifiés.* Thèse de doctorat: Université de Franche Comté, Besançon, FRANCE.
15. Richard, F. (1999). MIC2M, modélisation et identification du comportement mécanique non linéaire des matériaux. http://mic2m.univ-fcomte.fr. Accessed 2017-03-17.
16. Bourmaud, A., Morvan, C., Bouali, A., Placet, V., Perré, P., & Baley, C. (2013). Relationships between micro-fibrillar angle, mechanical properties and biochemical composition of flax fibers. *Industrial Crops and Products, 44,* 343.
17. Del Masto, A., Trivaudey, F., Guicheret-Retel, V.,Placet, V., & Boubakar, L. (2017). Nonlinear tensile behaviour of elementary hemp fibres: a numerical investigation of the relationships between 3D geometry and tensile behaviour. *Journal of Materials Science.*
18. Baets, J., Plastria, D., Ivens, J., & Verpoest, I. (2014). Determination of the optimal flax fibre preparation for use in unidirectional flax–epoxy composites. *Journal of Reinforced Plastics Composites, 33,* 493.
19. Lineo—Flax Fiber Impregnation. URL http://www.lineo.eu/. Accessed 2016-01-24.
20. Placet, P. (2009). Characterization of the thermo-mechanical behaviour of Hemp fibres intended for the manufacturing of high performance composites. *Composites Part A-Applied Science, 40,* 1111.

Investigating the Transient Response of Hybrid Composite Materials Reinforced with Flax and Glass Fibres

Mehmet Cihan, James I.R. Blake and Adam Sobey

Abstract Considerable attention has been devoted to the integration of natural-derived composites into the composite market. In this regard flax fibre can be considered as the most promising natural reinforcement among other natural fibres due to its high specific mechanical properties although these properties fall behind the mechanical properties of conventional fibres. Notwithstanding the downside of flax fibres against conventional fibres in terms of mechanical properties, their inherent viscoelastic characteristics bring an advantage of high vibration damping property. Evaluating the behaviour of flax fibre composites, it is also required to have an understanding of such composite material's response to dynamic loads as structures are exposed not only to static loads but also short duration loads. This paper presents an experimental methodology for investigating the transient response of flax and glass fibre reinforced composites, as well as hybrid of the two, to impulsive loads.

Keywords Flax composites · Hybrid composites · Transient response

Introduction

The commercial success of synthetic polymers reflects their demand such that the total production of synthetic polymers exceeds the sum of the production of all metals over the last few decades [1]. On the other hand, the increasing global environmental consciousness and social awareness, new environmental regulations

M. Cihan (✉) · J.I.R. Blake · A. Sobey
Fluid Structure Interactions Group, University of Southampton,
SO17 1BJ Southampton, UK
e-mail: mc3g14@soton.ac.uk

J.I.R. Blake
e-mail: J.I.R.Blake@soton.ac.uk

A. Sobey
e-mail: ajs502@soton.ac.uk

R. Fangueiro and S. Rana (eds.), *Advances in Natural Fibre Composites*,
https://doi.org/10.1007/978-3-319-64641-1_8

and high rate of depletion of petroleum have together provoked the composite industries to find greener solutions [2].

In parallel with these developments, natural fibre reinforced composites have received substantial attention due to their advantages over conventional materials, with perceived but not proven advantages in carbon footprint. Today, natural fibre reinforced composites have widely been used in many applications including structural, construction and automotive. The substitution of synthetic fibre composites with flax reinforced composites has already commenced with some interior items of cars including door panels and parcel shelves [3, 4].

The mechanical properties of flax fibres and their composites have already been studied by [5–9] while there is a less of an understanding of the materials response subjected to dynamic loads.

Designing a safe composite structure requires an understanding of material behaviour not only under static loads but also subjected to dynamic loads. Predicting the dynamic responses of composite materials is not straightforward as there are many concerns regarding the characterization of such materials. More issues arise when hybridizing conventional composites with flax fibres when there is such disparity in mechanical properties. The lack of literature on the effect of stacking sequences of flax/glass fibres on the response of such composite materials constitutes a challenge to characterize these materials.

Unlike the number of shock tube experimental studies on conventional materials there is only one shock tube experiment on flax fibre composites and no shock tube experiment on the hybrid of flax and glass fibres. The only shock tube experiment on flax composites was carried out by Huang et al. [10] who investigated the damage mechanism of both unidirectional and cross-ply flax fibres reinforced epoxy and polypropylene composite samples employing a helium gas driven shock tube. It was reported that the dynamic load resistance of cross-ply flax/epoxy samples is markedly superior to that of unidirectional flax/epoxy samples.

This paper presents preliminary results in the authors investigation into the transient response of flax and glass composite materials as well as their hybrid composites employing a vertical water shock tube.

Experimental

Presentation of the Materials Involved

In this study flax fibre and E-glass fibre were chosen as the reinforcements and epoxy resin as the matrix. Flax fibre (Flaxply BL200, woven cross-ply [0°/90°], 222.1 g/m^2) was purchased from Lineo, Belgium. Glass fibre (woven cross-ply [0°/90°], 590 g/m^2) was purchased from Gurit. The matrix system was Gurit Prime 20 LV Epoxy resin and Prime hardener.

Fabrication Process and the Volume Fraction Determination

Three different types of laminates were fabricated using a vacuum assisted resin infusion method, namely: $[F_6]$, $[G_6]$ and $[G_2/F]_s$. Once the resin infusion process was completed the laminate was left to cure at ambient temperature (8–15 °C) for 24 h. This curing process was followed by the recommended post curing schedule at which the laminates were placed into an oven at 65 °C for 7 h. The stacking sequences and the physical properties of the composite specimens were given in Table 1.

Due to the strong correlation between the fibre volume fraction and the mechanical properties, it is important to quantify the constituents of the composites. The flax fibre content in the composite samples was theoretically determined whilst the glass content was experimentally obtained as per ASTM 3171-15. The volume fractions of such composite samples can be seen in Table 2.

Shock Tube Setup

A shock tube experiment reproduces fluid structure interactions (FSI) that might be experienced in numerous applications from high speed craft planing to offshore structures. One of the advantages of using a shock tube is that shock tubes generate planar wavefronts regardless of the cross section of the tube [11].

In this study a vertical shock tube seen in Fig. 1 consists of a stainless steel tube with an outer diameter of 88.9 mm and a wall thickness of 5.49 mm. A piston is positioned at the top of the water column whilst the specimens were clamped at the bottom end of the tube, thus sealing the arrangement.

A free falling projectile mass is dropped at a desired height which impacts the piston. The impacted piston pressurises the water and generates an exponentially decaying pressure history in the tube at whose bottom end the sample is fastened. When the impulsive load is imparted to the specimen, the specimen undergoes a rapid escalation in pressure. The shock tube has been equipped with pressure

Table 1 Stacking sequences and the physical properties of the composites samples

Laminates	Stacking sequence	Laminate Thickness (mm)	Weight (g)
$[F_6]$	FFFFFF	3.40	46.1
$[G_6]$	GGGGGG	3.03	62.8
$[G_2/F]_s$	GGFFGG	3.35	58.5

Table 2 Fibre volume fractions of the composite samples

V_f (%)	$[F_6]$	$[G_6]$	$[G_2/F]_s$
V_{fflax}	36.46	–	14.09
V_{fglass}	–	47.42	26.07

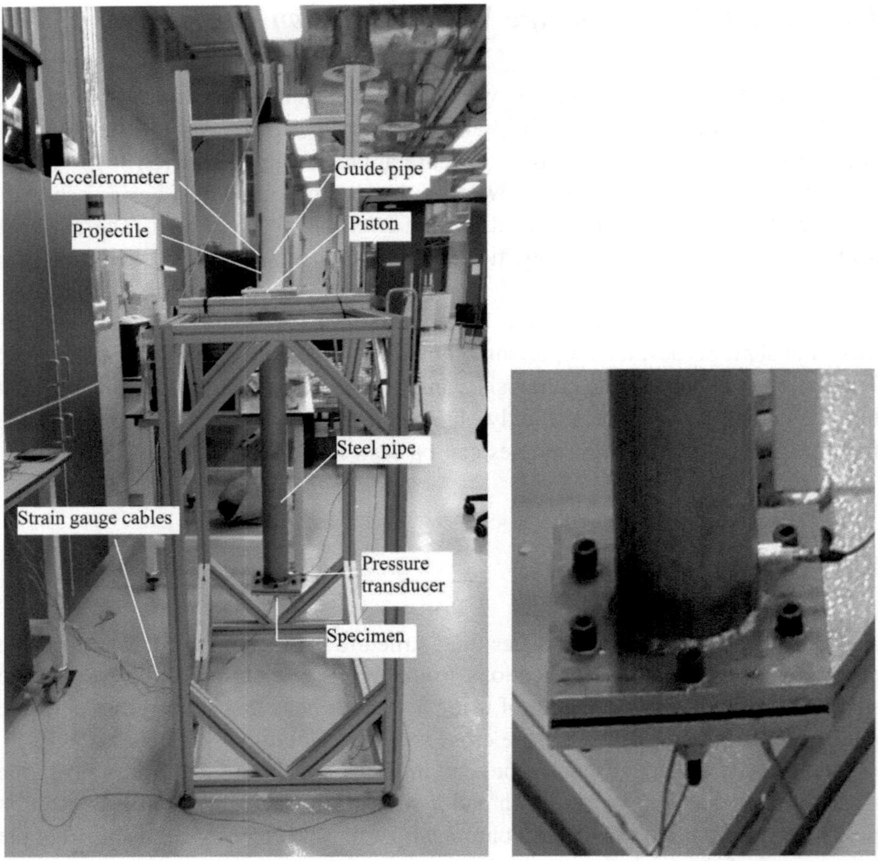

(a) Shock tube rig

(b) Position of the pressure transducer

Fig. 1 The shock tube setup with the measurement transducers

measurements to be able to measure the pressure near the specimen. It should be noted that the pressure mentioned here is the pressure read by the pressure transducer which is placed 30 mm above the specimen. The dynamic response of the samples was measured by means of a strain gauge bonded on the outer face of every sample.

The repeatability of the shock tube was assessed through some additional experiments prior to the actual experiments using a rigid stainless steel plate. A projectile mass was dropped from the same height several times. The pressure measurements were recorded and showed that the repeatability of the shock tube was good and the pressure measurement was reliable.

Once the generated impulse arrives, the pressure increases instantaneously to the peak value. Immediately afterwards, the pressure decreases at a nearly exponential

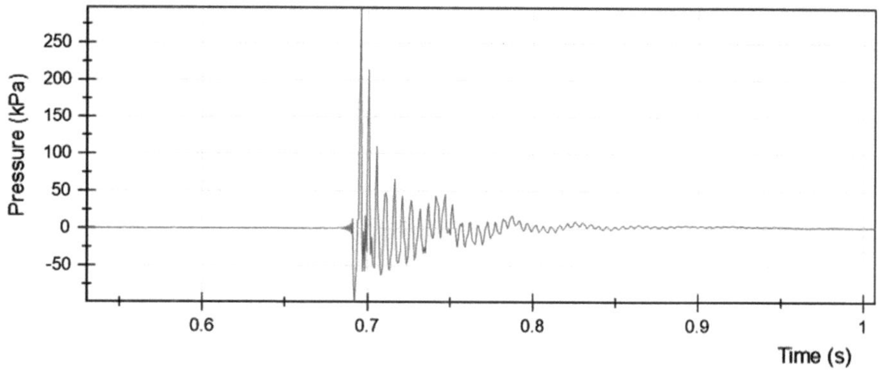

Fig. 2 The pressure profile generated by the shock tube

rate [12]. Consequently, the expected pressure profile was an exponentially decaying pressure with time which can be expressed as follows:

$$p = p_0 e^{-t/t_0} \tag{1}$$

where, p_0 is the initial peak and the t_0 is the decay time. The pressure history obtained using a rigid plate in this experiment is in Fig. 2.

The dynamic load impinged onto the clamped samples was generated through a water filled shock tube apparatus. A 3 kg of projectile was dropped free falling at a height of 30 mm for each sample for 6 times. The in-plane strain measurements experienced by the specimens were measured via the strain gauge bonded at the centre of the rear side of each sample. The strain measurement was acquired by National Instruments NI cDAQ-9135 data acquisition system and the sampling rate was 10 kHz.

Results and Discussion

The peak pressures for each run and the dynamic strain values at the centre of each composite specimen are given in Table 3. As can be seen, the pressure values measured for each composite material specimen was different.

Due to hydroelastic effects, it is clear that there is some coupled interaction between the fluid and the structure which is affecting for each specimen, under notionally the same imparted force on the piston, the perceived pressure. To be able to compare the dynamic responses of the composite specimens at the same pressure load the 3 kg projectile was dropped at different heights for each sample. The peak pressure was intended to be kept at around 270 kPa for each sample (therefore within the pressure transducer limit) as shown in Fig. 3.

Table 3 Peak in-plane strain values of the samples at the centre of the samples as well as the statistical values are reported

6 Runs	GGGGGG		FFFFFF		GGFFGG	
	Pressure (kPa)	Strain	Pressure (kPa)	Strain	Pressure (kPa)	Strain
1	330.99	1.57	227.08	2.64	292.35	1.34
2	342.74	1.64	223.60	2.63	300.54	1.39
3	324.61	1.53	216.90	2.57	309.94	1.42
4	315.78	1.49	229.59	2.71	310.24	1.43
5	330.09	1.56	226.28	3.19	305.64	1.41
6	344.25	1.62	230.07	3.24	297.29	1.36
Mean	331.41	1.57	225.59	2.83	302.67	1.39
S.D.	9.88	0.00006	4.44	0.00028	6.57	0.00003
CoV (%)	2.98	3.82	1.968	9.89	2.171	2.16

(Strain values multiplied by 10^3)

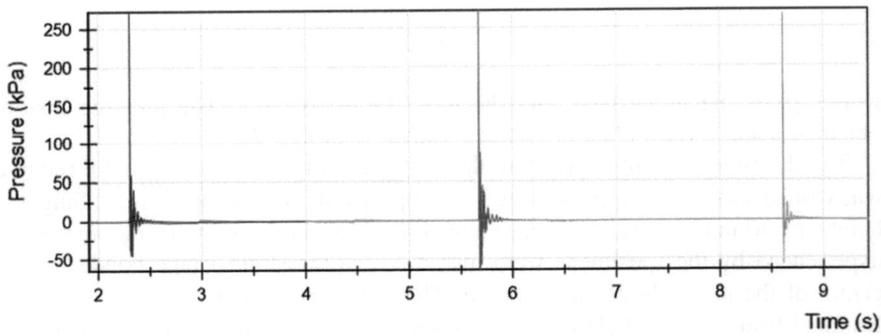

Fig. 3 The peak pressure was intended to be kept at 270 kPa to observe the response of the composite specimens at the same pressure load, *red*: glass, *green*: flax and *blue*: hybrid (Color figure online)

The in-plane strain that the flax sample experienced was more than twice that of glass sample as presented in Fig. 4.

It can be observed that hybridizing glass fibre with flax fibre in the configuration of GGFFGG did not actually show much difference in terms of in-plane strain in comparison to that of the glass sample. A similar transient response of flax/glass composite sample was observed by Morye et al. [13] where the behaviour GFG composite sample was attributed to the energy absorption characteristics of flax fibre when employed as a core in a composite.

Although the pressure history of the tube shows a similar trend which is an instant pressure increase followed by an exponential decay of the pressure, the pressure decayed much quicker in the flax sample, shown in Fig. 5.

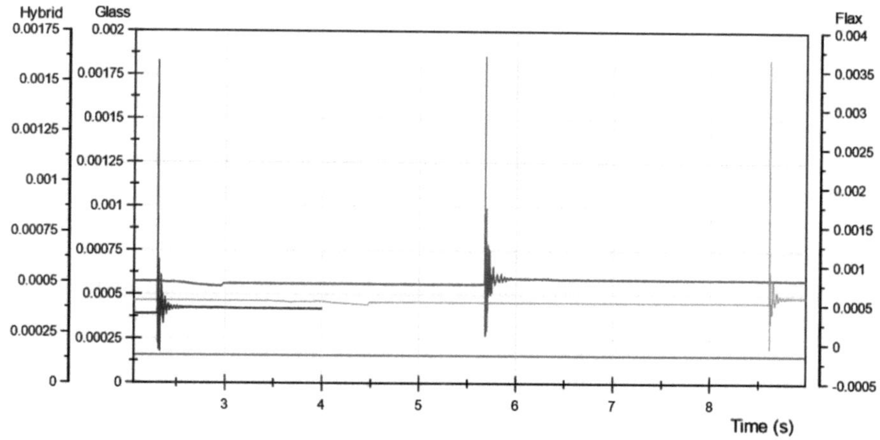

Fig. 4 Dynamic peak strain values of the composite specimens measured by the strain gauges bonded at the centre of the specimens, *red*: glass, *green*: flax and *blue*: hybrid (Color figure online)

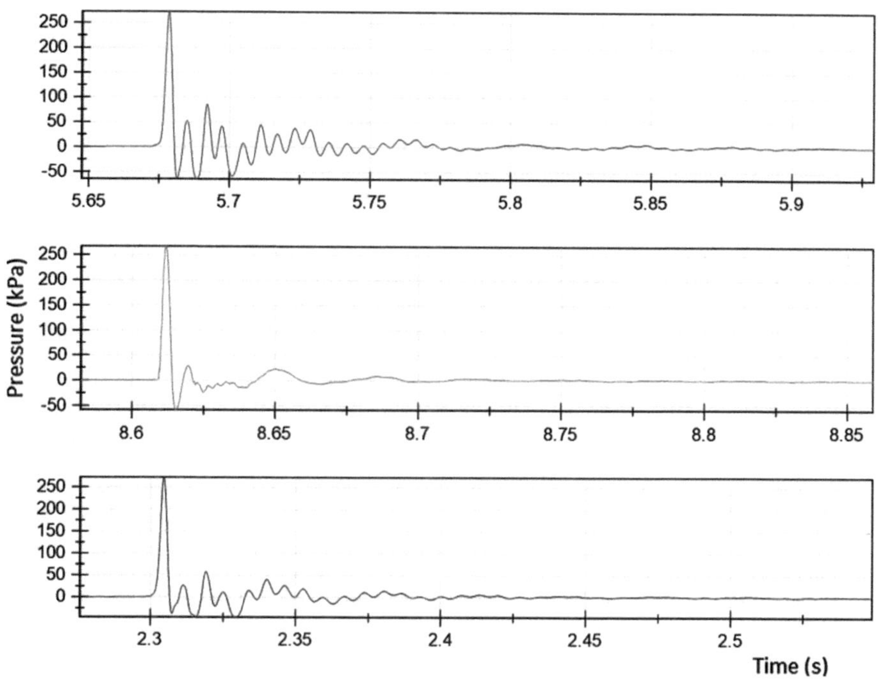

Fig. 5 Pressure profiles of the composite specimens over time, *red*: glass, *green*: flax and *blue*: hybrid (Color figure online)

Conclusion

An experimental study which includes the effect of fluid structure interaction was conducted to evaluate the dynamic response of three different laminate configurations under impulsive loading conditions. A shock tube was used to produce a dynamic pressure similar in profile to those encountered at sea. As expected, fluid structure interaction was observed as a difference between the measured pressures from material type to material type for the same drop height.

The results of this study show that GGFFGG composite specimens experienced slightly less in-plane strain than the GGGGGG composite sample while the FFFFFF composite sample's response was notably different. Consequently, sandwiching two plies of flax fibres between four plies of glass fibres enhanced the performance of the sample against the imparted impulsive load.

References

1. Brinson, H., & Brinson, L. C. (2008). *Polymer Engineering Science and Viscoelasticity*. Berlin: Springer.
2. Netravali, A. N., & Chabba, S. (2003). Composites get greener. *Materials today, 6*(4), 22–29.
3. Marsh, G. (2003). Next step for automotive materials. *Materials Today, 6*(4), 36–43.
4. Yan, L., Chouw, N., & Jayaraman, K. (2014). Flax fibre and its composites–a review. *Composites Part B Engineering, 56,* 296–317.
5. Baley, C., Perrot, Y., Busnel, F., Guezenoc, H., & Davies, P. (2006). Transverse tensile behaviour of unidirectional plies reinforced with flax fibres. *Materials Letters, 60*(24), 2984–2987.
6. Bax, B., & Müssig, J. (2008). Impact and tensile properties of pla/cordenka and pla/flax composites. *Composites Science and Technology, 68*(7), 1601–1607.
7. Adekunle, K., Cho, S. W., Ketzscher, R., & Skrifvars, M. (2012). Mechanical properties of natural fiber hybrid composites based on renewable thermoset resins derived from soybean oil, for use in technical applications. *Journal of Applied Polymer Science, 124*(6), 4530–4541.
8. Oksman, K. (2001). High quality flax fibre composites manufactured by the resin transfer moulding process. *Journal of Reinforced Plastics and Composites, 20*(7), 621–627.
9. Åkesson, D., Skrifvars, M., Seppälä, J., & Turunen, M. (2011). Thermoset lactic acid-based resin as a matrix for flax fibers. *Journal of Applied Polymer Science, 119*(5), 3004–3009.
10. Huang, K., Rammohan, A. V., Kureemun, U., Teo, W. S., Lee, H. P., et al. (2016). Shock wave impact behavior of flax fiber reinforced polymer composites. *Composites Part B Engineering, 102,* 78–85.
11. Bigler, B. R., Yu, A. W., & Bass, C. R. Planar wave propagation in shock tubes for replicating blast injury.
12. Swisdak, M. M Jr. (1978). Explosion effects and properties. Part ii. Explosion effects in water. Technical report, DTIC Document.
13. Morye, S. S., & Wool, R. P. (2005). Mechanical properties of glass/flax hybrid composites based on a novel modified soybean oil matrix material. *Polymer Composites, 26*(4), 407–416.

The Response of *Manicaria saccifera* Natural Fabric Reinforced PLA Composites to Impact by Fragment Simulating Projectiles

Sebastian Quintero, Alicia Porras, Camilo Hernandez and Alejandro Maranon

Abstract This chapter presents the impact behavior of a recently developed green composite material made of *Manicaria saccifera* natural fabric reinforced Poly-Lactic Acid (PLA). Composite coupons made of PLA and *Manicaria saccifera* fabric were produced by compression molding using the film stacking method. The composite ballistic limit (V_{50}) was determined by subjecting PLA/Manicaria coupons, of varying lay-ups and thicknesses, to ballistic impact loading using fragment simulating projectiles (FSPs) according to the MIL-STD-662F standard. It was found that coupons with areal densities between 0.2 and 0.3 g/cm^2 displayed a V_{50} between 50 and 70 m/s. Also, it was found that the V_{50} increased nonlinearly as a function of coupon thickness, but it does not depend on the composite stacking sequence. Finally, the energy absorbed by the material at impact on complete penetrations is uniform and independent of the striking velocity, whereas for partial penetrations increases exponentially.

Keywords *Manicaria saccifera* · Natural fiber composite
Fragment simulating projectile · Ballistic limit

S. Quintero · A. Porras (✉) · A. Maranon
Structural Integrity Research Group, Mechanical Engineering Department,
Universidad de los Andes, CR 1 ESTE 19A 40, 111711 Bogota, Colombia
e-mail: n-porras@uniandes.edu.co

A. Maranon
e-mail: emaranon@uniandes.edu.co

C. Hernandez
Sustainable Design in Mechanical Engineering Research Group (DSIM), Mechanical
Engineering Department, Escuela Colombiana de Ingeniería Julio Garavito,
AK 45 205-59, 111166 Bogotá, Colombia

R. Fangueiro and S. Rana (eds.), *Advances in Natural Fibre Composites*,
https://doi.org/10.1007/978-3-319-64641-1_9

Introduction

Environmental issues and regulations have promoted the use of natural fiber composites in diverse applications over many industries. This is the case for the construction industry where current demand of both energy-saving, and environmentally friendly buildings has encouraged the use of novel composite materials [1, 2]. Therefore, natural fiber composites have found an important space due to their environmental benefits combined with competitive structural and non-structural properties [3].

Most common natural fibers used as a reinforcement in composites for construction applications are flax, hemp, and jute. For example, composite panels made of plant oil-based resin and natural fiber mats made of flax, cellulose, pulp and hemp were found to have mechanical properties suitable for applications in housing construction materials [4]. Likewise, cellular beams and plates made from industrial hemp and flax fibers with unsaturated polyester resin showed potential to serve as primary load bearing components with competitive mechanical performance against conventional materials [5]. Also, jute/polypropylene laminate composites were studied for their potential application in structural insulated panels in construction as replacement of traditional Oriented Strand Board (OSB) laminates. This study showed a significant improvement in the energy absorption and the mechanical properties of the natural fiber composite laminates compared to traditional OSB laminates [6]. In overall, natural fibers have gained interest as reinforcing material in composites to produce new building materials for the construction industry. Main applications currently used for natural fiber composites include: building panels, roofing sheets, boards partitions, doors and windows, tiles, etc. [7].

Although the use of natural composite materials for construction applications has expanded and its main properties have been studied, there is a quest to develop even lighter materials with additional damage tolerance [8]. For instance, there is an interest to investigate the impact response of natural fiber composites. In particular, the response of architectonical panel against impacts in the ballistic range that would represent the effect of secondary blast, hurricane, tornado and foreign object debris from roads and runways [9].

The response of synthetic composite materials to ballistic impact has been investigated by many researchers. Studies have identified that fiber properties, fabric structure, number of fabric layers, areal density, projectile parameters and impact parameters are the major factors in energy absorption capability and ballistic impact resistance of composite systems [10, 11]. Most of the high performance synthetic fibers and their composites studied are Kevlar [12, 13], Ultra-High Molecular Weight Polyethylene [14, 15], and glass fiber composite panels [16, 17]. However, studies examining ballistic impact responses of natural fiber composites are limited. Wambua et al. [18] studied the ballistic properties of flax, hemp and jute fabric reinforced polypropylene composites processed using hot compression molding. Cassiano et al. studied the ballistic impact performance of Epoxy Composite Reinforced with Malva natural fibers [19].

In this direction, the research presented in this chapter investigates the ballistic impact behavior and tensile properties of a novel composite material made of Polylactic Acid (PLA) matrix reinforced with a natural fabric extracted from the *Manicaria saccifera* Palm using standard fragment-simulating projectiles (FSPs). The influence of composite lay-up configuration and thickness were evaluated.

Materials and Methods

Composite coupons used to characterize the impact behavior by fragment simulating projectiles were manufactured from *Manicaria saccifera* natural fabric, as reinforcement, and polylactic acid (PLA) as matrix. *Manicaria saccifera* is a palm that grows in lowland forests of the tropical and subtropical regions of South America. Portions of the palm are used by local population as sources for house thatches and sails (leaves), food (fruits and seedlings), hats (spathes), sago (trunk), and medicines (fruits and leaves). Among these, pouch-like spathes, with a typical length between 0.8 and 1.2 m, consist of a brown fibrous material constituted of numerous inter-crosslinked fibers with a closely woven texture, resembling a natural fabric. A previous study [20], found that Manicaria's specific fracture stress and specific modulus are 72.09 ± 15.73 MPa/g cm^{-3} and 2.20 ± 0.44 GPa/g cm^{-3}, respectively. Also, the Manicaria fabric displays a strength between 254.88 and 391.08 cN/m, which is equal or better than some synthetic fabrics, such as fiberglass fabrics, which is used in a wide range of industrial applications as reinforcement for polymers. Typically, to preprocess the natural fabric, both ends of the spathe are removed, and a longitudinal cut is made to open it in half.

PLA is a versatile thermoplastic produced from lactic acid monomers sourced mainly from the fermentation of corn, potatoes, sugar beets, and sugar cane [21]. Its main advantages are its fast bio-degradability, its mechanical properties, and its processing methods. In contrast to conventional polymers, PLA degrades under natural conditions in a short period of time after disposal [22]. PLA's mechanical properties, in terms of its tensile strength and elastic modulus, are superior to those of similar bio-degradable materials; and comparable to those of polystyrene. Finally, PLA is easily processed using conventional manufacturing methods as its melting temperature, of around 150 °C, facilitates the manufacturing of composites using standard processing equipment at temperatures below natural fiber degradation [23]. In this work, PLA was provided by Nature Works LLC, under the trade name PLA2003D. Granulates of PLA were converted into a thin sheet using a Brabender Plasticorder 331 extruder at a processing temperature between 180 and 210 °C.

Prior to composite manufacturing, *Manicaria saccifera* fabrics and PLA were dried in oven at 110 °C for 1 h. Composite coupons of approximately 12 cm by 12 cm were produced by the compression molding process using the film stacking method with the following stacking sequences $[0^P/0^M]_s$, $[0^P/90^M/0^M/0^P]$, $[0^P/0^M_2/0^P]_s$, and $[0^P/90^M/0^M/0^P]_s$, where superscripts P and M denote PLA and *Manicaria saccifera*, respectively. PLA and *Manicaria saccifera* fabrics were heat compressed in a LabTech Molding Press using the processing parameters described in [24]. Quasi-static tensile properties of composites for every stacking sequence were obtained using the ASTM D3039-08 standard using rectangular samples 15 mm wide and 250 mm long. Tests were carried out using a universal Instron 3367 testing machine, with a crosshead speed of 5 mm/min and a gauge length of 50 mm, using an extensometer device. Four specimens were tested for every stacking sequence.

Impact testing on PLA/Manicaria coupons were performed in a purposely built gas gun capable of delivering fragment simulating projectiles (FSPs) at speeds up to 250 m/s [25]. Fragment simulating projectiles (FSPs) with a straight tail, 0.22 cal (5.46 mm in diameter, 1.1 ± 0.1 g in weight), were manufactured according to the MIL-DTL-46593B standard [26] from a 4340H steel bar. Black nylon sabots, 10 mm in diameter, were used to both center fragments, and reduce the friction between fragments and the gas gun muzzle [27]. Composite coupons were clamped in place using a metallic frame resulting in an unclamped area of 10 cm by 10 cm. Figure 1 shows the experimental setup used in this study.

To characterize the V_{50} ballistic limit for the different PLA/Manicaria composite stacking sequences, the MIL-STD-662F standard procedures were applied. Specifically, the velocities of five partial and complete FSP penetrations were averaged, within a velocity spread between 36 and 48 m/s. A witness plate was not employed during experimentation, and as such, the types of penetrations are defined as follows: complete penetration (CP), when the projectile fully penetrates the material, and goes through it completely; and partial penetration (PP), when the projectile either partially penetrates the material (does not go through it completely)

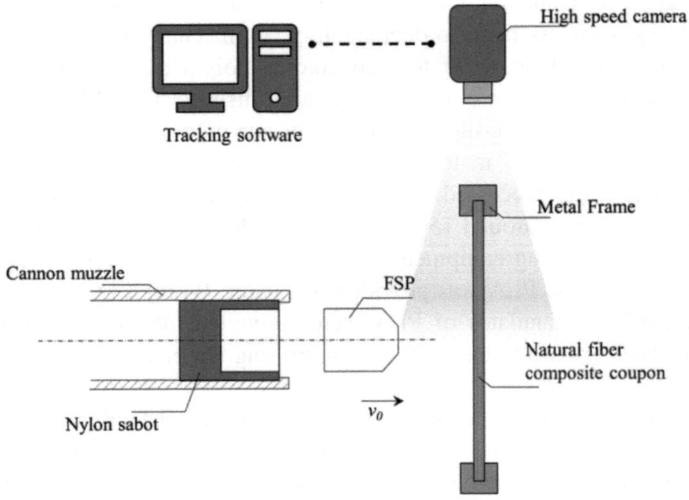

Fig. 1 Experimental setup for impact loading by FSPs

or bounces back without noticeably indenting the composite coupon. Also, the Up-and-Down method [28] was used during the V_{50} characterization: if a CP was achieved, the FSP striking velocity was reduced by approximately 15 m/s to yield a PP. If a PP was achieved, then the FSP striking velocity was increased by approximately 15 m/s to yield a CP. This process was repeated until a spread velocity between 36 and 48 m/s was achieved. In these conditions, the final striking velocity is the estimation of the V_{50} for the composite material. This method is valid if the impact tests can be represented as a cumulative normal (Gaussian) distribution.

To calculate the energy absorbed by the composite material during a FSP impact test (E_A), a simple energy balance approach [29–31], was employed, as shown in Eq. 1.

$$E_A = E_S - E_R = \frac{1}{2} m_{FSP} \left(V_S^2 - V_R^2 \right), \tag{1}$$

where E_S is the kinetic striking energy of the FSP, E_R is the FSP residual kinetic energy, m_{FSP} is the mass of the FSP, V_S is the FSP striking velocity, and V_R is the residual velocity (FSP velocity after impact, for a CP or PP alike). Since the distance from the metallic frame, where the material was clamped, to the cannon muzzle was very small, the effect of wind drag was not considered. Finally, it was assumed that a composite coupon absorbed as much energy, through damage mechanisms and elastic deformation, as the difference between the initial and the residual FSP kinetic energy. No other forms of energy dissipation were considered, although some energy was lost in the forms of heat and most noticeably, sound.

Results and Discussion

Figure 2 and Table 1 shows the physical and tensile properties of PLA/Manicaria coupons for the different stacking sequences studied. Given the orthotropy of the *Manicaria saccifera* fabric, unidirectional coupons $[0^P/0^M]_s$ and $[0^P/0^M_2/0^P]_s$ display better mechanical properties compared to the cross-ply laminates. Compared with previous studies, unidirectional coupons have comparable tensile strength and Young's modulus. For example, Porras et al. [32], found tensile strengths of about 68.5 ± 5.8 MPa, and Young's modulus of 4.9 ± 0.3 GPa. Also, in Porras et al. [24], authors reported tensile strengths of 123.36 ± 6.42 MPa, and Young's modulus of 7.61 ± 0.48 GPa.

Figures 3 and 4 show examples of high speed recordings for partial and complete penetrations of FSP into PLA/Manicaria coupons captured at 33,000 fps. Postmortem inspection of composite coupons revealed that most of them experienced delamination and fracture along the principal direction in the outer layers of the laminate. Internally, it was observed that the main energy absorption damage mechanisms were interlaminar delamination and fiber rupture.

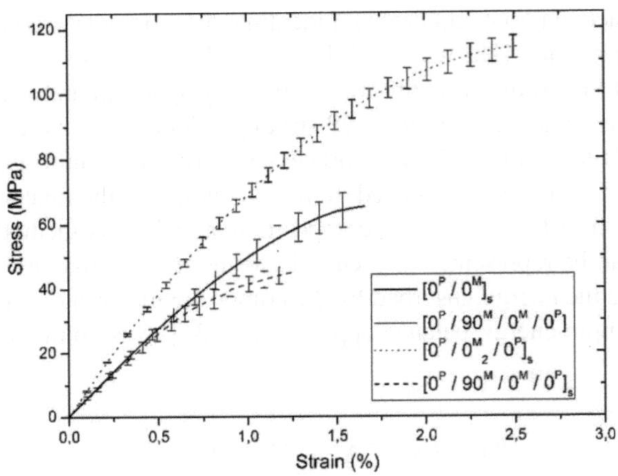

Fig. 2 Tensile stress-strain curves for PLA/Manicaria coupons

Table 1 Mechanical properties of PLA/Manicaria coupons

Stacking sequence	Mass (g)	Thickness (mm)	Tensile strength (MPa)	Tensile elongation (%)	Young's modulus (GPa)
$[0^P/0^M]_s$	27.3 ± 2.5	1.75 ± 0.3	67.14 ± 7.4	1.87 ± 0.1	5.66 ± 0.4
$[0^P/90^M/0^M/0^P]$	30.2 ± 2.3	1.80 ± 0.1	31.85 ± 6.0	0.69 ± 0.1	5.02 ± 0.3
$[0^P/0^M_2/0^P]_s$	43.8 ± 3.1	2.47 ± 0.1	114.83 ± 3.2	2.5 ± 0.3	7.57 ± 0.4
$[0^P/90^M/0^M/0^P]_s$	41.5 ± 3.9	2.65 ± 0.2	43.83 ± 5.9	1.26 ± 0.3	5.46 ± 0.2

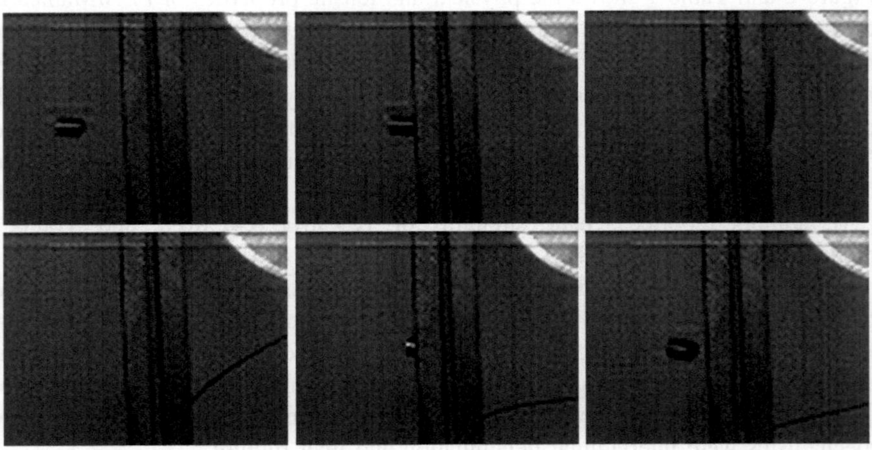

Fig. 3 Sequence of a partial penetration of a FSP at 48.1 m/s into a $[0^P/0^M]_s$ coupon

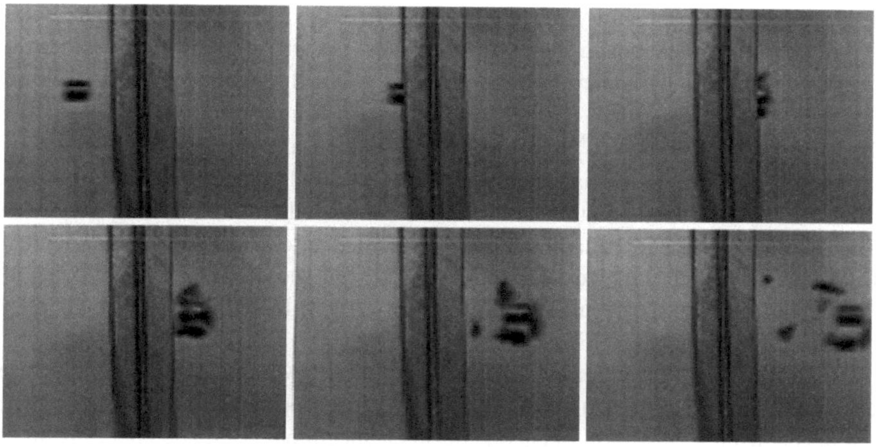

Fig. 4 Sequence of a complete penetration of a FSP at 69.9 m/s into a $[0^P/90^M/0^M/0^P]$ coupon

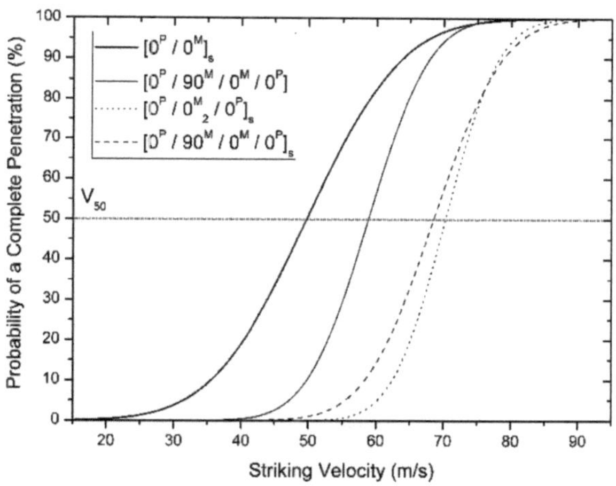

Fig. 5 Ballistic limit curves for of PLA/Manicaria coupons for 0.22 caliber FSP

Using the Tracker Video Analysis software, the FSP striking and residual velocities were measured for each impact, PP and CP alike. The software locates the position of the FSP on each individual frame and, using a reference distance, calculates the velocity per frame. Since various data points were recorded per flight, it was observed that the velocity of such points was distributed around a mean. Hence these velocities were averaged to obtain a single measurement of velocity (striking or residual) and the respective measure of error.

Figure 5 shows the ballistic limit curves for PLA/Manicaria composite coupons. Since V_{50} is a probabilistic property, for each stacking sequence the velocities are plotted against a standard cumulative normal distribution. From this plot is possible to determine the probability that, at a given velocity and for a given panel configuration, a complete penetration will take place. For this study, V_{50} ballistic limits for PLA/Manicaria laminates, as a function of the stacking sequence, are as follows: 49.71 ± 10.67 m/s for $[0^P/0^M]_s$, 58.59 ± 6.47 m/s for $[0^P/90^M/0^M/0^P]$, 70.33 ± 5.42 m/s for $[0^P/0^M_2/0^P]_s$, and 68.57 ± 7.43 m/s for $[0^P/90^{M/}0^M/0^P]_s$.

These results showed that there is no statistical difference of the V_{50} ballistic limit between coupons with similar thickness and different stacking sequence. Therefore, the ballistic behavior of the PLA/Manicaria composite is independent from configuration. However, the V_{50} ballistic limit showed a non-linear behavior with thickness. Similar results were previously found by Wambua et al. for other natural fiber composites [18].

Regarding the ballistic energy absorbed by the PLA/Manicaria coupons, it was calculated from the residual velocities. It was assumed that the projectile's mass remains constant and is homogeneous between specimens. Ballistic energy resulted uniform through the spectra of striking velocity and for all coupons configuration to an average of 2.0 J. This result implies that the energy dissipated by the composite material is independent from the striking velocity, and coupon configuration for the case of complete penetrations.

Conclusions

This study shows that the *Manicaria saccifera* natural fabric reinforced PLA composite could potentially be used for shielding against secondary debris from secondary blasts or debris from diverse sources.

The mechanical tensile properties and low range ballistic properties of the PLA/Manicaria were analyzed varying the thickness and stacking sequence. Different behavior was found for tensile quasi-static properties and ballistic properties. While for tensile mechanical properties, significant differences in the tensile strength were found when stacking configuration was varied, ballistic properties remained statistically constant for different configurations.

Ballistic limit for the PLA/Manicaria composite showed a nonlinear relation as a function of coupon thickness similarly to other tested natural fiber composites. However, the stacking configuration showed insignificance improvement on the V_{50} ballistic limit. Finally, the energy absorbed by the material at impact on complete penetrations is uniform and independent of the striking velocity and configuration. For the analyzed material thickness range, configurations and striking velocities of the present study, the material configuration does not affect the amount of energy that the panel is able to absorb for complete penetration impacts.

References

1. Laborel-Préneron, A., et al. (2016). Plant aggregates and fibers in earth construction materials: A review. *Construction and Building Materials, 111,* 719–734.
2. Korjenic, A., Zach, J., & Hroudová, J. (2016). The use of insulating materials based on natural fibers in combination with plant facades in building constructions. *Energy and Buildings, 116,* 45–58.
3. Mohanty, A., Misra, M., & Drzal, L. (2002). Sustainable bio-composites from renewable resources: Opportunities and challenges in the green materials world. *Journal of Polymers and the Environment, 10*(1), 19–26.
4. O'donnell, A., Dweib, M., & Wool, R. (2004). Natural fiber composites with plant oil-based resin. *Composites Science and Technology, 64*(9), 1135–1145.
5. Burgueño, R., et al. (2004). Load-bearing natural fiber composite cellular beams and panels. *Composites Part A Applied Science and Manufacturing, 35*(6), 645–656.
6. Kalyankar, R., & Uddin, N. (2012). Structural characterization of natural fiber reinforced polymeric (NFRP) laminates for building construction. *Journal of Polymers and the Environment, 20*(1), 224–229.
7. Singh, B., et al. (2011). Natural fiber-based composite building materials. In *Cellulose fibers: Bio-and nano-polymer composites* (pp. 701–720). Berlin: Springer.
8. Faruk, O., et al. (2012). Biocomposites reinforced with natural fibers: 2000–2010. *Progress in Polymer Science, 37*(11), 1552–1596.
9. Jover, N., Shafiq, B., & Vaidya, U. (2014). Ballistic impact analysis of balsa core sandwich composites. *Composites Part B Engineering, 67,* 160–169.
10. Ahmad, M. R., et al. (2008). Effect of fabric stitching on ballistic impact resistance of natural rubber coated fabric systems. *Materials and Design, 29*(7), 1353–1358.
11. Cheeseman, B. A., & Bogetti, T. A. (2003). Ballistic impact into fabric and compliant composite laminates. *Composite Structures, 61*(1), 161–173.
12. Lee, Y. S., Wetzel, E. D., & Wagner, N. J. (2003). The ballistic impact characteristics of Kevlar® woven fabrics impregnated with a colloidal shear thickening fluid. *Journal of Materials Science, 38*(13), 2825–2833.
13. Yahaya, R., et al. (2016). Investigating ballistic impact properties of woven kenaf-aramid hybrid composites. *Fibers and Polymers, 17*(2), 275–281.
14. Nguyen, L., et al. (2016). The efficiency of ultra-high molecular weight polyethylene composite against fragment impact. *Experimental Mechanics, 56*(4), 595–605.
15. Morye, S., et al. (2000). Modelling of the energy absorption by polymer composites upon ballistic impact. *Composites Science and Technology, 60*(14), 2631–2642.
16. Nunes, L., Paciornik, S., & d'Almeida, J. (2004). Evaluation of the damaged area of glass-fiber-reinforced epoxy-matrix composite materials submitted to ballistic impacts. *Composites Science and Technology, 64*(7), 945–954.
17. Deka, L., Bartus, S., & Vaidya, U. (2008). Damage evolution and energy absorption of E-glass/polypropylene laminates subjected to ballistic impact. *Journal of materials science, 43*(13), 4399.
18. Wambua, P., et al. (2007). The response of natural fibre composites to ballistic impact by fragment simulating projectiles. *Composite Structures, 77*(2), 232–240.
19. Nascimento, L. F. C., et al. (2017). Evaluation of ballistic armor behavior with epoxy composite reinforced with Malva fibers. In *Characterization of Minerals, Metals, and Materials* (pp. 647–655). Berlin: Springer.
20. Porras, A., Maranon, A., & Ashcroft, I. (2015). Characterization of a novel natural cellulose fabric from *Manicaria saccifera* palm as possible reinforcement of composite materials. *Composites Part B Engineering, 74,* 66–73.
21. Gupta, A., & Kumar, V. (2007). New emerging trends in synthetic biodegradable polymers–polylactide: A critique. *European Polymer Journal, 43*(10), 4053–4074.

22. Shih, Y.-F., Huang, C.-C., & Chen, P.-W. (2010). Biodegradable green composites reinforced by the fiber recycling from disposable chopsticks. *Materials Science and Engineering A, 527* (6), 1516–1521.
23. Mohanty, A. K., Misra, M., & Drzal, L. T. (2005). *Natural fibers, biopolymers, and biocomposites*. Boca Raton: CRC Press.
24. Porras, A., Maranon, A., & Ashcroft, I. (2016). Optimal tensile properties of a Manicaria-based biocomposite by the Taguchi method. *Composite Structures, 140,* 692–701.
25. Buchely, M. F., Maranon, A., & Silberschmidt, V. V. (2016). Material model for modeling clay at high strain rates. *International Journal of Impact Engineering, 90,* 1–11.
26. MIL-DTL-46593B—Projectile, Calibers 0.22, 0.30, 0.50, and 20 MM Fragment Simulating, 2006.
27. Stilp, A. J. (2005). Sabot designs for launching penetrators and projectiles. In *High-pressure shock compression of solids VIII* (pp. 201–225). Berlin: Springer.
28. MIL-STD-662F, Military standard V50 ballistic test for armor, 1997.
29. Abrate, S. (2007). Ballistic impact on composites. In *16th International Conference on Composite Materials*.
30. Hsieh, A. J., et al. (2004). *The effects of PMMA on ballistic impact performance of hybrid hard/ductile all-plastic-and glass-plastic-based composites*, DTIC Document.
31. Nilakantan, G., & Nutt, S. (2014). Effects of fabric target shape and size on the V50 ballistic impact response of soft body armor. *Composite Structures, 116,* 661–669.
32. Porras, A., Maranon, A., & Ashcroft, I. (2016). Thermo-mechanical characterization of *Manicaria saccifera* natural fabric reinforced poly-lactic acid composite lamina. *Composites Part A Applied Science and Manufacturing, 81,* 105–110.

Advances in Natural Fibre Reinforced Thermoplastic Composite Manufacturing: Effect of Interface and Hybrid Yarn Structure on Composite Properties

Mahadev Bar, R. Alagirusamy and Apurba Das

Abstract Natural fibre reinforced thermoplastic composite materials are becoming very popular in the material community due to several advantages of natural fibre and thermoplastic polymer. The demand of stronger and better composite than the existing ones is also increasing simultaneously with their growing popularity. However, natural fibre reinforced thermoplastics have some disadvantages associate with the poor fibre-matrix interaction, short length of the natural fibres and high melt viscosity of thermoplastic resins. All these factors decrease the ultimate mechanical properties of the natural fibre composites. However, the surface treatment of natural fibres, use of low twisted yarn and hybrid yarn during composite manufacturing significantly improve the mechanical properties of the natural fibre composites. The scope of all three approaches in determining composites mechanical properties have been reviewed here. Finally, the combined effects of interface and DREF spun hybrid yarn structure on the tensile and flexural properties of flax-PP based unidirectional composite specimen have been discussed in this chapter.

Keywords Natural fibre · Interface · Biocomposite · Hybrid yarn

Introduction

Reduction of carbon footprint, global warming and climate changing are the major concern of the present world. Governments are encouraging the researchers as well as industries to invent and implement the environment friendly, ecologically sustainable development mechanisms to overcome these major climate related issues. Use of natural resources and recyclable thermoplastic matrices for composite manufacturing, fit well into this picture [1–9]. Beside this, use of natural fibre as

M. Bar (✉) · R. Alagirusamy · A. Das
Department of Textile Technology, Indian Institute of Technology Delhi, New Delhi, India
e-mail: mahadevbar07@gmail.com

R. Fangueiro and S. Rana (eds.), *Advances in Natural Fibre Composites*,
https://doi.org/10.1007/978-3-319-64641-1_10

composite reinforcing material has few more advantages over the glass and other synthetic fibres. For instance, natural fibres are abundantly available in every corner of the world at reasonable price. They have low density, high specific strength and stiffness with compare to glass and other synthetic fibres. Natural fibres also exhibit unique acoustic and thermal insulating property due to their hollow cellulosic structure. The production of natural fibre requires very less energy and it involves CO_2 absorption whilst returning oxygen to the environment. In other words, natural fibres are CO_2 neutral i.e. the combustion of natural fibre does not return excess CO_2 to the atmosphere. The production of natural fibre does not depend on the fossil resources and it does not promote the activities like sand mining (sand is the major raw material for glass fibre) which is one of the main reason of soil erosions. Moreover, the use of natural fibre develops a non-food agricultural based economy and creates job opportunities in the rural areas [4–10].

Like natural fibres, recyclable thermoplastic polymers are also getting more attention of the composite researchers as well as composite manufacturing industries. This is mainly due to several advantages of thermoplastic polymers over the thermoset polymers. For instance, thermoplastic polymers are non-corrosive in nature, have long self-life and shorter curing cycle. They are ductile in nature and can be repaired, recycled and reused as per requirement. In room temperature, thermoplastic polymers remain in solid form. Hence, during processing it does not contaminate the machine parts [6–10]. The pre-mentioned advantages of natural fibre and thermoplastic resin find several uses of natural fibre reinforced thermoplastic composite in different areas such as automobiles, constructions, households, sporting goods, packaging etc. [8–15].

Since, last few decades lot of research have been conducted on natural fibre composite and many research articles have been published addressing various challenges. Besides research institutions, many industries such as Libeco Lagae (Belgium), Lineo (Belgium) and NPSP (Netherlands) have shown their interest over natural fibre reinforced thermoplastic composites. In a recent survey, it is predicted that the market size for natural fibre composites will reach to $5.83 billion by 2019 and it will continue with a compounded annual growth rate of 12.3% [16]. Along with this increasing market growth the demand of natural fibre based strong composites are also increasing exponentially [1–6].

However, natural fibre reinforced thermoplastics have some drawbacks. For instance, the poor fibre-matrix interaction which restricts the full strength utilization of natural fibre to the final composites. Secondly, natural fibres have short fibre length which make it difficult to control the fibre alignment into the composite structure. Unlike thermoset resins, thermoplastic resins have very high melt viscosity which restricts the even resin distribution into the composite structure and ultimately resulted in a composite with lot of void. The high viscous thermoplastic resin also promotes the fibre misalignment into the composite structure. All these factors diminish the mechanical performance of natural fibre composite [17–21]. Surface modification of natural fibre, use of low twist yarn and hybrid yarn during thermoplastic composite manufacturing can overcome these pre-mentioned shortcomings. Different shortcomings of natural fibre composites and their remedies are discussed in this chapter.

Interface Modification and Its Effect on Composite Properties

Natural fibres are hydrophilic in nature while thermoplastic matrices are hydrophobic in nature. Hence, the fibre and matrices are not compatible to each other. On the other hand, fibre-matrix interface plays an important role in determining the mechanical properties of the fibre reinforced composites. Because, the stress transfer from matrix to fibre is taking place across the interface. The poor or imperfect interface acts as a stress concentrator during mechanical testing [19–22]. Interfacial bonding between fibre and matrix can occur by means of four mechanism which are (i) mechanical bonding, (ii) electrostatic bonding, (iii) chemical bonding and (iv) reaction or inter-diffusion. In a good fibre-polymer interface, more than one force is acted at the same time and same interface [23].

Extensive research has been carried out, nevertheless research is still going on to improve the fibre-matrix bonding behavior of the natural fibre reinforced composite system [24–27]. The interfacial strength of a natural fibre composite system can be improved either by modifying the fibre surface or by modifying the matrix or by modifying the both. Among these methods, fibre surface modification is more effective than the matrix modification approach. Fibre surface modification to improve the fibre-matrix interaction can be largely divided into chemical approaches, physical approaches and biological approaches or enzyme treatment [1, 6–9]. The advantages and disadvantages of all three approaches have been discussed below.

Surface modification of natural fibre through chemical treatment is a very effective way of improving fibre-matrix interaction. There are several chemical treatment techniques available which can effectively modify the natural fibre surfaces. Among them, natural fibres surface treatment with alkali, MAgPP, saline coupling agent, benzyl, acryl, permanganate, peroxide, isocyanate are very well-known [1, 6–9, 19–22]. Alkali treatment removes the fat, wax, lignin and hemicellulose from the natural fibre surfaces. This makes the fibre surface rough and exposes the cellulose for interfacial bonding. Mild alkali treatment also improves the crystallinity and the breaking strength of the natural fibre [28–30]. While, the other chemicals except alkali are behaved as coupling agent. The one end of such chemicals is polar in nature while the other end is non-polar in nature. The polar end of the coupling agent reacts with the polar functional groups of the natural fibre while the non-polar end is entangled with the thermoplastic polymer chains. This phenomenon enhances the fibre-matrix interaction of the natural fibre composite system [25–28]. For example, MAgPP which can react with the hydroxyl groups present on the natural fibre surface through covalent bond

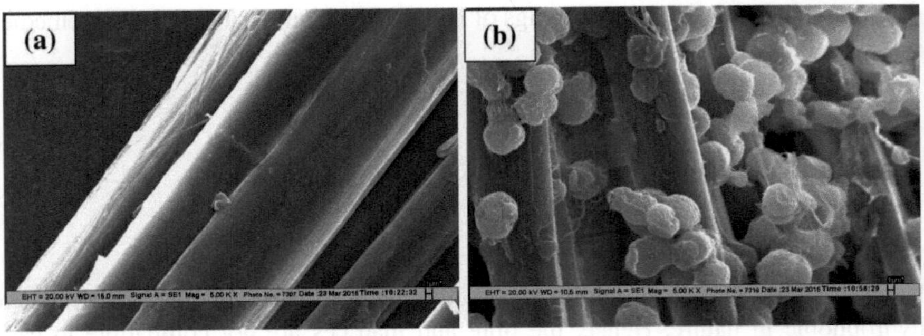

Fig. 1 SEM images of a untreated flax fibres, b MAgPP treated flax fibres

formation and forms a continuous layer of MAgPP around the natural fibre. Figure 1 shows the SEM images of untreated and MAgPP treated flax fibre. During composite manufacturing, the PP chains present on the fibre surface, adhere with the thermoplastic matrices.

It has been observed that the chemical treatment of natural fibre improves the tensile, flexural and other properties of the natural fibre composite. It has also been observed that among all methods of improving interfacial bond strength, the natural fibre treatment with MAgPP could be regard as most successful. It has been shown to give almost twice the composite strength as obtained with silane treatment [27].

Surface modification of natural fibre through chemical means successfully improves the interfacial bonding strength of natural fibre composite system. But, there are unresolved pollution problem related to disposal of chemicals after treatment. The cost of these chemicals is very high and most of them are health hazardous in nature [1, 6]. However, the physical approach of natural fibre surface modification does not have the pre-mentioned problems. Physical approach includes plasma treatment, corona discharge, electronic beam radiation, IR treatment, fibre beating etc. [1, 6–9, 24, 31]. Plasma is known as forth state of matter. It can be defined as partially ionized gases that have a collective ionized behavior. The main advantage of the plasma treatment is that it confines to the fibre surface only. It does not change the bulk properties of the material. The proper selection of starting compounds and external plasma parameters such as pressure, power and treatment time can create desire compound on the fibre surface [24]. Yuan et al. [32] have subjected wood fibre to the cold-plasma and Ar-Plasma treatment and produced wood-PP composite using the same. It is observed that the plasma treatment enhance the hydrophobicity and roughness of the wood fibre which significantly improves the tensile and flexural properties of the resultant composite. Corona is defined as a luminous, audible discharge that occurs due to inhomogeneous electrode geometries such as point electrode and plane. Compared to plasma, the corona discharges are relatively low power electrical discharge that takes place at atmospheric pressure. The corona discharge brings the chemical and physical changes of fibres including increasing surface polarity and roughness of the fibre. However, corona discharge is not so effective on three dimensional surfaces such as textile fabric [31, 32]. Belgacem et al. [33] have studied the effect of corona discharge on Cellulose-PP composite. It is observed that the composite strength improves when either one or both components are modified by corona discharge pretreatment. Electron beam radiation is another way of fibre surface modification. It improves the interfacial bonding between natural fibres and thermoplastic polymer by producing free radicals that encourage crosslinking [34]. The next physical approach of natural fibre surface modification is ultrasound treatment. Ultrasound is defined as very high frequencies of sound, above 20 kHz, generally used for medical and diagnostic purposes. Ultrasound treatment of natural fibre clean the fibre surface and make it rough, as a result it improves the fibre matrix interaction [31, 35].

Surface modification of natural fibre through biological means is another way of improving fibre-matrix interaction. Compared to chemical and physical approaches of natural fibre surface modification, the biological approach is quite new and advantageous [1, 6, 9]. In biological approaches natural fibres are treated with enzymes or fungi's. The biological treatment is environment friendly and have a focused performance. In general, the enzyme treatment removes hemi-cellulose and lignin from the fibre and makes the fibre surface rough. In addition, it also create holes on the fibre surface which helps in better interlocking with the matrix. Bledzki et al. [36] have studied mechanical properties of PP composite reinforced with enzyme treated abaca fibre. It is observed that the enzyme treatment enhances the fibre surface roughness as a result up to 45% improvement in composite tensile strength is observed. But, the enzyme treatment is time consuming and very specific to temperature and pH [1, 9].

Effect of Yarn Twist on Composite Properties

Composite materials are anisotropic in nature and exhibit maximum mechanical properties along the fibre direction [13]. Goutianos et al. [37] have reported that a woven fabric reinforced composite shows at least 3–4 time better mechanical properties than a nonwoven mat reinforced composite. Hence, it can be concluded that a composite will exhibit highest mechanical properties when the reinforcing fibres are completely aligned to the direction of applied load [38–40]. Effect of fibre orientation on composite properties is expressed by the rule of mixture equation:

$$E_c = \eta_0 \eta_l V_f E_f + \left(1 - V_f\right) E_m \tag{1}$$

where, E_f, E_m and E_c are the mechanical properties (modulus or failure strength) of the fibre, matrix and composite respectively, η_l is a factor related to fibre length, and η_0 is related to fibre alignment.

Unlike synthetic fibres, natural fibres have short fibre length. Hence, it is difficult to control these fibre alignment in a composite structure [41–43]. Textile yarns or yarn based textile structures can control the fibre orientation into the composite.

Yarn is a structure in which short staple fibres are twisted together. Twist enhances the cohesion between the fibres and holds the fibre together into the yarn structure. On the other hand, twist diminishes the maximum fibre strength utilization due to obliquity effect. At the same time, twist reduces the rate of production which enhances the ultimate cost of the yarn [43, 44]. In a composite structure fibres are held together by means of resin. Hence, once the composite is made, the yarn twist has no such importance. Goutianos et al. [37] have observed that low twisted yarn has low strength thus, it cannot be used in processes such as pultrusion or textile manufacturing routes like knitting or weaving. Therefore, it is concluded that the natural fibre based reinforcing yarn should have minimum level of twist which should allow it to process on textile machines.

Ma et al. [43] have manufactured sisal yarn reinforced phenol composite using sisal yarn of different twist level and have reported the effect of yarn twist on different composite properties. It is observed that high twisting level of the reinforcing yarn diminishes the mechanical properties of the resultant composite. A modified rule of mixture equation has been used in the same study to calculate the ultimate stress (σ_1) of the composite samples which are further compared with the experimental results. The modified rule of mixture equation is mentioned below.

$$\sigma 1 = \frac{(\mathrm{Vf} * \sigma \mathrm{f} * \mathrm{N} * \mathrm{Af} * \cos\theta\, \mathrm{mean}\ )}{\mathrm{Ay}} + \mathrm{Vm}\, \sigma \mathrm{m} \qquad (2)$$

where, σ_f is the axial tensile strength of the yarn A_f and A_y is the cross sectional area of sisal fibers and sisal yarns respectively. N is the no of fibre in the yarn cross-section. θ_{mean} is a function of average twist angle at the yarn surface. V_f and V_m are the fibre and matrix volume fraction respectively. σ_m is the axial tensile strength of the matrix.

Hybrid Yarn and Its Effect on Composite Properties

In the arena of natural fibre composite, thermoplastic polymers are getting more interest due to their several advantages which are already discussed earlier. However, unlike thermoset resins thermoplastic polymer melts have very high viscosity. This high melt viscosity makes the even distribution of thermoplastic resin in the composite structure very difficult. This problem increases the void content in the resultant composite and ultimately leading to inferior composite properties [9, 18, 45]. Use of hybrid yarn during thermoplastic composite manufacturing can solve the above mentioned problems. Hybrid yarns are the yarn having more than one fibre component in its structure. The hybrid yarn used for thermoplastic composite manufacturing has both the reinforcing as well as matrix forming fibre components.

Based on yarn structure, hybrid yarns are of three types namely wrapped yarn, core spun yarn and commingled yarn [18]. In a wrapped yarn structure, thermoplastic filaments are wrapped around a core of reinforcing fibers while in core spun yarn structure, thermoplastic staple fibres are wrapped around the core of reinforcing fibre. Commingled yarns are produced via commingling process in which the reinforcing and matrix forming filaments are passed through a compressed air nozzle and mixed together. Among these hybrid yarn structures, commingled yarn provides the high potential for through mixing of matrix and reinforcing components. However, natural fibres have short fibre length thus, it is not suitable for commingling [46]. Natural fibre based core spun or wrapped yarn structure can be manufactured through different spinning techniques such as ring spinning, rotor spinning, DREF spinning, wrap spinning, micro-braiding etc. Alagirusamy et al. [18] have summarized different methods of hybrid yarn manufacturing and have explained their effectiveness on composite properties.

In general, hybrid yarn reduces the effective resin flow distance as a result the thermoplastic resin distribution into the composite structure is improved. George et al. [47] have developed a bio-commingled composite using PP filament and twisted jute yarn. Bio-commingling improves the wetting of reinforced natural fibre which ultimately results in superior composite property. A similar kind of result is observed in case of micro-braided hybrid yarn reinforced composite studies where the micro-braided yarn is prepared using twisted jute yarn as core and PP filament as sheath [48]. Although, micro-braiding and bio-commingling enhance the natural fibre reinforced composite properties by lowering effective resin flow distance but these methods are not commercially viable. Because, the rate of production of micro-braided yarn is very low and the bio-commingling method is suitable for producing small and simple structures only.

Composites with Low Twisted Core Based Hybrid Yarn with Modified Interface

So far, the hybrid yarns for natural fibre composite reinforcement have a twisted core made of natural fibres. In most of these studies, the core natural fibres are not subjected to any surface modification. In this part of the present chapter, the combined effect of natural fibre surface modification and the hybrid yarn structure on composites tensile and flexural properties have been discussed. Hence, Flax-PP based core-sheath structured hybrid yarns are manufactured through DREF spinning method. During DREF spinning, yarn parameters such as core yarn twist and sheath percentage are varied at three different levels i.e. 0.72, 4.42, 8.12 TPI; and 40, 50, 60% respectively. Before hybridization, flax yarns are treated with MAgPP and depending on the experimental design MAgPP treated or untreated flax yarns are fed as core during DREF spinning. In this way, total 10 sets of hybrid yarn have been manufactured. The details of the produced hybrid yarn parameters are tabulated in Table 1.

Table 1 The details of hybrid yarns used for composite manufacturing

Untreated flax core		MAgPP treated flax core	
Core twist (twists/in.)	Natural fibre percentage	Core twist (twists/in.)	Natural fibre percentage
0.72	40	0.72	40
	50		50
	60		60
4.42	60	4.42	60
8.12	60	8.12	60

These hybrid yarns are further consolidated in a compression molding machine to manufacture unidirectional composite samples. These composite samples are then tested for tensile and flexural performances. The test results are analyzed critically and are reported below accordingly.

The stress-strain curves of different untreated as well as MAgPP treated flax reinforced composite samples are presented in Figs. 2 and 3 respectively. The photographs of tensile failure surfaces of untreated and MAgPP treated flax core based hybrid yarn reinforced unidirectional composite specimens are shown in Fig. 4 and in Fig. 5 respectively. The photographs reveal the manner in which composite specimens fail in tensile tests. In case of untreated flax yarn reinforced composite specimens and twisted, MAgPP treated flax yarn reinforced composite specimens, the failure is rather dominated by shear due to poor interfacial bonding between the fibres and matrices. It can also be predicted that in the above mentioned

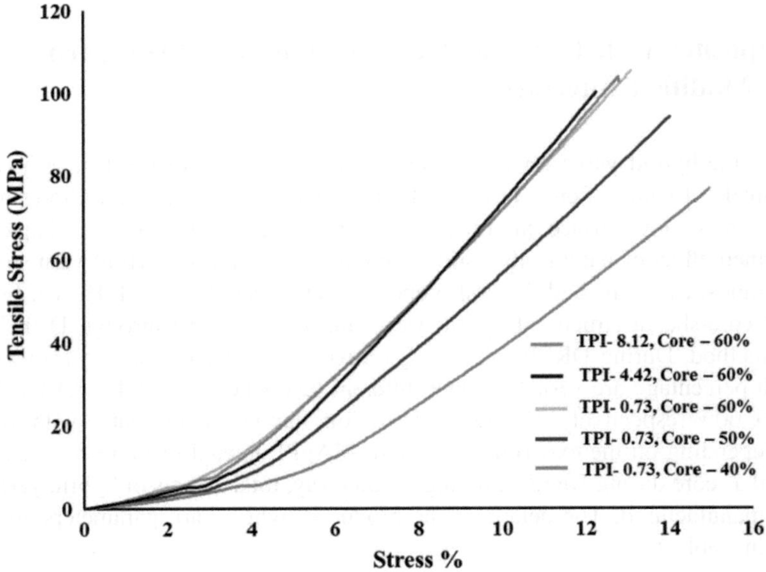

Fig. 2 Stress-strain curves of untreated flax core based hybrid yarn reinforced composite samples

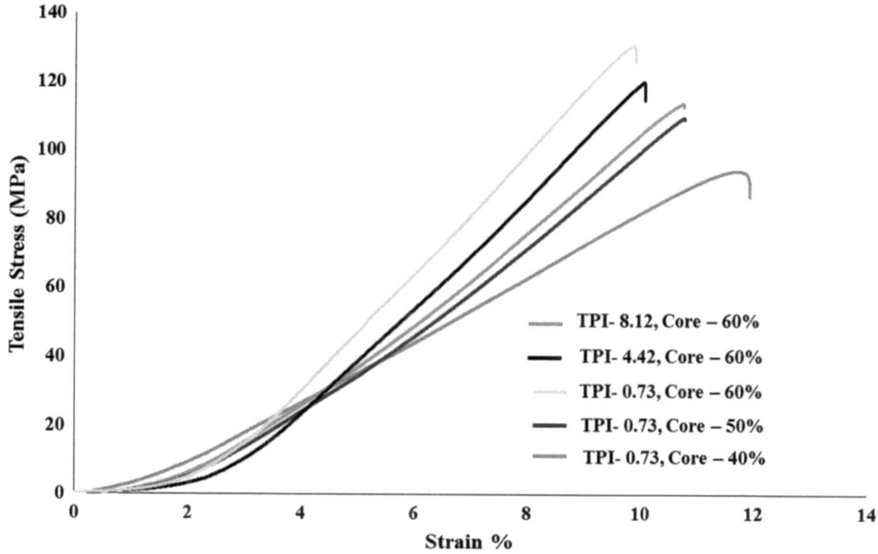

Fig. 3 Stress-strain curves of MAgPP treated flax core based hybrid yarn reinforced composite samples

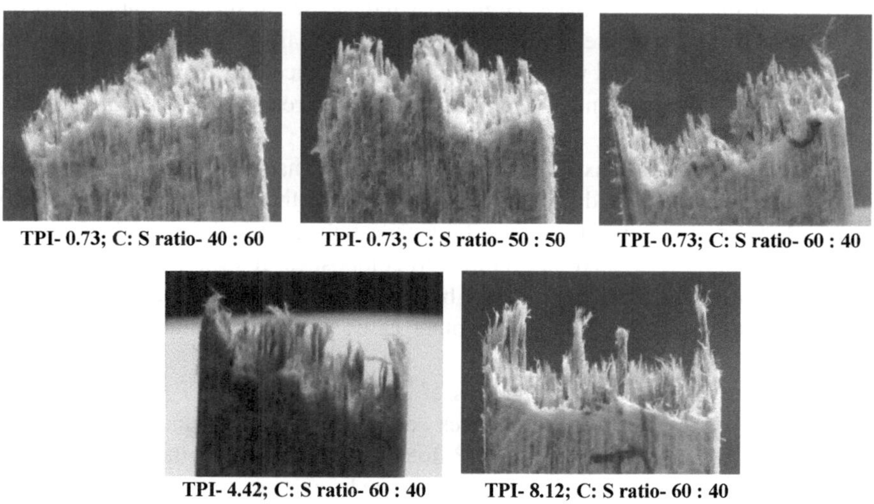

Fig. 4 Photographs of tensile failure ends of untreated flax core based hybrid yarn reinforced unidirectional composites

cases, the matrix failure occurs first then fibre slippage become more dominant. This resulted in tensile failure composite ends with lot of pulled out fibres. The photographs of tensile failure composite samples also state that the fibre pullout length decreased significantly with decreasing flax yarn twist and with decreasing

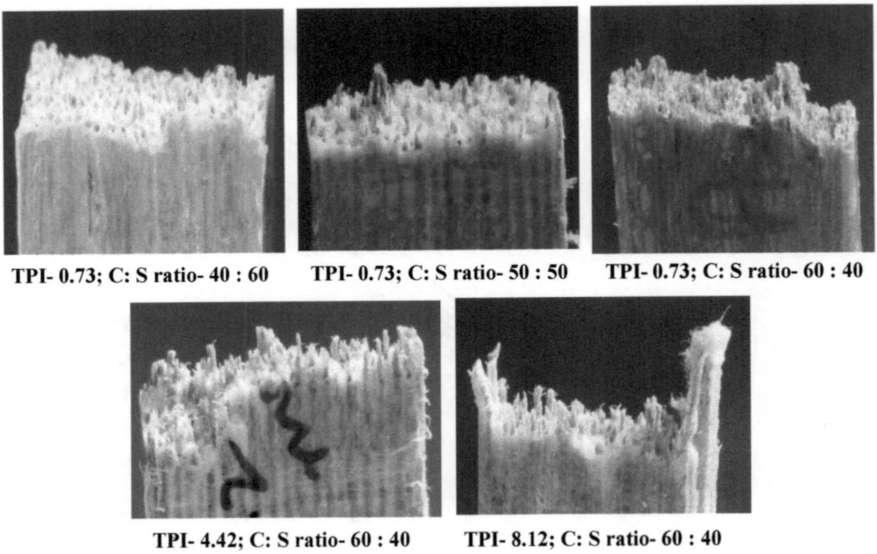

Fig. 5 Photographs of tensile failure ends of MAgPP treated flax core based hybrid yarn reinforced unidirectional composites

flax content into the composite. High twist reduces the resin penetration into the structure of yarn. The molten resin penetration into the fibre bundle is less when the composite is manufactured with low resin content. Hence, the resin distribution into the composite structure hinders significantly with increasing flax content and flax yarn twist.

MAgPP treatment of flax fibres improves the fibre-matrix bonding as a result the length of pulled out fibres decreases significantly after the MAgPP treatment of flax fibres. However, no fibre pull out at the tensile failure end is observed, in case of low twisted, MAgPP treated flax reinforced composites. It seems that in the above mentioned cases the fibre and matrix both behave as one unit and fail together almost in a straight line. While in all other cases the line of failure is an inclined curvy line.

SEM images of tensile fracture ends of untreated and MAgPP treated flax core based hybrid yarn reinforced composites are shown in Figs. 6 and 7 respectively. SEM images reveal that the molten PP does not penetrate into the highly twisted flax yarn structure. In case of partially untwisted flax yarn reinforced composites, the resin penetrates into the yarn surface layer only while the yarn core remains dry. Complete wetting of flax yarns are observed in case of low twisted flax yarn reinforced composite. Lot of fibre pull out is observed in the SEM images of untreated flax yarn reinforced composites and MAgPP treated, twisted flax yarn reinforced composites. While, in case of MAgPP treated, untwisted flax yarn reinforced composites no fibre pullout is observed due complete wetting of thermoplastic fibres and improved bonding between the fibres and matrices.

TPI- 0.73; C: S ratio- 60 : 40

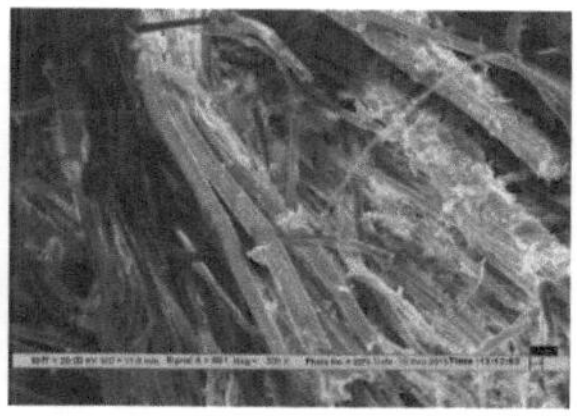
TPI- 0.73; C: S ratio- 50 : 50

TPI- 0.73; C: S ratio- 40 : 60

TPI- 4.42; C: S ratio- 40 : 60

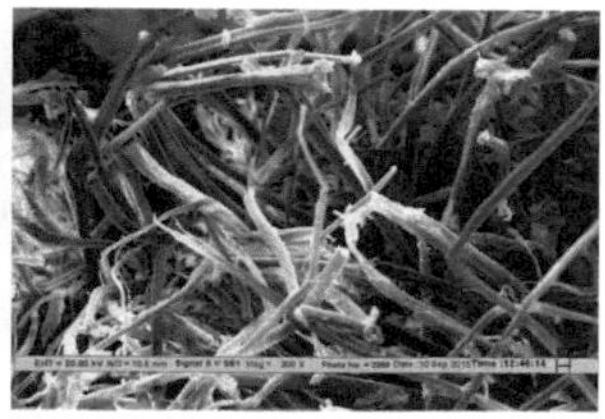
TPI- 8.12; C: S ratio- 40 : 60

Fig. 6 SEM images of untreated flax core based hybrid yarn reinforced, tensile fracture composite ends

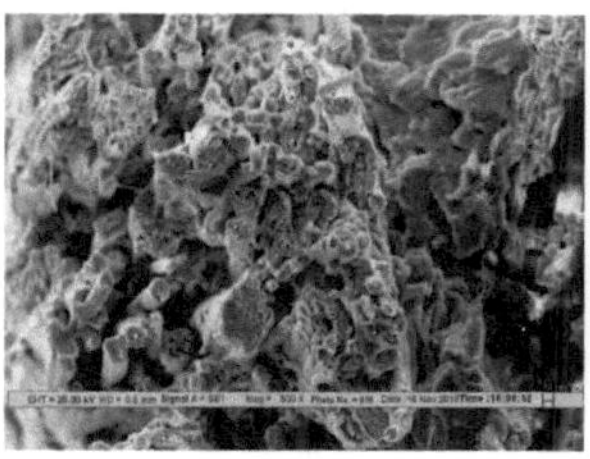
TPI- 0.73; C: S ratio- 60 : 40

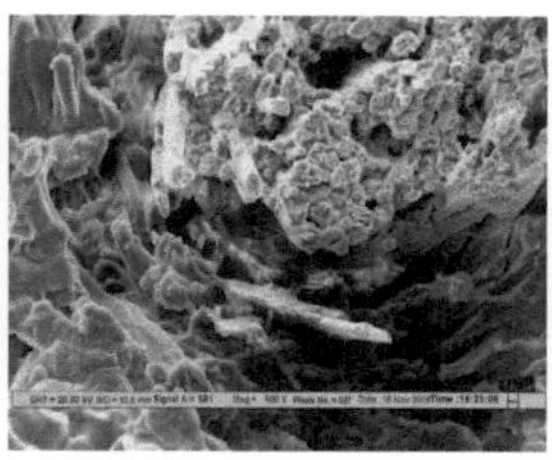
TPI- 0.73; C: S ratio- 50 : 50

TPI- 0.73; C: S ratio- 40 : 60

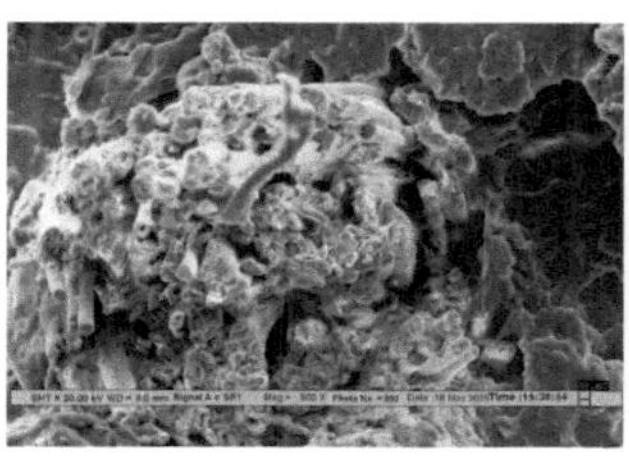
TPI- 4.42; C: S ratio- 40 : 60

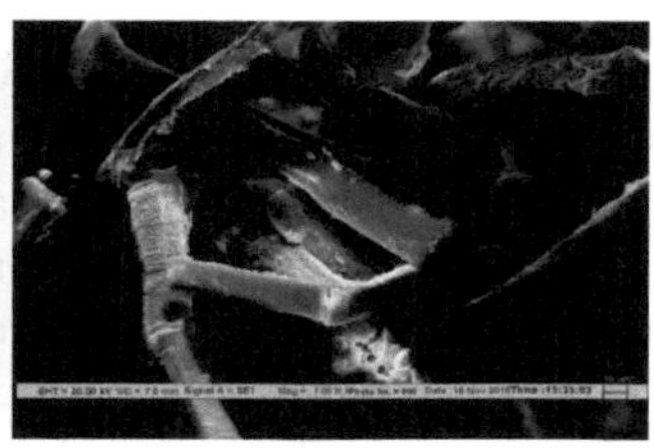
TPI- 8.12; C: S ratio- 40 : 60

Fig. 7 SEM images of MAgPP treated flax core based hybrid yarn reinforced, tensile fracture composite ends

Maximum tensile stress and the modulus of the MAgPP treated and untreated flax core based hybrid reinforced unidirectional composite samples are graphically presented in Fig. 8 and in Fig. 9 respectively. Flax fibres are the prime load bearing components of the flax-PP based hybrid yarn reinforced unidirectional composite. The fibre-matrix interface becomes stronger after the MAgPP treatment of flax

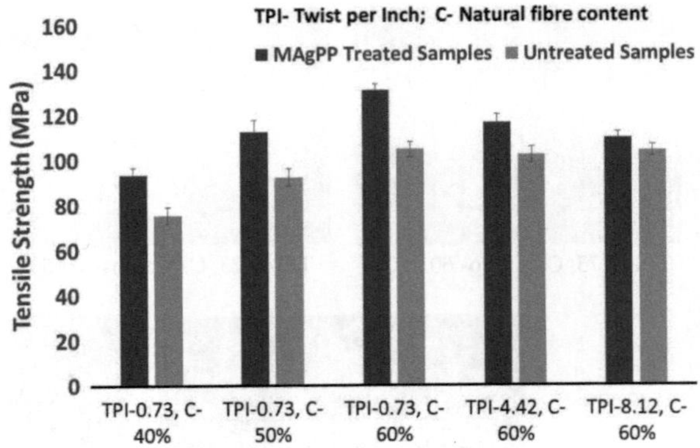

Fig. 8 Maximum tensile stress of the hybrid reinforced unidirectional composites

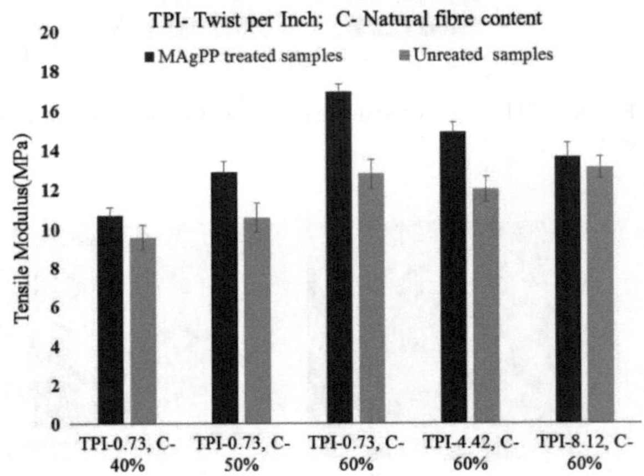

Fig. 9 Tensile modulus of the hybrid reinforced unidirectional composites

fibre [20, 21]. Hence, up to 25% improvement in the tensile strength and up to 33% improvement in tensile modulus of the unidirectional composites are observed after the MAgPP treatment of reinforcing flax yarns. Similar results are experienced in case of increasing flax fibre content from 40 to 60% in the composite structure. Low yarn twist enhances the fibre orientation in the composite structure. However, it does not exhibit any significant influence on the tensile strength and modulus of the untreated flax yarn reinforced unidirectional composites due to poor interfacial performance. On the other hand, around 20% improvement in tensile strength and 25% improvement in tensile modulus are observed while the reinforcing untwisted yarns are treated with MAgPP. This is mainly the result of improved fibre-matrix interaction and improved resin distribution into the composite structure.

Flexural strength and modulus of the flax-PP based hybrid yarn reinforced composite samples are tested according to three point bending test method. During three point bending test, one side of the composite specimen is subjected to a compressive force while the other side is subjected to the tensile force. The photographs of tensile and compression side of the untreated and MAgPP treated flax core based hybrid yarn reinforced unidirectional composite specimens are presented in Figs. 10 and 11 respectively. The photographs show that, after 3-point bending tests, PP matrix is accumulated on the compression side and no fracture is observed on the tensile side of the high twisted, untreated flax core based hybrid yarn reinforced unidirectional composite.

Few kinks on the compression side and some line of fracture on the tensile side are observed in case of low twisted, untreated flax reinforced composite and highly twisted, MAgPP treated flax reinforced composites. However, no matrix accumulation or kinks on the compression side and a sharp line of fracture on the tensile side of the MAgPP treated, low twisted flax reinforced composite are observed. The flexural stress-deformation curves of untreated and MAgPP treated composite samples are shown in Figs. 12 and 13 respectively. It is observed that the MAgPP treated composite samples have reached to a maximum stress level, then it experienced a sharp fall. After that, it continues to take load for a certain deflection and then it fails. On the other hand, untreated, twisted flax yarn reinforced composite reaches to a maximum stress level and then gradually fails after experiencing some deflection. However, a sharp fall after reaching the maximum stress is observed in case of low twisted, untreated flax yarn reinforced composite. It can be concluded that the interfacial strength is better in case of MAgPP treated flax yarn reinforced samples than the untreated one. Hence, the MAgPP treated flax yarn reinforced composites and low twisted, untreated flax yarn reinforced composites show a sharp fall in the flexural stress-deformation curves due to better interphase. On the other hand, due to poor interface no such sharp fall in the flexural stress-deformation curve of the untreated, twisted flax yarn reinforced composite is observed.

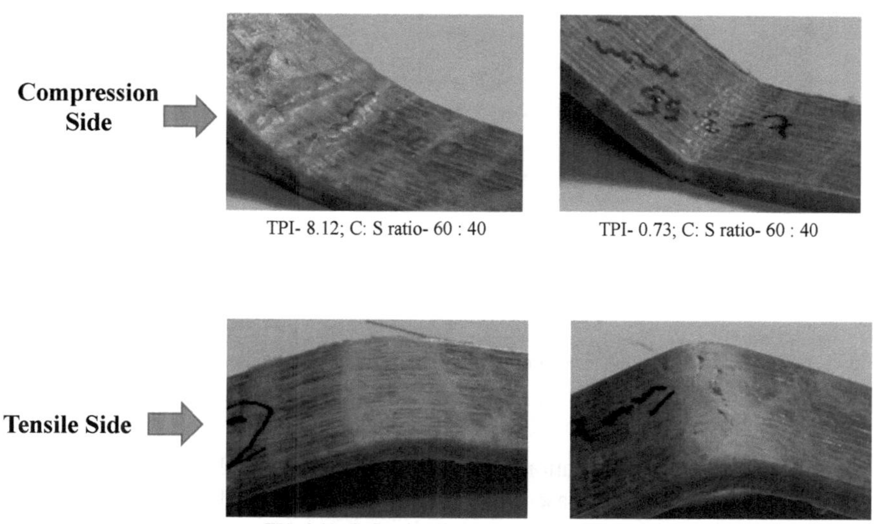

Fig. 10 The photographs of untreated flax core based hybrid yarn reinforced, 3-point bending tested unidirectional composite specimens

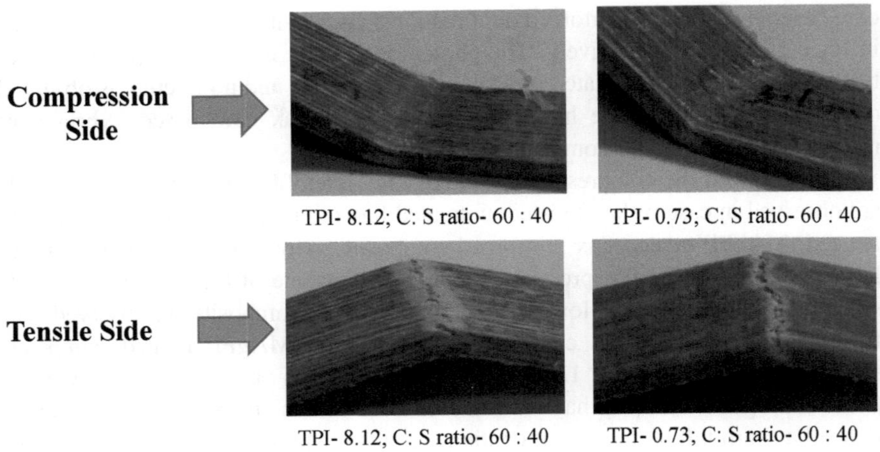

Fig. 11 The photographs of MAgPP treated flax core based hybrid yarn reinforced, 3-point bending tested unidirectional composite specimens

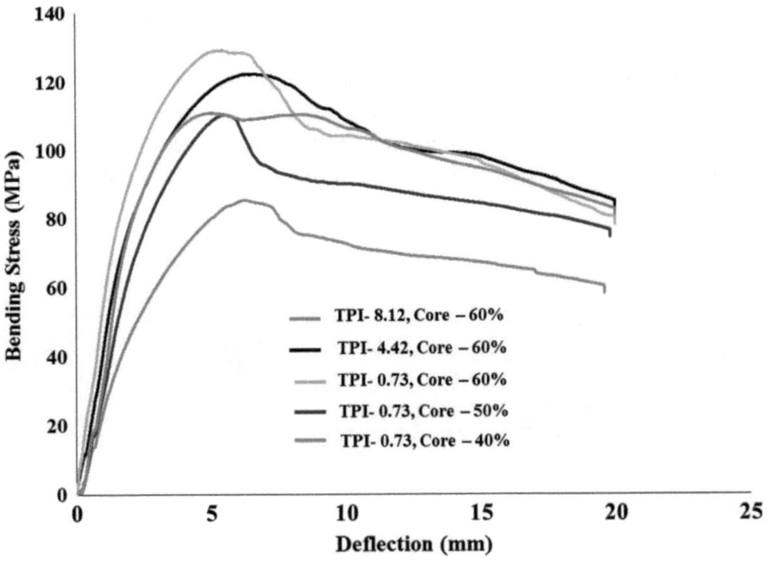

Fig. 12 Bending stress-deflection curves of untreated flax core based hybrid yarn reinforced composite samples

From the previous observations, it can also be concluded that the failure of high twisted, untreated flax core based hybrid yarn reinforced unidirectional composites is mainly dominated by the compressive force during 3 point bending tests. The combined action of tensile and compressive forces dominates in case of low twisted, untreated flax core based hybrid yarn reinforced composites and high

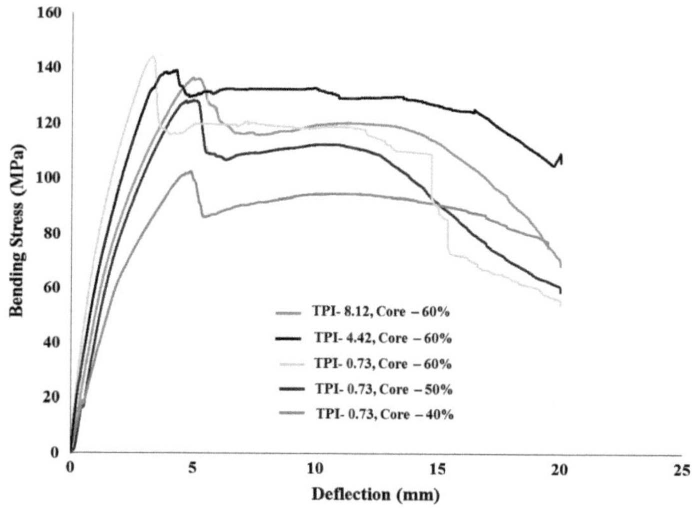

Fig. 13 Bending stress-deflection curves of MAgPP treated flax core based hybrid yarn reinforced composite samples

twisted, MAgPP treated flax core based composites. Tensile forces mainly dominate during the 3-point bending failure of low twisted, MAgPP treated flax reinforced composite samples.

Figures 14 and 15 represent the flexural strength and modulus values of the hybrid yarn reinforced unidirectional composite samples. It is observed that 40% improvement in tensile strength and 70% improvement in tensile modulus is experienced while the flax fibre content in the composite increases from 40 to 60%. However, up to 20% improvement in flexural modulus and 7% improvement in flexural strength are observed after the MAgPP treatment of core flax yarn. Similar

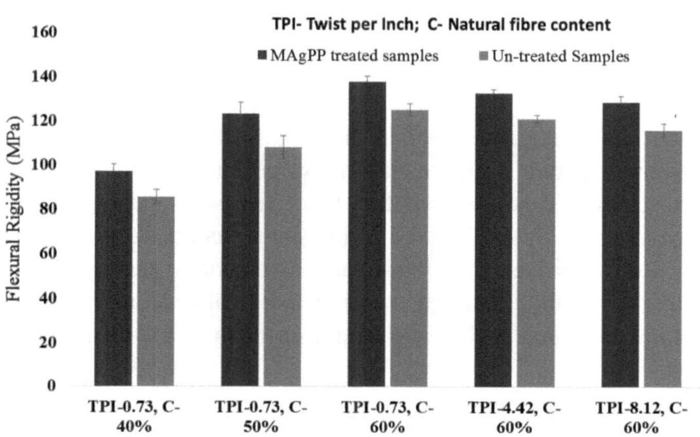

Fig. 14 Flexural rigidity of the hybrid reinforced unidirectional composites

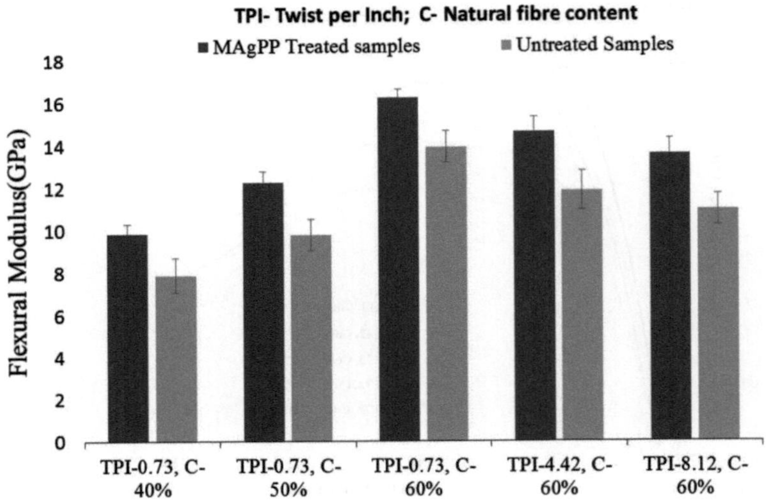

Fig. 15 Flexural modulus of the hybrid reinforced unidirectional composites

changes are experienced while the flax yarn twist is decreased from 8.12 TPI to 0.73 TPI. In this present composite system, flax fibres are the main load bearing components and after the MAgPP treatment of flax fibres the fibre-matrix interface become stronger. Hence, the flexural strength and modulus of the composite samples increase after MAgPP treatment of flax yarn and with increasing flax fibre content into the composite structure. Yarn twist of the reinforced flax yarn decreases the alignment of the reinforced flax fibres into the composite structure and it also restricts the PP resin distribution into the composite. Hence, flexural strength and modulus of the composite samples increase with decreasing reinforced flax yarn twist.

Conclusions

In this chapter, the individual as well as the combined effect of interface, yarn twist and hybrid yarn structure on the natural fibre reinforced thermoplastic composite have been discussed. It is observed that low yarn twist and high matrix content improve the resin distribution into the composite structure. Surface modification of the natural fibre improves the fibre-matrix interaction. Hence, the tensile and flexural properties of the unidirectional composite enhance after the surface treatment of natural fibre and with decreasing yarn twist and with increasing natural fibre content.

References

1. Pickering, K. L., Efendy, M. G. A., & Le, T. M. (2016). A review of recent developments in natural fibre composites and their mechanical performance. *Composites Part A, 83,* 98–112.
2. Netravali, A. N., & Chabba, S. (2003). Composites get greener. *Materials Today, 6*(4), 22–29.
3. Pandey, J. K., Nagarajan, V., Mohanty, A. K., et al. (2015). Commercial potential and competitiveness of natural fiber composites. In M. Misra, J. K. Pandey, & A. K. Mohanty (Eds.), *Biocomposites: Design and mechanical performance* (pp. 1–15). Cambridge, England: Woodhead Publishing.
4. Joshi, S. V., Drzal, L. T., Mohanty, A. K., et al. (2004). Are natural fiber composites environmentally superior to glass fiber reinforced composites? *Composites Part A, 35,* 371–376.
5. Wambua, P., Ivens, J., & Verpoest, I. (2003). Natural fibres: Can they replace glass in fibre reinforced plastics? *Composites Science and Technology, 63,* 1259–1264.
6. John, M. J., & Thomas, S. (2008). Biofibres and biocomposites. *Carbohydrate Polymers, 71,* 343–364.
7. Gurunathan, T., Mohanty, S., & Nayak, S. K. (2015). A review of the recent developments in biocomposites based on natural fibres and their application perspectives. *Composites Part A, 77,* 1–25.
8. Bar, M., Alagirusamy, R., & Das, A. (2015). Flame retardant polymer composite. *Fibres and Polymers, 16*(4), 705–715.
9. Faruk, O., Bledzki, A. K., Fink, H. P., et al. (2012). Biocomposites reinforced with natural fibers: 2000–2010. *Progress in Polymer Science, 37,* 1552–1596.
10. Dittenber, D. B., & GangaRao, H. V. S. (2012). Critical review of recent publications on use of natural composites in infrastructure. *Composites Part A, 43,* 1419–1429.
11. Pickering, K., Beckermann, G., Alam, S., et al. (2007). Optimizing industrial hemp fibre for composites. *Composites Part A, 38,* 461–468.
12. Pandey, J. K., Ahn, S. H., Lee, C. S., et al. (2010). Recent advances in the application of natural fiber based composites. *Macromolecular Materials and Engineering, 295*(11), 975–989.
13. Alves, C., Ferrao, P. M. C., Silva, A. J., et al. (2010). Ecodesign of automotive components making use of natural jute fiber composites. *Journal of Cleaner Production, 18,* 313–327.
14. Koronis, G., Silva, A., & Fontul, M. (2013). Green composites: A review of adequate materials for automotive applications. *Composites Part B, 44,* 120–127.
15. Davoodi, M. M., Sapuan, S. M., Ahmad, D., et al. (2010). Mechanical properties of hybrid kenaf/glass reinforced epoxy composite for passenger car bumper beam. *Materials and Design, 31*(10), 4927–4932.
16. Ali, A., Shaker, K., Nawab, Y., et al. (2016). Hydrophobic treatment of natural fibers and their composites—A review. *Journal of Industrial Textiles,* 1–33 doi:10.1177/1528083716654468.
17. Mohanty, A. K., Misra, M., & Drzal, L. T. (2002). Sustainable bio-composites from renewable resources: Opportunities and challenges in the green materials world. *Journal of Polymers and the Environment, 10,* 19–26.
18. Alagirusamy, R., Fangueiro, R., & Ogale, V. (2006). Hybrid yarns and textile preforming for thermoplastic composites. *Textile Progress, 38*(4), 1–71.
19. Torres, F. G., & Cubillas, M. L. (2005). Study of the interfacial properties of natural fibre reinforced polyethylene. *Polymer Testing, 24,* 694–698.
20. Zafeiropoulos, N. E., Baillie, C. A., & Hodgkinson, J. M. (2002). Engineering and characterisation of the interface in flax fibre/polypropylene composite materials. Part II. The effect of surface treatments on the interface. *Composites: Part A, 33,* 1185–1190.
21. Lu, J. Z., Wu, Q., & McNabb, H. S. (2000). Chemical coupling in wood fiber and polymer composites: A review of coupling agents and treatments. *Wood and Fiber Science, 32*(1), 88–104.

22. Chen, P., Lu, C., & Yu, Q. (2006). Influence of fiber wettability on the interfacial adhesion of continuous fiber-reinforced PPESK composite. *Journal of Applied Polymer Science, 102*(3), 2544–2551.
23. Matthews, F. L., & Rawlings, R. D. (1999). *Composite materials: Engineering and science*. Cambridge, England: Woodhead Publishing.
24. de Farias, J. G. G., Cavalcante, R. C., & Canabarro, B. R. (2017). Surface lignin removal on coir fibers by plasma treatment for improved adhesion in thermoplastic starch composites. *Carbohydrate Polymers, 165,* 429–436.
25. Yanjun, X., Callum, A. S. H., Zefang, X., et al. (2010). Silane coupling agents used for natural fiber/polymer composites: A review. *Composites Part A, 41,* 806–819.
26. Mohanty, A. K., Khan, M. A., & Hinrichsen, G. (2000). Surface modification of jute and its influence on performance of biodegradable jute-fabric/biopol composites. *Composites Science and Technology, 60*(11), 15–24.
27. Bera, M., Alagirusamy, R., & Das, A. (2010). A study on interfacial properties of jute-PP composites. *Journal of Reinforced Plastics and Composites, 29,* 3155–3161.
28. Kabir, M. M., Wang, H., & Lau, K. T. (2012). Chemical treatments on plant-based natural fibre reinforced polymer composites: An overview. *Composites Part B, 43*(7), 2883–2892.
29. Beckermann, G. W., & Pickering, K. L. (2008). Engineering and evaluation of hemp fibre reinforced polypropylene composites: Fibre treatment and matrix modification. *Composites Part A, 39*(6), 979–988.
30. Gomes, A., Matsuo, T., Goda, K., et al. (2007). Development and effect of alkali treatment on tensile properties of curaua fiber green composites. *Composites Part A, 38*(8), 1811–1820.
31. Mukhopadhyay, S., & Fangueiro, R. (2009). Physical modification of natural fibers and thermoplastic films for composites—A review. *Journal of Thermoplastic Composite Materials, 22,* 135–163.
32. Yuan, X., Jayaraman, K., & Bhattacharyya, D. (2004). Effect of plasma treatment in enhancing the performance of woodfibre-Polypropylene composite. *Composites Part A, 35,* 1363–1374.
33. Belgacem, M. N., Bataille, P., & Sapieha, S. (1994). Effect of corona modification on cellulose/PP composites. *Journal of Applied Polymer Science, 53,* 379–385.
34. Huber, T., Biedermann, U., & Muessig, J. (2010). Enhancing the fibre matrix adhesion of natural fibre reinforced polypropylene by electron radiation analyzed with the single fibre fragmentation test. *Composite Interfaces, 17*(4), 371–381.
35. Laine, J. E., & Goring, D. A. I. (1977). Influence of ultrasonic irradiation on the properties of cellulosic fibers. *Cellulose Chemistry and Technology, 11*(5), 561–567.
36. Bledzki, A. K., Mamun, A. A., & Jaszkiewicz, A. (2010). Polypropylene composites with enzyme modified abaca fibre. *Composites Science and Technology, 70,* 854–860.
37. Goutianos, S., Peijs, T., & Nystrom, B. (2006). Development of flax fibre based textile reinforcements for composite applications. *Applied Composite Materials, 13*(4), 199–215.
38. Shah, D. U. (2016). Damage in biocomposites: Stiffness evolution of aligned plant fibre composites during monotonic and cyclic fatigue loading. *Composites Part A, 83,* 160–168.
39. Amor, I. B., Rekik, H., & Kaddami, H. (2010). Effect of Palm tree fiber orientation on electrical properties of palm tree fiber-reinforced polyester composites. *Journal of Composite Materials, 44*(13), 1553–1568.
40. Norman, D. A., & Robertson, R. E. (2003). The effect of fiber orientation on the toughening of short fiber-reinforced polymers. *Journal of Applied Polymer Science, 90*(10), 2740–2751.
41. Santamala, H., Livingston, R., Sixta, H., et al. (2016). Advantages of regenerated cellulose fibres as compared to flax fibres in the processability and mechanical performance of thermoset composites. *Composites Part A, 84,* 377–385.
42. Kobayashi, S., & Takada, K. (2013). Processing of unidirectional hemp fiber reinforced composites with micro-braiding technique. *Composites Part A, 46,* 173–179.
43. Ma, H., Li, Y., & Luo, Y. (2011). The effect of fiber twist on the mechanical properties natural fiber reinforced composites. In *18th International Conference on Composite Materials.*

44. Omrani, F., Wang, P., Soulat, D., et al. (2017). Mechanical properties of flax-fibre-reinforced preforms and composites: Influence of the type of yarns on multi-scale characterization. *Composites Part A, 93,* 72–81.
45. Bar, M., Das, A., & Alagirusamy, R. (2017). Studies on flax-polypropylene based low-twist hybrid yarns for thermoplastic composite reinforcement. *Journal of Reinforced Plastics and Composites*. doi:10.1177/0731684417693428.
46. Ye, L., Friedrich, K., Kastel, J., et al. (1995). Consolidation of unidirectional CF/PEEK composites from commingled yarn prepreg. *Composites Science and Technology, 54*(4), 349–358.
47. George, G., Jose, E. T., Jayanarayanan, K., et al. (2012). Novel bio-commingled composites based on jute/polypropylene yarns: Effect of chemical treatments on the mechanical properties. *Composites Part A, 43,* 219–230.
48. Khondker, O. A., Ishiaku, U. S., Nakai, A., et al. (2006). A novel processing technique for thermoplastic manufacturing of unidirectional composites reinforced with jute yarns. *Composites Part A, 37,* 2274–2284.

Environmental Friendly Thermoplastic Composite Laminates Reinforced with Jute Fabric

Pietro Russo, Giorgio Simeoli, Valentina Lopresto, Antonio Langella and Ilaria Papa

Abstract A commercial jute fabric was used as the reinforcement of two thermoplastic matrices: polypropylene (PP) and poly(lactic acid) (PLA) having glass transition temperatures equal to 8 and 60 °C respectively. The latter were evaluated by previous dynamic-mechanical tests. Plates prepared by conventional film stacking and compression molding procedures have been systematically subjected to low-velocity impact tests at room temperature. The comparison between the two kind of samples, reported not only in terms of load-displacement and energy-time curves but also in light of morphological observations, is discussed taking into account, among others, the rubbery behavior of PP based samples and the glassy nature shown by the fully-biodegradable system PLA/jute at the considered test temperature.

Keywords Thermoplastic composite laminates · Jute fabric
Low velocity impact

Introduction

In the last decades, stringent environmental legislation and consumer awareness have driven the researcher community to develop new products from renewable feedstock, independent of fossil resources. About composite systems, although glass and carbon fibers reinforced materials have somewhat superior mechanical properties, their life cycle performance is very questionable. In fact, the manufac-

P. Russo (✉)
Institute for Polymers, Composites and Biomaterials,
National Council of Research, Pozzuoli (Na), Italy
e-mail: pietro.russo@unina.it

G. Simeoli
CRdC Tecnologie Scarl, Naples, Italy

V. Lopresto · A. Langella · I. Papa
Department of Chemical, Materials and Industrial Production,
University of Naples Federico II, Naples, Italy

R. Fangueiro and S. Rana (eds.), *Advances in Natural Fibre Composites*,
https://doi.org/10.1007/978-3-319-64641-1_11

turing of these products not only consume huge energy but their disposal at the end of their useful life is also very difficult since difficulties related to their recycling/reuse. In this frame, a lot of efforts have been already spent to highlight valuable potentials of natural fibers as the lignocellulosic ones or, more in general, coming from vegetative sources as flax, hemp, cotton, jute, kenaf, sisal, banana, bamboo and so on. These fibers offer some advantages over man-made fibers being highly available and renewable, with low density, non-abrasive to process equipments and, mainly, very safe during handling, processing and use [1]. However, some drawbacks as lack of consistency of fibre quality, high level of variability in fibre properties depending on source and cultivars, poor compatibility between fibres and matrix, high moisture absorption, problems of storing for extended times, low resistance to ultraviolet exposition and fluctuation in price depending upon the global demand and production still restrict their applications. Thus, focusing the interest mainly on the moisture absorption and the poor wettability of natural fibres, various strategies for surface modifications of fibers [2, 3] or matrices [4] and the use of suitable coupling agents [5, 6] have been widely reported to improve the interfacial adhesion in polymer matrices. Alternatively, again with the aim to enhance the ultimate properties of products, attention has been also paid towards hybridization approaches consisting in the simultaneous use of at least two different natural fibers or in the combination of natural fibers with synthetic ones [7–10].

Among all the natural reinforcing materials, the jute, obtained from inner bast tissues of the plant stem, appears to be a promising fiber due to its high toughness and aspect ratio with respect to other reinforcements and commercially available in the required form, including bi-directional fabrics. Jute fibers are characterized by a high cellulose content and a low microfibril angle ensuring moderate tensile and flexural properties of polymer composites with respect to systems reinforced with other natural fibers [1, 11–14].

Recently, thermoplastic composites have been introduced in the automotive industry [15] and as structural materials for aerospace applications to replace thermoset prepregs, mostly due to their potential for rapid or cost-effective processing [16].

Among thermoplastics, polypropylene (PP) is one of the most extensively used in industrial applications due to its high chemical and wear resistance, easy processability, low cost and excellent mechanical properties. While, regarding bio-based thermoplastic matrices, a relevant role is played by the poly(lactic acid) (PLA): a versatile compostable polymer that is produced from renewable resources such as corn. The ever growing interest toward this resin is related to the fact that PLA offers high mechanical properties and, mainly, can degrade into natural products in just a few years [17].

Berhanu et al. [18] examined the effect of jute fabric reinforcement content on tensile and flexural properties of polypropylene based composites. Authors demonstrated that while tensile modulus of samples uniformly increased with the content of the reinforcement, the tensile and flexural strength followed a non-monotonous trend with the highest values assumed for samples with 40% by weight of jute fibers. The flexural modulus, instead, linearly increased up to 40% by weight of reinforcement, remaining almost constant for further increase of the jute fiber content.

Russo et al. [19] investigated the mechanical properties of laminates based on three different polyolefin matrices: one virgin polypropylene resin and two polyolefin fractions coming from end-of-life bumpers and packaging wastes, respectively; all reinforced with layers of a commercial jute fabric. In case of virgin PP and scraps coming from the bumpers recycling, the effect of a common coupling agent, polypropylene grafted with 1 wt% of maleic anhydride (PP-g-MA), on static flexural and dynamic low-velocity and Charpy impact properties has been also considered. The experimental findings, always supported by morphological observations, demonstrated that the addition of the coupling agent increases flexural parameters, impact initiation energy and Charpy breaking load of virgin PP based samples as a sign of a good load transfer at the matrix-fiber interface. On the contrary, for laminates based on recycled polyolefins, the relatively poor adhesion at the interface gave rise to a ductile behavior of products showing higher impact propagation energies and higher resilience with respect to ones based on the neat matrix.

Khan et al. [20] studied PLA composites reinforced with woven and non-woven jute fabric and subjected to static tensile and flexural tests and Charpy impact loading. Results highlighted, among others, higher tensile, flexural and impact strength of PLA/woven jute fabric composites in the warp direction than in the weft one.

Papa et al. [21] analyzed internal damages, coming from low-velocity impact events carried out up to penetration, with the help of an ultrasonic NDI technique and by the visual inspection of impacted PLA/jute woven fabric composite laminates. Results revealed the occurrence of both delamination and fibre breakage phenomena, even at relatively low energy levels.

In this work, a commercial jute fabric was used as the reinforcement of two thermoplastic matrices: polypropylene (PP) and poly(lactic acid) (PLA) having glass transition temperatures equal to 8 and 60 °C, as evaluated by previous dynamic-mechanical tests. Plates, prepared by conventional film stacking and compression molding procedures have been systematically subjected to low-velocity impact tests at room temperature up to complete perforation. The comparison between the two kind of samples, reported not only in terms of load-displacement and energy-time curves but also in light of morphological observations, is discussed taking into account, among others, the rubbery behavior of PP based samples and the glassy nature shown by the fully-biodegradable system PLA/jute at the considered test temperature.

Experimental

Materials

Investigated systems are based on a virgin polypropylene PP MA712 supplied by Unipetrol (MFI@230 °C/2.16 kg = 12 g/10 min) and a poly(lactic acid) resin furnished by Nature Work under the trade name Ingeo 7001D (ρ = 1.24 g/cc, MFI@210 °C/2.16 kg = 6 g/10 min, T_g = 55–60 °C) as matrices and a commercial plain wave type jute fabric with a specific mass of 250 g/m^2 from Deyute (Alicante, Spain) as the reinforcement.

Laminate Preparation

Layers of jute fabric, predried in a vacuum oven at 70 °C for 4 h, were alternated with films of neat matrix using conventional film stacking and compression molding techniques. Operating in this way PP/jute and PLA/jute laminated structures consisting of 8 balanced fabric layers 0°/90° $[(0/90)_4]_s$, respectively, symmetrically stacked with respect to the middle plane of the laminate, were fabricated. Each sample have an average thickness of 3 mm for PP based laminates and 3.5 mm for PLA based ones and a volumetric fraction of reinforcement approximately equal to 50 and 45%, respectively.

Low-Velocity Impact Tests

Impact measurements were conducted on square samples located on a steel cylindrical support with an internal hole 80 mm in diameter, according to the ASTM 3029, using an instrumented free-fall drop dart testing machine (CEAST, mod. "MK16"). The instrumented impactor, cylindrical in shape, was equipped with a hemispherical head having a radius of 19.8 mm and a minimum weight of 3.6 kg. Additional masses is possible to add to reach, together with the height of drop, higher impact energies. Samples were always clamped between the support and a steel plate by compressed air and impacted in the centre up to the complete penetration obtained, on the studied laminates, with an energy level equal to 15.98 and 20 J for PP/jute and PLA/jute laminates, respectively. The impact velocity was 4 m/s. In each case, a minimum of 4 samples were tested, reporting the average data in terms of typical load-displacement and energy-time diagrams. The first curve is very important since it represents a map of the behaviour of the material under dynamic load.

Results and Discussion

Figure 1 reports load-deflection curves for impacted PP and PLA woven jute samples. In both cases, the open type curve testifies the successful penetration of the reference laminates. In general, the shape of the curve depends on many issues mainly including the intrinsic structural features of the investigated materials and the impact conditions. The fact that the two curves are completely different in shape and values, as it is clearly visible by the figure, denotes a completely different impact response of the laminates. First of all, the contact loads obtained on poly lactic acid are more than double respect to the poliolefin, as well as the displacement in correspondence of the maximum load is higher in the first case and it was reached after a series of load drops meaning a series of internal damages, probably due to fibre failures. Soon after the maximum value, the load suddenly drops up to 0 denoting a catastrophic damage. The laminates made of poliolefin, on the other side, showed a less disturbed shape of the curve that gradually goes to 0 after reaching the maximum value. A complete failure of the reinforcement occurs in correspondence of the maximum value of the load which, in turn, begins to decrease so much faster, the lower the ductility of the matrix. Future experimental tests, carried out at increasing impact energy values, are necessary to investigate the damage start and propagation and the different failure mechanisms interactions.

Curves of energy necessary to penetrate the laminate, called penetration energy (Up), evaluated as the area under the curves from the above discussed load-deflection ones, are compared in Fig. 2. The picture clearly shows, as expected by the observation of the load displacement curves, that the amount of energy transferred from the impactor to the investigated biocomposite laminates, at the end

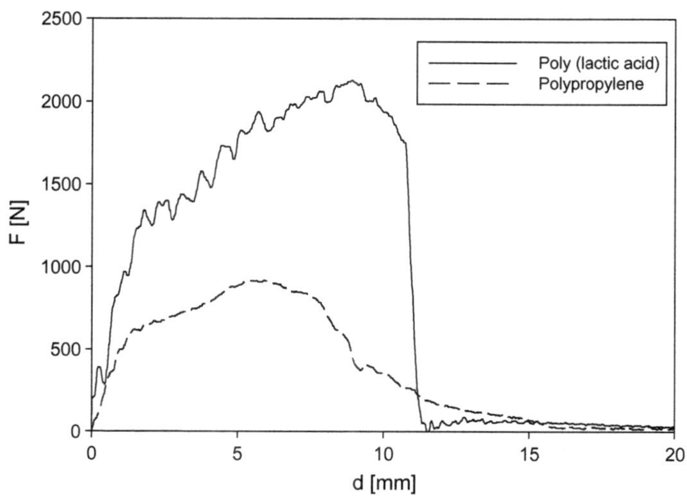

Fig. 1 Load–displacement curves up to complete penetration

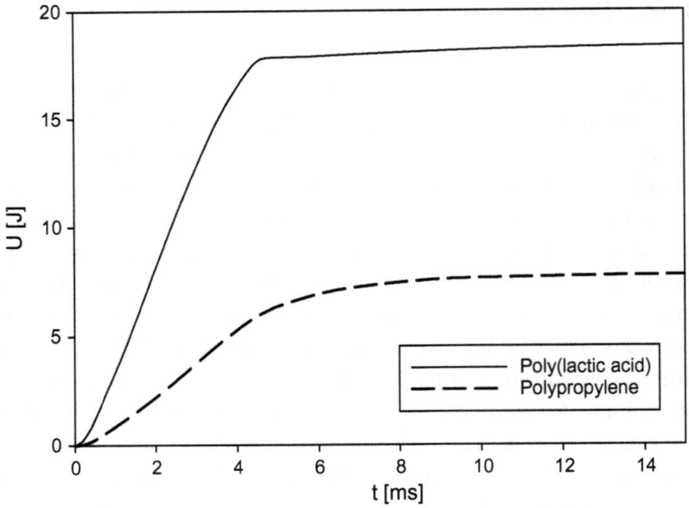

Fig. 2 Energy-time curves of composites from the low velocity impact tests

Table 1 Impact parameters

	Umax [J]	Utot [J]	Fmax [N]	Upeak [J]	DI
Polypropylene/jute	7.1	7.76	920	3.96	0.96
Poly(lactic acid)/jute	18.44	8.25	2128	5.25	0.57

of an impact events, increases with the use of the poly lacticid matrix, as highlighted in Table 1. It is so, necessary a larger amount of energy to penetrate the Poly Lactic Acid. In Table 1, the impact parameters as the maximum energy, the total amount of energy that is absorbed by the samples, the peak load, the energy at peak load and the ductility index DI soon hereafter defined, are collected.

The ductility index (DI), evaluated by taking the ratio between the energy absorbed after the force peak ($E_{tot} - E_{peak}$) and the energy absorbed up to the peak force (E_{peak}), reflects the ductility of the material, therefore, the results reported in Table 1, indicate a higher plastic deformation ability of PP laminates compared with PLA based ones. This behavior is reasonable taking into account that tests are conducted at room temperature at which the polypropylene matrix, being above its Tg, behave as a rubbery material while the PLA, being below its Tg, behaves as a glass.

Conclusions

This research compared two thermoplastic laminated structures reinforced with a commercial jute fabric in terms of low-velocity impact behavior. Specifically, samples $[(0/90)_4]_s$ based on a polypropylene (PP) and a poly(lactic acid) (PLA) commercial grade, respectively, were impacted at room condition.

Expectedly, collected results, reported in terms of typical load-deflection and energy-time curves, and further emphasized by the evaluation of some impact parameters, reflected the actual structural state of the matrix at test temperature. Indeed, the characteristic glass transition temperatures (T_g) of PP and PLA matrices approximately equal to 0° and 60 °C, respectively, can properly justify the more pronounced ductility detected for the former composite laminates with respect to the fully biodegradable ones.

References

1. Bogoeva-Gaceva, G., Avella, M., Malinconico, M., Buzarovska, A., Grozdanov, A., Gentile, G., et al. (2007). Natural fiber eco-composites. *Polymer Composites, 28,* 98–107.
2. Mohanty, A. K., Misra, M., & Drzal, L. T. (2001). Surface modifications of natural fibers and performance of the resulting biocomposites: An overview. *Composite Interfaces, 8,* 313–343.
3. Kabir, M. M., Wang, H., Lau, K. T., & Cardona, F. (2012). Chemical treatments on plant-based natural fibre reinforced polymer composites: An overview. *Composites Part B Engineering, 43,* 2883–2892.
4. Aziz, S. H., Ansell, M. P., & Clarke, S. J. (2005). Modified polyester resins for natural fibre composites. *Composite Science and Technology, 65,* 525–535.
5. Xie, Y., Hill, C. A. S., Xiao, Z., Militz, H., & Mai, C. (2010). Silane coupling agents used for natural fiber/polymer composites: A review. *Composites Part A Applied Science and Manufacturing, 41,* 806–819.
6. El-Sabbagh, A. (2014). Effect of coupling agent on natural fibre in natural fibre/polypropylene composites on mechanical and thermal behaviour. *Composites Part B Engineering, 57,* 126–135.
7. Boopalan, M., Niranjanaa, N., & Umapathy, M. J. (2013). Study on the mechanical properties and thermal properties of jute and banana fiber reinforced epoxy hybrid composites. *Composites: Part B, 51,* 54–57.
8. Ramesh, M., Palanikumar, K., & Reddy, K. H. (2009). *Composites: Part B, 48,* 1–9.
9. Santulli, C., Sarasini, F., Tirillò, J., Valente, T., Valente, M., Caruso, A. P., et al. (2013). Mechanical behaviour of jute cloth/wool felts hybrid laminates. *Material and Design, 50,* 309–321.
10. Petrucci, R., Santulli, C., Puglia, D., Sarasini, F., Torre, L., & Kenny, J. M. (2013). Mechanical characterisation of hybrid composite laminates based on basalt fibres in combination with flax, hemp and glass fibres manufactured by vacuum infusion. *Material & Design, 49,* 728–735.
11. Acha, B. A., Marcovich, N. E., & Reboredo, M. M. (2005). Physical and mechanical characterization of jute fabric composites. *Journal of Applied Polymer Science, 98,* 639–650.
12. Puglia, D., Biagiotti, J., & Keny L. M. (2005). A review on natural fibre-based composites—Part II: Application of natural reinforcements in composite materials for automotive industry. *Journal of Natural Fibers, 1,* 23–65.

13. Mohanty, A. K., Misra, M., Drzal, L. T. (2002). Sustainable bio-composites from renewable resources: Opportunities and challenges in the green materials world. *Journal of Polymers and the Environment, 10*, 19–26.
14. Khan, M. A., Hassan, M. M., Drzal, L. T. (2005). Effect of 2-hydroxyethyl methacrylate (HEMA) on the mechanical and thermal properties of jute-polycarbonate composite. *Composites A, 36*, 71–81.
15. Wakeman, M. D., Rud, C. D., Cain, T. A., Brooks, R., & Long, A. C. (2000). Compression moulding of glass and polypropylene composites for optimised macro-and micro-mechanical properties. 4: Technology demonstrator—A door cassette structure. *Composites Science and Technologies, 60,* 1901–1918.
16. Hou, M. (1997). Stamp forming of continuous glass fibre reinforced polypropylene. *Composites A, 28,* 695–702.
17. Wang, K. H., Wu, T. M., Shih, Y. F., & Huang, C. M. (2008). Water bamboo husk reinforced poly (lactic acid) green composites. *Polymer Engineering & Science, 48,* 1833–1839.
18. Berhanu, T., Kumar, P., & Singh, I. (2014). In *Proceedings of the AIMTDR 2014*, Assam, India, December 12–14, 2014.
19. Russo, P., Simeoli, G., Acierno, D., & Lopresto, V. (2015). Mechanical properties of virgin and recycled polyolefin-based composite laminates reinforced with jute fabric. *Polymer Composites, 36,* 2022–2029.
20. Khan, G. M. A., Terano, M., Gafur, M. A., & Alam, M. S. (2016). Studies on the mechanical properties of woven jute fabric reinforced poly (l-lactic acid) composites. *Journal of King Saud University—Engineering Sciences, 28,* 69–74.
21. Papa, I., Lopresto, V., Simeoli, G., Langella, A., & Russo, P. (2016). *Composites Part B*, (in press).

Selection Chart of Flame Retardants for Natural Fiber Polymer Composites

A. Elsabbagh, A. Ramzy, T. Attia and G. Ziegmann

Abstract Increasing the share of natural fiber thermoplastic composites NFTC in the market of engineering market demand to guarantee flame retardant properties. On the other side, addition of flame retardants FRs affects the mechanical properties negatively specially the whole tensile strength. Polypropylene PP reinforced with 30% natural fibers (flax, jute, hemp and sisal) are prepared using kneading and different FRs are mixed together. The effect of the different FRs on the flame retardance level, namely UL94, as well as the mechanical properties are studied. Hence, a selection chart for FRs is established. The studied FRs materials in this work are mineral, halogenated, halogen-free intumescent. Also, the effect of synergism with nanoclays and antimony tri oxide is considered. As a result of this experimental work regarding both the mechanical properties and the flame retardance levels, a material selection chart is built considering the tensile strength and flame retardance test UL94. This chart is enriched by other literature results. Also, the concept of this chart can be extended to deal with another mechanical property or another flame retardance test.

Keywords Natural fibers · Flame retardance · Composites · Material selection

Abbreviation List

NFTC	Natural fiber thermoplastic composites
FRs	Flame retardants
PP	Polypropylene
UL94	Flame retardance level
MAPP	Maleated anhydride grafted polypropylene
DECA	Decabromodiphenyl oxide

A. Elsabbagh (✉) · T. Attia
Faculty of Engineering, Ain Shams University, Cairo, Egypt
e-mail: elsabbagh.ahmed@eng.asu.edu.eg

A. Ramzy · G. Ziegmann
Institute of Polymer Materials and Plastics Engineering,
Clausthal University of Technology, Clausthal-Zellerfeld, Germany

R. Fangueiro and S. Rana (eds.), *Advances in Natural Fibre Composites*,
https://doi.org/10.1007/978-3-319-64641-1_12

UTS Ultimate tensile strength
ATH Aluminum trihydrate

Introduction

NFTC are recently applied to high added value applications in electronic and automobile industries [2]. The ability of NFTC to burn plays a key role in the application of composites [9], [8]. Lignin, cellulose and hemicellulose are main constituents of natural fibers. The flammability of the fibers is controlled by these constituents. The flammability of fibers is affected by two opposite factors. Firstly, lignin contains aromatic rings which hinders the oxidation of fibers. [3, 7, 10, 13]. Secondly, the cellulose supports the fiber burning as it imparts volatile flammable materials [3, 5]. This can be rearticulated as following: The increase of cellulose percent helps the flammability and on the other side, lignin induces forming of insulating char [1, 4, 11].

Experimental Work

Flax, jute, hemp and sisal fibers were supplied by Sachsenleinen-Germany. The fibers are subjected to a treatment of NaOH 10% solution at 90 °C/10 min then NaOH 5% solution for 24 h and finally neutralized. The fibers are then left overnight in air and afterwards for 24 h at 90° in an adapted atmosphere oven. Polypropylene DOW H 734 52 RNA (Dow Customer Information Group, Edegem, Belgium) is used as matrix. Melt flow index of this homopolymer at 230 °C and 2.16 kg is 52 g/10 min, so it is suitable for extreme flow levels in injection moulding purposes. Maleated anhydride grafted polypropylene matrix MAPP of 9100 g/mol molecular weight is supplied by Sigma Aldrich. The used FRs in this study are inorganic metal hydrate $Mg(OH)_2$, The synergetic effect of nano clay (Sodium montmorillonite) and intumescent FR as ammonium polyphosphate Exolit AP-766, supplied by Clariant, is investigated. Decabromodiphenyl oxide DECA alone and with Antimony trioxide Sb_2O_3 is also studied. Table 1 shows the compositions of the trials in this study. All FRs except AP-766 are supplied by Sigma Aldrich.

Compounding is carried out using 600P kneader Haake Rheomix at 180 °C/50 rpm for 15 min. The compounded material is cut by a mechanical shredder to small granules which are then dried overnight and then injected for tensile samples of 1BB size according to ISO 527-2 and impact samples of according to ISO 179-1. Samples of mechanical testing are conditioned at 23 °C/50% relative humidity for

Table 1 Plan of work

Group	Fiber	Effect of:	Parameter wt%
A	30 and 50% flax	$Mg(OH)_2$	0, 20, 30
B	30% flax	AP-766	20, 22, 24, 26, 28
C	30% flax	AP-766+ Nanoclay	20 0, 1, 2, 3, 4, 5
D	30% flax	DECA+ Sb_2O_3	20, 22, 24, 26, 28 6
E	30% flax	DECA+ Sb_2O_3	24 3, 6, 9, 12
F	30% (flax, hemp, jute, sisal)	$Mg(OH)_2$	30%
G	30% (flax, hemp, jute, sisal)	AP-766	24, 26, 28

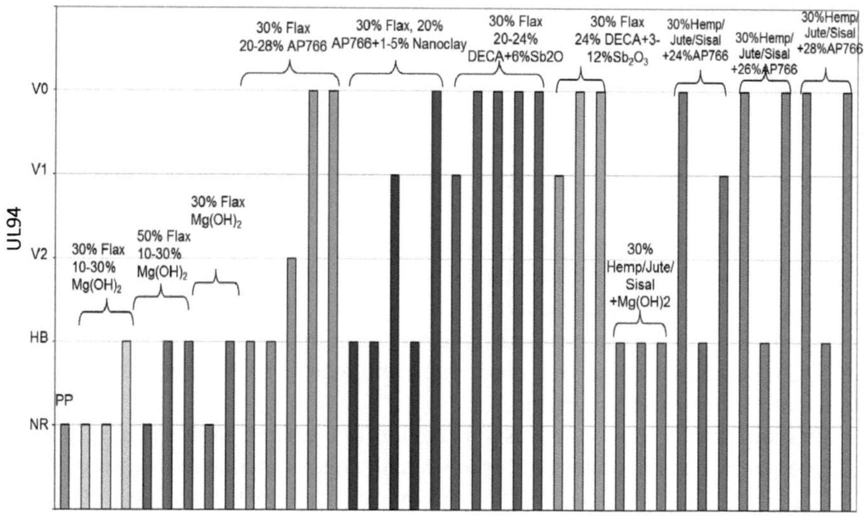

Fig. 1 UL94 results of all groups

at least 88 h per ISO 291 for test room conditions. Samples for flammability test UL94 are also injection molded.

Results and Conclusions

Figure 1 illustrates the UL94 values of all groups listed in Table 1. It is obvious that $Mg(OH)_2$ is not sufficient till 50 wt% to reach any V-class. However, it helps in reaching HB class. Even only 20% of $Mg(OH)_2$ is enough for 50% flax samples to get the HB class. Nanoclays do not show significant synergism within the qualitative test of UL94.

Intumescent AP-766 proved its efficiency starting from 24% level to get V2 class till V0 at 26%. Smoke evolves slightly with 20% AP-766 which is not the case with 28%. No reddish glowing points remain obvious after the test.

Nanoclay shows significance with AP-766. 3 and 5% clay with 20% AP 766 make the rating jump to V1 and V0. Results of nanoclay drop at 4% clay. This is probably due to poor mixing quality which fails to ensure good dispersion/exfoliation of the nanoclay in the matrix/fibre compound. This can be attributed to low applied shear rate. On the other side, high shear rates cause fiber damage and reduce the fibre aspect ratio which is the responsible of load transfer efficiency. Therefore, nanoclays are added before fibers' addition. This strategy represents a challenge for the already mixed commercial natural fiber polymer granules. In other words, an optimized mixing strategy is required for good mix without fiber damaging.

DECA also performs successfully with 30% flax at all tried contents in synergism with 6% Sb_2O_3. However, the synergism of Sb_2O_3 is obvious because the UL94 class drops to V1 when only 3% of Sb_2O_3 is applied. Effect of using different fibers shows insignificance. It is worthily noted that reddish glowing point remains for seconds after finishing the test in case of DECA application. Not only this, but evolving smoke is also observed.

The type of fiber shows significance where flax and hemp fibers have higher class. Jute has failed to pass the V0 class and sticks to HB level. Sisal, the leaf fiber, shows also a slightly better behavior than jute but worse than flax and hemp. Flax and hemp show the fastest distinguishing time after the removal of the flame source. $Mg(OH)_2$, does not show the fiber type significance because its amount is not sufficient even for flax and hemp.

Similarly, Fig. 2 shows the tensile behavior for the same groups named in Fig. 1. The inorganic metal hydrates affect the strength negatively due to its heavy loading. 32% drop in strength for the 50% flax group is measured whereas only 10% drop in UTS is found in 30% flax composites. The effect of AP-766 on the strength of the 30% Flax composite. The drop in UTS is close to that resulted by $Mg(OH)_2$%. The drop is almost 25% at 20% AP-766. However, more AP-766 is not following a decreasing linear trend line. On the other side, nanoclays show a significant reducing effect for composite UTS in an approximate linear way. 5% nanoclays lead to almost 29% loss in UTS. The same discussion can imply on group H for nanoclays with $Mg(OH)_2$%.

The effect of DECA is too small. UTS decreases only 14% at maximum level of DECA used which is 28%. Also, the effect of the synergist antimony trioxide is insignificant. However, the healthy restrictions (2002/95/EC) on the assumed toxic nature of DECA should be considered. The effect of fiber type at different FRs on UTS. Only sisal is a little bit less than the other fiber types of flax, hemp and jute. Jute composites have higher UTS than the others. The gradual increase of AP-766 shows no stable negative effect in dependence on fibers except for jute. Considering the reference UTS of composites (without FR), the effect of AP-766 on all fiber

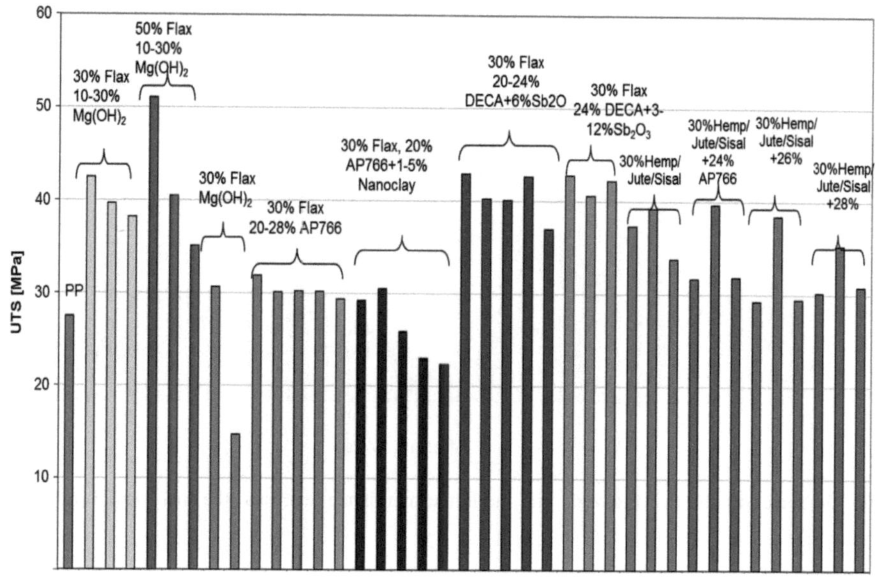

Fig. 2 Ultimate tensile strength (UTS) of all groups

types is the same. That is different from what noticed with magnesium hydroxide, where only sisal and jute are heavily affected more than flax and hemp.

To summarize both the mechanical and flame retardance properties, a material selection chart was established based on the results of Figs. 1 and 2. Figure 3 illustrates the mutual effect of flame retardant materials on reducing the mechanical properties. Besides, the effect of FR type on the strength of the NFTC is shown. This chart in Fig. 3 is a tool in FR selection of NFTC. For the sake of illustration, main points considering different FRs are given in the selection chart. The following remarks can be inferred:

- DECA or DECA/Sb_2O_3 show the best result of flame retardance. But considering the healthy regulations,
- AP-766 has the next best performance starting from 26% to reach V0 class but with 29% loss in UTS,
- Nanoclays affect significantly where 3% of nanoclays with 20% AP766 reaches the V0 class,
- The mineral FR Magnesium hydroxide, with its high thermal stability and good ecological impact, needs thorough studying to attain better classification. However, for HB-class, it can be considered with a loss in UTS around 30%.

For the sake of comparison with other literature, it is found that 30% of jute with PP gives almost the same UL94 rank as 30% flax [9]. The property profile of cordenka fibers reinforced PP composites can be tailored by a partial replacement of the cordenka by natural fibers such as jute. This imparts an improvement in tensile

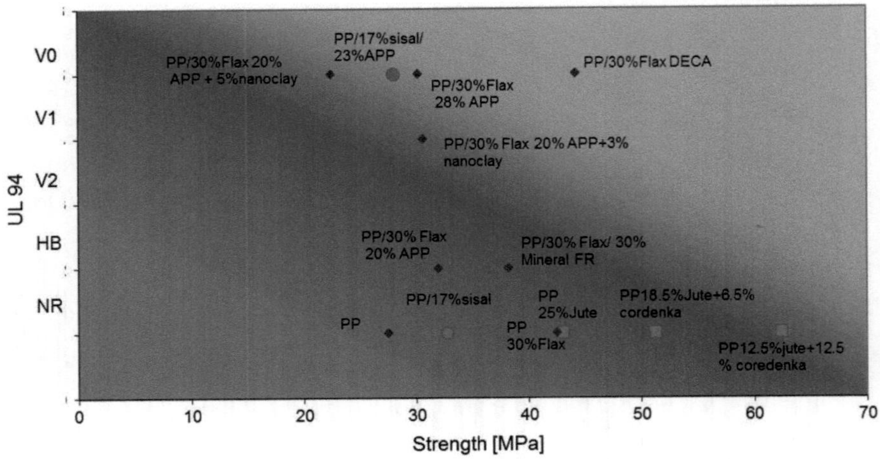

Fig. 3 Selection chart of FR regarding the flame retardance level and the mechanical property

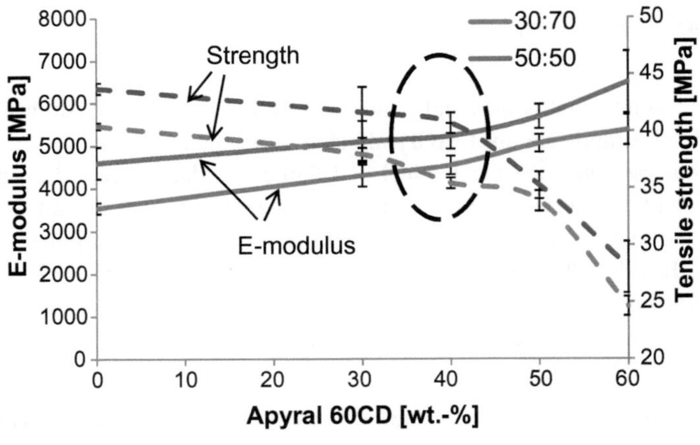

Fig. 4 Impact/strength of 30:70 and 50:50 (NF:PP) at different FR %

strength with slight decrease in UL94 values as shown in Fig. 3 (PP18.5% Jute + 6.5% cordenka and PP12.5%jute + 12.5% coredenka).

Adding sisal instead of flax to PP [8] have almost the same effect on the tensile strength. To improve the UL94 rank by adding APP is also like the PP + 30% flax + 28%APP or PP + 27%sisal + 23%APP Fig. 3

Although the low UL94 results shown by mineral FR like $Mg(OH)_2$ (alone or in synergism with ammonium polyphosphate), it would be attractive to optimize the behavior of this type of FR. That is attributed to the environmental friendly nature of FR. That is why a separate study is carried on to study the effect of mineral FR

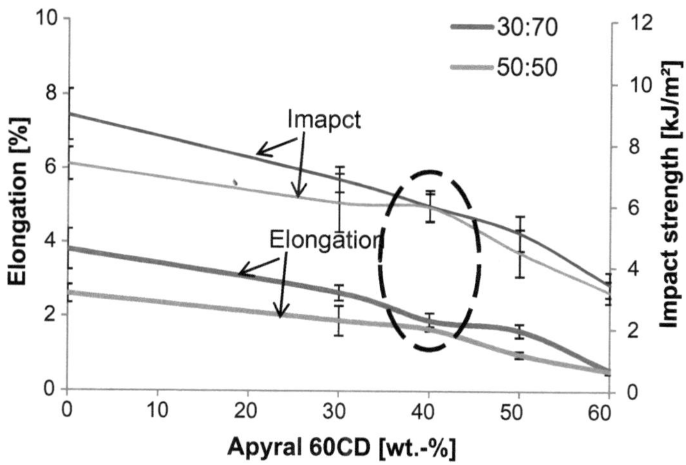

Fig. 5 Elongation/impact strength of 30:70 and 50:50 (NF:PP) at different FR %

content on the tensile strength of NFTC. A commercial mineral FR is selected for this further study.

Aluminium trihydrate ATH (Apyral 60 CD of average 6 m^2/g and median size of 1 μm is supplied by Nabaltec, germany). According to the supplier datasheet and [12], 60% of ATH should be added to reach V-0 level of UL94 test. A spectrum of lower FR content is applied starting from 30 to 60 wt% with 10% intervals is tried. Figures 4 and 5, on the contrary, show that 40% is enough with natural fibers to keep the V-0 retardance without much loss of mechanical properties [6].

Conclusion

A survey for different FR on different fiber types of NFTC is carried out. Flame resistance as well as the resulting mechanical properties are considered. A material selection chart is developed to help in FR selection with the new categories of engineering NFTC. Also an optimization study is carried out to minimize the required mineral FR keeping high flame retardance level and mechanical properties as well.

References

1. Azwa, Z. N., et al. (2013). A review on the degradability of polymeric composites based on natural fibres. *Materials & Design, 47,* 424–442.
2. Bledzki, A. K., Zhang, W., & Chate, A. (2001). Natural-fibre-reinforced polyurethane microfoams. *Composites Science and Technology, 61,* 2405–2411.

3. Chapple, S., & Anandjiwala, R. (2010). Flammability of natural fiber-reinforced composites and strategies for fire retardancy: A review. *Journal of Thermoplastic Composite Materials, 23,* 871–893.
4. Dittenber, D. B., & Gangarao, H. V. (2012). Critical review of recent publications on use of natural composites in infrastructure. *Composites Part A: Applied Science and Manufacturing, 43,* 1419–1429.
5. El-Sabbagh, A., Steuernagel, L., & Ziegmann, G. (2013). Low combustible polypropylene/flax/magnesium hydroxide composites: Mechanical, flame retardation characterization and recycling effect. Mechanical, flame retardation characterization and recycling effect. *Journal of Reinforced Plastics and Composites, 32,* 1030–1043.
6. El-Sabbagh, A., et al. (2016). Optimization of flame retardant content with respect to mechanical properties of natural fiber polymer composites. Case study of polypropylene/flax/aluminum trihydroxide. *Polymer Composites, 37,* 3310–3325.
7. Ferdous, D., et al. (2002). Pyrolysis of lignins: Experimental and kinetics studies. *Energy & Fuels, 16,* 1405–1412.
8. Jeencham, R., Suppakarn, N., & Jarakumjorn, K. (2010). Flammability and mechanical properties of sisal fiber/propylene composites: Effect of combination of flame retardants. *Advanced Materials Research, 123–125,* 85–88.
9. Khan, M. A., Ganster, J., & Fink, H. P. (2007). Natural and man-made cellulose fiber reinforced hybrid polypropylene composites: Effect of fire retardants. *Advanced Materials Research, 29–30,* 341–344.
10. Kozłowski, R., & Władyka-Przybylak, M. (2008). Flammability and fire resistance of composites reinforced by natural fibers. *Polymers for Advanced Technologies, 19,* 446–453.
11. Suardana, N. P. G., Ku, M. S., & Lim, J. K. (2011). Effects of diammonium phosphate on the flammability and mechanical properties of bio-composites. *Materials & Design, 32,* 1990–1999.
12. Troitzsch, J. (Ed.). (2004). *Plastics flammability handbook. Principles, regulations, testing, and approval*. Munich: Hanser.
13. Virk, A. P., Sharma, P., & Capalash, N. (2012). Use of laccase in pulp and paper industry. *Biotechnology Progress, 28,* 21–32.

Mechanical Properties of Raffia Fibres Reinforced Geopolymer Composites

Kinga Korniejenko, Michał Łach and Janusz Mikuła

Abstract Geopolymer composites are "green" alternative to the traditional cementitious materials. They have good compressive strength, durability and other properties such as highly resistant to flame and heat and corrosion resistance. However, these composites have relatively low tensile and flexural strength, which limits their use in many areas, especially in construction industry. This paper describes possibilities to improve this mechanical properties by fibre addition. The study is intended to analyse the influence of addition of various raffia fibres on the mechanical properties of the geopolymer based on fly ash. The empirical part of the research was based on the compressive strength tests, flexural strength tests and detailed microstructure examination. The samples were prepared using sodium promoter and raffia fibres (the 1% by mass of the composite). The research involved the samples reinforced by raffia and artificial fibre—PP (polyprophylene) for comparison. PP is the traditional additive for building materials such a cement. The results show the possibility to produce the composites of reasonable properties from the industrial wastes (fly ash) and renewable resources—raffia fibres, which makes them a new class of environmentally friendly materials.

Keywords Geopolymer composites · Raffia · Fly-ash based geopolymers

Introduction

Cements have been reinforced with natural fibers for many years, particularly in developing countries that have used local materials such as bamboo, sisal, jute and coir with some success [1]. There are also same investigation with using the raffia fibres [2, 3]. It is well established that the choice of fibers used to reinforce concrete

K. Korniejenko (✉) · M. Łach · J. Mikuła
Institute of Material Engineering, Faculty of Mechanical Engineering,
Cracow University of Technology, Cracow, Poland
e-mail: kinga.korniejenko@mech.pk.edu.pl

R. Fangueiro and S. Rana (eds.), *Advances in Natural Fibre Composites*,
https://doi.org/10.1007/978-3-319-64641-1_13

can affect its mechanical properties. The type of fibers, its form, surface properties and matrix properties, all need to be considered [1].

Geopolymer composites reinforced with the chosen natural fibres have relatively short history. They are interesting and eco—friendly alternative to the traditional construction materials such as Portland cement [4]. Natural fibers have special advantages when compared to their synthetic counterparts, where the former represents an environmentally friendly alternative, with lower density, lower cost, non-toxicity, ease of processing, renewability and recyclability, ready availability high strength-to-mass ratio and good tensile strength [5]. These properties make them attractive alternatives to the synthetic fiber composites used in more industrialized countries [1]. In addition, the use of natural fibres in geopolymer composites has the potential to produce materials with higher specific strength and specific modulus, due to their lower density [5, 4]. So far as a reinforcement of geopolymers were used and researched a following fibers:

- Cotton fibre; The use of cotton fibers has several advantages which include low cost, renewability, and low weight when compared to synthetic fibres, good intrinsic mechanical properties [1]. Results show that the enhancement of mechanical properties was achieved at an optimum fibre content of 2.1 wt%. Results of thermal analysis show that fly-ash based geopolymer can prevent the degradation of cotton fabric at elevated temperatures [5]. Results show that the appropriate addition of cotton fibres can improve the mechanical properties of geopolymer composites. In particular, the flexural strength and the fracture toughness increase at an optimum fibre content of 0.5 wt%. However, as the fibre content increases, the density of geopolymer composites decreases due to an increase in porosity and tendency of fibre agglomeration [1].
- Sisal fibre. The results of the research show that the appropriate addition of sisal fibres can improve the mechanical properties of geopolymer composites, especially flexible strength. Sisal fibres has good coherences with geopolymer matrix [4].
- Flex fibres. The mechanical properties of the geopolymers reinforced by fibre-reinforced composites improve with increasing fibre content. This represents a significant improvement on the flexural strength of the unreinforced geopolymer matrix, and all the composites show graceful failure, unlike the brittle failure of the matrix. Despite the formation of microcracks due to water loss from the geopolymer matrix, the fibres are thermally protected by the matrix up to 400 °C. The flax fibres do not appear to be compromised by the alkaline environment of the matrix, suggesting new possible applications for these low-cost simply prepared construction materials [6].
- Coir (fibres from coconut). The results of the research show that the appropriate addition of coir can improve the mechanical properties of geopolymer composites, especially flexible strength [4, 7].
- Sorghum fiber. The results indicate that the unit weight of the sweet sorghum fiber–geopolymer composite decreases with higher fiber content. Although the inclusion of sweet sorghum fiber slightly decreases the unconfined compressive

strength, the splitting tensile, and flexural strengths as well as the post-peak toughness increase with the fiber content up to 2% and then start to decrease. The splitting tensile tests also clearly show the transition from the brittle failure of the plain geopolymer specimen to the "ductile" failure of the geopolymer specimen containing sweet sorghum fiber [8].

- Wool fibres. The composites reinforced with natural protein-based (wool) have approximately 40% improvement in ultimate flexural strength compared with the unreinforced geopolymer matrix [9, 10].
- Other fibres such as jute, corn husk, pineapple leaf fibre and cellulose [11].

The Raffia palms (called also Raphia) are a genus of twenty species of palms [12] native to tropical regions of Africa, especially Madagascar, with one species also occurring in Central and South America. The raffia is especially popular in: the Province of Bohol in the Philippines, Kuba of Democratic Republic of the Congo, Nso of Cameroon, the Igbo and Ibibio/Annang of South-Eastern, the Urhobo and Ijaw people of Niger delta Nigeria and the Yoruba of South-Western Nigeria, and among several other West African ethnic nations [2, 13]. The raffia palm tree belongs to the multifunctional plant family. It is the largest palm and one of the most useful economically [14, 15]. In their local environments, it is used for ropes, sticks to tying up plants that require support and supporting beams, including binding together vegetables to be marketed, weaving baskets, and various roof (branches as well as leaves). Traditionally, it is used also to beds, shoes, handbags, carpets, hats and other native cloth, mats and food (sap/wine) and also to traditional art such as crafts, decoration, braces and sculptures [12, 16]. Its nuts are source of food and cosmetic oil, whereas the petioles and raw leaves are used as construction materials [17]. Contemporary, the fibres are used to reinforce composites, clay bricks and concrete, or to make panels and geotextiles [17–19]. It is sometimes combining with polymer materials [20, 21], including composites for automotive industry [22] and it is solicited in the protection of the environment (water and soil conservation) and nutrition [23]. The plant can be also potential source for bioethanol production [24].

The raffia palms grow in the swampy and semi-swampy areas of the equatorial rainforest or derived savannah [14, 25]. The plants are monocarpic or hapaxanthic, usually they flowering once and individual stems dying after fruiting but the root system remaining alive and sending up new stems. The palms grow up to 16 m tall and are remarkable for their compound pinnate leaves. The raw fibres are extracted from the upper surface of the leaflets [17, 3]. The raffia palm is crowned with enormous leaves that may be up to 25 m long, up 3 m wide (and composed of 80 to 100 leaflets [2]. The raffia fibres are made from leafs. The membrane on the underside of each individual frond leaf is taken off to create a long thin fiber which can be dyed and woven as a textile. Fiber from these leaves is torn into small strips and dried in the sun. The fibers are soft, pliable, and have "strong nature". Mechanical properties of fibres: Young's modulus approx. 30 GPa, tensile strength approx. 500 MPa, total elongation between 2 and 4% and density 0.75 [17, 26]. The raw fibre has semi-crystalline and that the allotrope form is I_b cellulose [17].

Additionally, the raffia has very good thermal properties. The research results which confirm its effectiveness as an insulation material having good heat storage capacity. The overall transverse thermal conductivity of anhydrous raffia bamboo gives an average of 0.07 W/m K [23]. This fact makes the raffia fibre attractive material for isolation.

Materials and Methods

The geopolymer matrix was made from fly ash from the CHP plant in Skawina (Poland). The fly ashes were thoroughly investigated as a possible raw material for the production of the geopolymer matrix being a base for various composites. SEM observations, EDS analysis and the previous research confirmed its suitability for such matrix. The chemical composition of the mentioned ash consists of approx. 56% SiO_2, 23.5% Al_2O_3, 6% Fe_2O_3, 3.5%, K_2O, below 3% CaO and MgO, and less than 1% of other components e.g., TiO_2, Na_2O, P_2O_5 and BaO [4]. High value of SiO_2 and Al_2O_3 is advantageous for creating geopolymers, additionally low amount CaO confirm the usability of this particular fly ashes for the prosecco alkali activation. The fly ash density amounted to 2.22 g/cm^3. The morphology of the particles of fly ash was typical of such by-products of coal combustion. Regarding the particle size distribution of the examined fly ash, the size of approx. 60% particles was <56 μm [4].

The matrix of the composites was 8 M sodium hydroxide solution combined with the sodium silicate solution (liquid glass at a ratio of 1:2, 5). In order to manufacture the composites flakes of technical sodium hydroxide were used and water solution of sodium silicate R–145 whose modulus was 2.5 and density 1.45 g/cm^3. Tap water was used as batched water instead of the distilled one. The alkaline solution was prepared by means of pouring the aqueous solution of sodium silicate and water over solid sodium hydroxide. The solution was mixed and left until its temperature became stable and the concentrations equalized.

The samples were prepared using sodium promoter, fly ash and raffia fibres (the 1% by mass of the composite). To reinforce the geopolymer matrix raffia fibres were used whose length was approx. 3 mm and diameter approx. 1 mm. The research involved the samples reinforced by raffia (Fig. 1) and artificial fibre—PP (polypropylene). PP is the traditional additive for building materials such a cement. The fly ash was mixed with fibres about 5 min by using mixing machine. Then, the solution was mixed with fly ash and raffia fibres about 15 min (to receive the homogeneous paste). Next, it was poured into two sets of plastic moulds. The first set consisted of the moulds dedicated to undergo compressive strength tests and the second set consisted of the moulds dedicated to undergo flexural strength tests. The samples were hand-formed and then subjected to vibratory removal of air bubbles. Tightly closed moulds were heated in the laboratory dryer for 24 h at 75 °C. Then, the samples were unmoulded. They were investigated after 28 days.

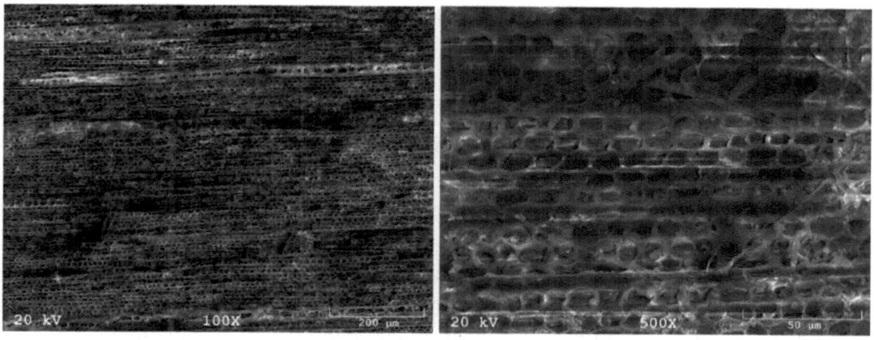

Fig. 1 Exemplary SEM images for raffia fibres

Microstructure research has been performed by means of scanning electron microscopy (SEM) type JEOL JSM 820 with EDS (The observations were made for raffia fibers and composites (geopolymer and fibres: raffia and PP). The investigation regarded the samples previously broken while compressive strength tests (in case of composities). The samples were sprinkled with a thin layer of gold with JEOL JEE-4X vacuum sputter. The observation was made at various magnifications (between 20–1000×). For the samples geopolymer—PP also the polished specimen were prepared.

Due to the lack of separate standards for geopolymer materials, the compressive strength tests were carried out according to the methodology described in the standard EN 12390-3. ('Testing hardened concrete. Compressive strength of test specimens'). Samples used in the flexural strength tests had cuboid shape and dimensions (approx.): 50 mm × 50 mm × 50 mm. The tests involved at least 6 samples. They were performed with a concrete press (MATEST).

Due to the lack of separate standards for geopolymer materials, the flexural strength tests were carried out according to the methodology described in the standard EN 12390-5 ('Testing hardened concrete. Flexural strength of testspecimens'). Samples used to the flexural strength test had cuboid shape and dimensions (approx.): 200 mm × 50 mm × 50 mm. Tests were performed with a universal testing machine—single-point load (Instron type 4465).

Results

The SEM observations were made form composites and raffia fibres. The investigations for raffia fibres confirm the literature data about these fibres [17]. There are composed of superimposing layers. The top layer is composed of strands, themselves constituted by scales aligned in the longitudinal direction of the fibre and the bottom one has an alveolus structure resembling a honeycomb [17] and Fig. 1. These structure could serve as mechanical bonding in a composite matrix [17].

The observation of microstructure for composities gives a preliminary information about the coherency of fibers (filler) and the geopolymer matrix as well as lets to evaluate of the fiber distribution.

The results from the scanning electron microscopy analysis are shown in Figs. 2 and 3. They presents the raffia fibres in geopolymer matrix. The investigation in low magnification 20× allow to estimate the fibres distribution (Fig. 2).

The investigation in high magnification 100–1000× allow to estimate the fibres coherency with matrix (Fig. 3).

The fibre distribution in the case of natural fibres as well as artificial is regular (Fig. 4). The fibre distribution in the matrix heavily influences the properties of the specific composite, as the fibres aggregation can decrease its mechanical properties. Here, in the same sample one could notice some fibre agglomerations and it was connected with a concomitant decrease of mechanical properties of the composite.

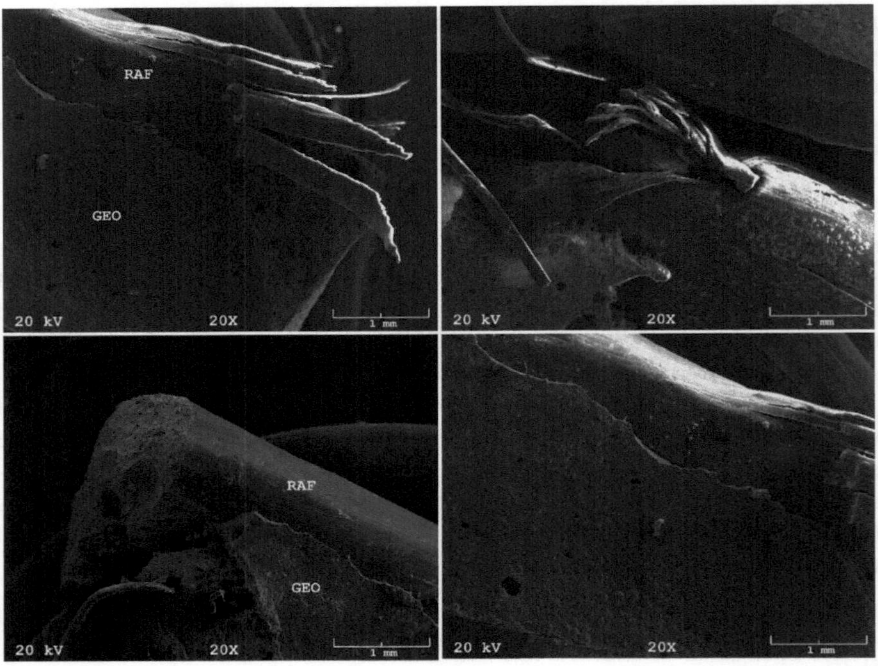

Fig. 2 Exemplary SEM images for composite: geopolymer matrix reinforced by raffia fibres in low magnification 20×

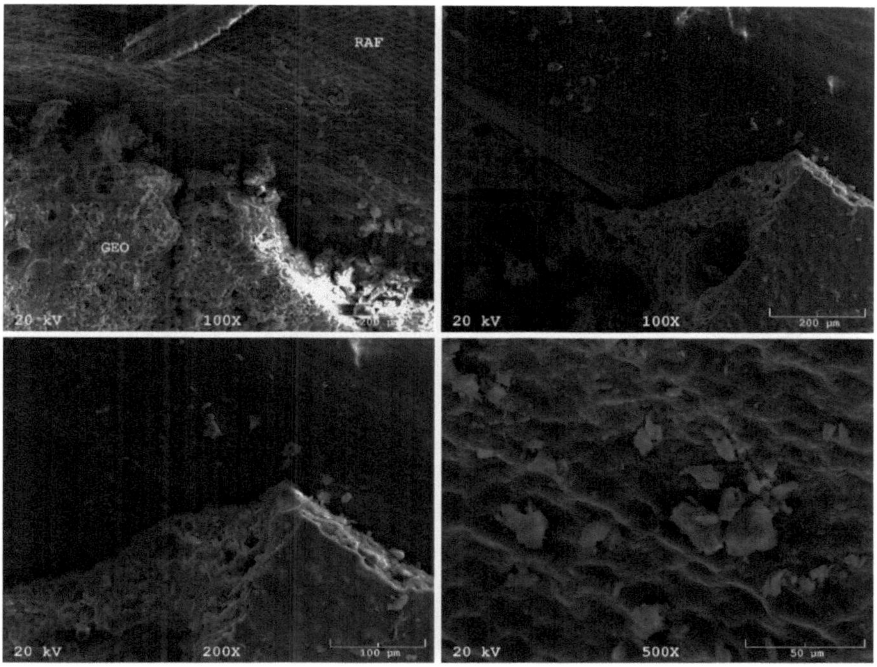

Fig. 3 Exemplary SEM images for composite: geopolymer matrix reinforced by raffia fibres in low magnification 100–500×

The coherency between the filler and the matrix not always exists. In the case of the composites reinforced with PP the observation confirms their cohesiveness (Fig. 4). But in the case of other natural fibres such as raffia the structure is not always coherent with the matrix: most fibres are surrounded by well-visible empty spaces. The exemplary ones are shown in the Fig. 3. The incoherent structure has a negative impact on the mechanical properties of the composites. However, even with such structure the presence of the fibres significantly reduces cracks propagation (SEM observation was made on the samples after the compressive strength test).

The results from the compressive strength test and flexural strength tests are shown in Tables 1 and 2.

The composites with raffia had worse mechanical properties as material with PP fibres additive, however they are still reasonable for some potential application. The addition of the raffia fibres decreased the mechanical properties of the samples comparison with other naturals fibres [4]. It is caused by the lack of cohesiveness between the fibres and the geopolymer matrix.

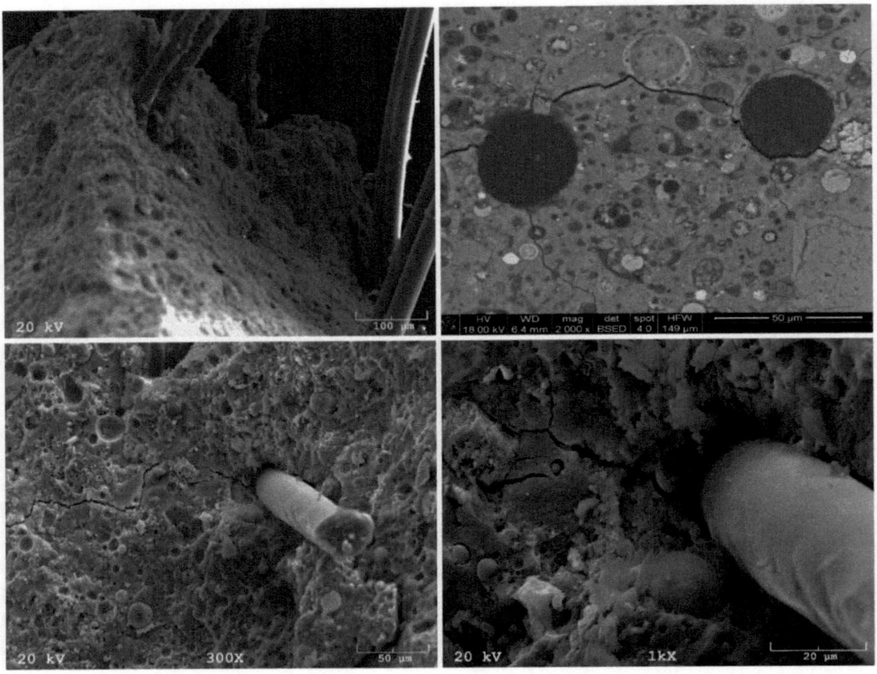

Fig. 4 Exemplary SEM images for composite: geopolymer matrix reinforced by PP fibres

Table 1 Compressive strength test after 28 days

Sample	MPa	StD
Geopolymer with raffia fibers (1%)	13.66	1.71
Geopolymer with PP fibers (1%)	16.14	3.52

Table 2 Flexural strength test after 28 days

Sample	MPa	StD
Geopolymer with raffia fibers (1%)	3.05	0.35
Geopolymer with PP fibers (1%)	3.95	0.21

Conclusions

Geopolymer composites reinforced with the raffia fibres and PP fibres have been produced and characterized. The samples were prepared using sodium promoter and four various types of fibre additive (each time it amounted to the 1% by mass of the composite (8 M)). The research involved both reinforced samples with natural and artificial fillers (for the sake of comparison).

The microstructure observations shows the difference between coherence natural and artificial fibres with geopolymer matrix that have an influence for mechanical properties. The composites with artificial fibers have better properties, however there are not such beneficial for environment. This study shows that it is possible to produce the composites of reasonable properties from the industrial wastes (fly ash) and renewable resources—raffia fibers, which makes them a new class of environmentally friendly materials. Further tests should be performed in order to optimize the composite properties.

Acknowledgements The authors gratefully acknowledge the funding by National Centre for Research and Development, Poland, under grants "Development of eco-friendly composite materials based on geopolymer matrix and reinforced with waste fibers" under the ERANet-LAC: Latin America, Caribbean and European Union funded by the European Commission, within the 7th Framework Programme for Research and Technology Development (FP7), Topic#02: Waste management, recycling and urban mining (project no. ELAC2015/T02-0721).

References

1. Alomayri, T., Shaikh, F. U. A., & Low, I. M. (2013). Characterisation of cotton fibre-reinforced geopolymer composites. *Composites: Part B, 50,* 1–6.
2. Odera, R. S., Onukwuli, O. D., & Atuanya, C. U. (2015). Characterization of the thermo-microstructural analysis of *Raffia palm* fibers proposed for roofing sheet production. *Journal of Minerals and Materials Characterization and Engineering, 3,* 335–343.
3. Odera, R. S., Onukwuli, O. D., & Osoka, E. C. (2011). Tensile and compressive strength characteristics of *Raffia palm* fibre-cement composites. *Journal of Emerging Trends in Engineering and Applied Sciences, 2*(2), 231–234.
4. Korniejenko, K., Frączek, E., Pytlak, E., & Adamski, M. (2016). Mechanical properties of geopolymer composites reinforced with natural fibers. *Procedia Engineering, 151,* 388–393.
5. Alomayri, T., Shaikh, F. U. A., & Low, I. M. (2013). Low thermal and mechanical properties of cotton fabric-reinforced geopolymer composites. *Journal Materials Science, 48,* 6746–6752.
6. Alzeer, M., & MacKenzie, K. (2013). Synthesis and mechanical properties of novel composites of inorganic polymers (geopolymers) with unidirectional natural flax fibres (phormium tenax). *Applied Clay Science, 75–76,* 148–152.
7. Patel, R., & Joshi, R. (2016). Analysis of development of porous fly ash-based geopolymer with low thermal conductivity. *International Advanced Research Journal in Science, Engineering and Technology, 3*(12), 171–178.
8. Chen, R., Ahmari, S., & Zhang, L. (2014). Utilization of sweet sorghum fiber to reinforce fly ash-based geopolymer. *Journal Materials Science, 49,* 2548–2558.
9. Alzeer, M., & MacKenzie, K. (2012). Synthesis and mechanical properties of new fibre-reinforced composites of inorganic polymers with natural wool fibres. *Journal Materials Science, 47,* 6958–6965.
10. Yliniemi, J., Nugteren, H., Illikainen, M., Tiainen, M., Weststrate, R., & Niinimäki, J. (2016). Lightweight aggregates produced by granulation of peatwood fly ash with alkali activator. *International Journal of Mineral Processing, 149,* 42–49.
11. Yan, L., Kasal, B., & Huang, L. (2016). A review of recent research on the use of cellulosic fibres, their fibre fabric reinforced cementitious, geo-polymer and polymer composites in civil engineering. *Composite Part B-Engineering, 92,* 94–132.
12. Olumide, B. O., & Bako, W. G. (2016). Woven sculptural piece as added dimension to textile design. *Mgbakoigba, Journal of African Studies, 5*(2), 123–138.
13. Ramakrishna, G., & Sundararajan, T. (2005). Impact strength of a few natural fibre reinforced cement mortar slabs: A comparative study. *Cement & Concrete Composites, 27,* 547–553.

14. Ewuim, S. C., Akunne, C. E., Anumba, A. I., & Etaga, H. O. (2011). Insects associated with wine from *Raffia palm* (Raphia Hookeri) in Alor, Nigeria. *Animal Research International, 8* (1), 1328–1336.
15. Obahiagbon, F. I., & Osagie, A. U. (2007). Sugar and macro-minerals composition of sap produced by *Raphia Hookeri palms*. *African Journal of Biotechnology, 6*(6), 744–750.
16. Sikame Tagne, N. R., Njeugna, E., Fogue, M., Drean, J. Y., Nzeukou, A., & Fokwa, D. (2014). Study of water absorption in *Raffia Vinifera* fibres from Bandjoun. *Cameroon. The Scientific World Journal, 2014,* 1–11.
17. Elenga, R. G., Dirras, G. F., Goma Maniongui, J., Djemia, P., & Biget, M. P. (2009). On the microstructure and physical properties of untreated Raffia textilis fiber. *Composites: Part A, 40,* 418–422.
18. Reddy, N., & Yang, Y. (2005). Biofibers from agricultural byproducts for industrial applications. *Trends in Biotechnology, 23*(1), 22–27.
19. Uke, C., & Gowripalan, N. (1989). Strength properties of *Raffia Bamboo. Construction and Building Materials, 3*(1), 49–52.
20. Tiaya Mbou, E., Njeugna, E., Kemajou, A., Tagne Sikame, N. R., & Ndapeu, D. (2017). Modelling of the water absorption kinetics and determination of the water diffusion coefficient in the pith of *Raffia vinifera* of bandjoun, Cameroon. Hindawi. *Advances in Materials Science and Engineering, 2017,* 1–12.
21. Sathish, P., Kesavan, R., & Ramnath, V. (2015). Experimental investigation on flexural property of Abaca and Raffia hybrid composites. *ARPN Journal of Engineering and Applied Sciences, 10*(13), 5534–5536.
22. Ramnath, B. V., Elanchezhian, C., Manickavasagam, V. M., Gowri Prasad, S., Arvindh Swamy, S., & Keshav, Raj R. (2016). Experimental investigation on shear and hardness of abaca based hybrid composites. *MATEC Web of Conferences, 74*(00039), 1–4.
23. Foadieng, E., Talla, P. K., Nkamgang, G. B., & Fogue, M. (2017). Study of the thermal properties of *Raffia bamboo vinifera* L. Arecaceae. Hindawi. *Advances in Materials Science and Engineering, 2017,* 1–10.
24. Van Nguyen, D., Rabemanolontsoa, H., & Saka, S. (2016). SAP from various palms as a renewable energy source for bioethanol production. *Chemical Industry & Chemical Engineering Quarterly, 22*(4), 355–373.
25. Obahiagbo, F. I. (2009). A review of the origin, morphology, cultivation, economic products, health and physiological implications of *Raphia palm. African Journal of Food Sciences, 3* (13), 447–453.
26. Elenga, R. G., Djemia, P., Tinguand, D., Chauveau, T., Goma Maniongui, J., & Dirras, G. F. (2013). Effects of alkali treatment on the microstructure, composition, and properties of the Raffia textilis fiber. *BioResources, 8*(2), 2934–2949.

Sustainable Composites Based on Pine Resin and Flax Fibre

R. Ribeiro, A.T. Marques and J.L. Alves

Abstract In this investigation, we assess the use of pine resin and flax fibre as raw materials for the formulation of a biocomposite material to determine its applicability and demonstrate the hidden potential that may presently be underexploited with these two resources. The formulated biocomposite was subjected to bending tests, comparing reinforced and non-reinforced matrices with flax fibre. Different additives of natural and synthetic origin were used to improve the fragility of the rosin, with EVA (ethylene vinyl acetate) being the material that presented greater potential, in the proportion of 30%. This formulation was used to manufacture innovative products. The results obtained are very promising and opens the opportunity to design and manufacture new sustainable products that can be used in different industrial areas, like furniture and lighting. The products hereby introduced envision to contribute for the rehabilitation of the rosin industry and revive the flax culture in Portugal.

Keywords Biocomposites · Industrial Design · Sustainable development Sustainable composites · Flax fiber · Pine resin

Introduction

Nowadays, one of the biggest problems facing society is the environmental pollution caused by the use and excessive disposal of products from the petroleum-based plastics industry. This raw material is transversally suitable for the

R. Ribeiro (✉) · J.L. Alves
Master in Product and Industrial Design, University of Porto, 4200-465 Porto, Portugal
e-mail: rubentmribeiro@gmail.com

A.T. Marques
Department of Mechanical Engineering (DEM), University of Porto, 4200-465 Porto, Portugal

J.L. Alves
INEGI, Faculty of Engineering of University of Porto, 4200-465 Porto, Portugal

R. Fangueiro and S. Rana (eds.), *Advances in Natural Fibre Composites*,
https://doi.org/10.1007/978-3-319-64641-1_14

whole industry, mainly for its ease of processing, low cost, and high durability [21]. It is also worth mentioning that, in many applications, life cycle analysis (LCA) is favourable to plastics [13]. However, their main drawbacks in ecological terms are known, from the time it takes them to decompose to the amount of resources they consume during processing [16]. A new paradigm has emerged due to these issues, which reflects the origin of the raw material as a way to achieve sustainable development. Biocomposites consist of technological materials made from renewable and nature-based raw materials. Natural fibres, available as agricultural and forestry resources, have been integrating polymer matrices to obtain more ecological composites than those reinforced with synthetic fibres, which are more difficult to recycle. The adoption of natural fibres allows not only the recycling of fibres, but also the natural degradation of fibres in the environment [13]. Also, the choice of the composite matrix can be crucial to define recyclability and degradation in the environment after a product's useful life has been reached. Although several biobased materials are already available, it is pertinent to look for more alternatives that promote local development. In this sense, there are several resources that are not being adequately valued by the industry. This research intends to study the potential of the Portuguese pine resin and flax fibre as raw materials in the development of new products that promote Eco-Conscious Design and sustainable development.

Biocomposites Definition and Market

The main differentiating element between a conventional composite and a biocomposite is based on issues of sustainability and origin, seeking to use natural resources from renewable sources. Although the "bio" prefix of biocomposites suggests that the material needs to be formulated with resources exclusively of natural origin, this type of composite can also seek synthetic resources for its formulation [16]. The biocomposite market is segmented by Natural Fibre Composites (NFC) and Wood Polymer Composites (WPC) [16]. In Europe, biocomposites currently account for about 10 to 15% of the whole composite market (Bioplastics Magazine [4]. The global market exclusively for NFC was estimated at around \$3.36 billion in 2015 and it is projected to reach \$6.5 billion by 2021 at a rate of growth of 11.68% between 2016 and 2021 (Markets and Markets [15].

Natural Fibres Reinforcement

In a global market heavily dominated by the use of synthetic fibres, the search for ecological and sustainable alternatives becomes increasingly relevant [16]. In this frame, natural fibres begin to attract attention in the composite industry, due to energy consumptions 5–10 times inferior to glass fibres during processing [13], and

Table 1 Properties of the most used natural fibers in biocomposites

	Flax	Hemp	Jute	Cotton	Coconut	Sisal
Density (g/cm^3)	1.5	1.47	1.3–1.49	1.5–1.6	1.15–1.46	1.45
Diameter (μm)	40–600	25–500	25–200	12–38	100–460	50–200
Tensile Strength (MPa)	345–1500	690	393–800	287–800	131–220	468–700
Young's Modulus (GPa)	27.6	70	13–26.5	5.5–12.6	4–6	9.4–22
Elongation at break (%)	2.7–3.2	1.6	1.16–1.5	7–8	15–40	3–7

Adapted from Mohanty, Misra e Drzal [16]

also their low cost and abundance [21]. Nowadays, the fibres with greater interest for the industry of the composites are the vegetal fibres, originating from the plants. Plants used as a source for obtaining fibre can be classified as primary or secondary. Primary plants are those in which fibre production is the main product to be obtained, while secondary plants are more suitable for fruit production and, in this case, fibre is considered a by-product [10]. Natural fibres generally grow in vertical bundles and can be obtained from many parts of a plant, namely: leaves (sisal, banana and abaca), seeds/fruit (cotton, coconut), bast (hemp, jute and flax), and the plant's core and sap. In general, natural fibres have half the fiberglass density for similar rigidity, causing them to have a higher specific stiffness [18, 14]. Table 1 compares the mechanical properties of the most commonly used natural fibres in the biocomposites industry, showing ranges of data that may include fibre modification.

Flax Fibre

Flax (*Linum usitatissimum* L.) represents the oldest fibre in the world, and its culture was originated in Egypt thousands of years ago [16]. This annual plant grows vertically in humid lands and temperate climates, especially in Europe, being able to reach 1.20 m in height. Normally, flax is easily recognized by the bluish or satiny hue of the flower, although there are dozens of subspecies in which the characteristics vary. The flax plantations are characterized by high turnover, being a resource that is renewed very quickly. The flax harvesting can be performed manually or mechanically. The plant is torn out by its roots to take advantage of the maximum length of the stem. After the fibres extraction, they are subjected to many transformation processes that allow the tissue to be obtained.

Data from Faostat [9], considering the period from 2007 to 2014, indicate that the world production of flax exclusively for fibre is estimated at around 350,000 tons per year and is dominated by Europe and China. In France, for example, the European Confederation of Flax and Hemp—CELC was established to demonstrate that industry and agriculture can work in symbiosis. This confederation intends to increase the interest of the industry, counting already with 10,000 companies, having into account the farmer's rights. The applicability of flax composites is very wide and can be used as structural material on automobile parts such as doors and decorative elements, and everyday objects such as furniture. More recent applications show that these types of composites have been implemented in sports objects, mobility, music, and the potential integration of these materials in the aerospace industry is being investigated [17].

Matrices for Biocomposite Formulation

Polymer matrices of natural fibre composites are mainly of two types: thermoplastic and thermosetting, which can be obtained from synthetic, biobased, or mixed resources [16]. Currently, most of the biocomposites available on the market still rely on the use of synthetic matrices, due to the balance between high durability and resistance, coupled with low processing costs [19]. Nonetheless, biobased matrices are beginning to raise the attention of new markets to respond to the calls of a more responsible and concerned society [22].

Biobased Matrices

Biobased matrices are those that can be formulated with content derived from renewable natural resources. Biobased plastics are called biopolymers and may come from a variety of resources, from starch to cellulose [19]. Biopolymers can also be obtained from microorganisms or monomers derived from biomass, such as the lactic acid-PLA [22]. PLA represents the most-known bioplastic, which has been used as a filament material for 3D printers. This material emits about 50–70% less carbon dioxide during processing compared to traditional polymers such as polyethylene terephthalate-PET [7]. Also, biopolymers represent a possible solution to problems related to the waste and disposal of plastics in the environment associated with synthetic plastics. However, not all biopolymers are biodegradable [16]. Biobased thermosets, although they belong to the class of bioplastics, are commonly called bioresins, and can be obtained from the oil of plants such as soy, for example. Many types of bioresins/biopolymers are being developed and currently available on market, as shown in Table 2. Still in the field of biobased matrices, there are other types of natural resins, less commercial and less used in composites, in the form of natural gums. Natural gums are natural polymers of vegetable origin characterized by low densities and reduced melt temperatures [5]. They can be obtained by tapping plants, which are transformed to remove impurities. Some natural gums include arabic gum, latex and rosin derived from pine resin [5].

Table 2 Biopolymers bioresins used in the manufacture of biocomposites

Resin Name	Commercial name	Producer	Type of resin	Resin base
Polylactic acid	PLA	Cargill Dow LLC	Thermoplastic	Lactic Acid
	Ingeo	NatureWorks LLC		Corn
Bio-epoxy	Vikoflex 7170	Atofina Chemicals Inc.	Thermoset	Soy-bean oil
	Vikoflex 7190	Atofina Chemicals Inc.		Linseed oil
	Sorona EP	DuPont Eng. Polymers		Corn
	Super SAP 100 epoxy	Entropy Resins		Pine & Vegetable oils
Bio-polyester	Envirez 5000	Ashland Chemical Co.	Thermoset	Soy

Adapted from Pilla [18]

Pine Resin

The pines represent the tree from which the resin is obtained. This tree derives from the genus *Pinus*, with several species, which vary depending on the climate and soil characteristics of each country [6]. The traditional pine-tapping is the most common process for obtaining rosin, in which the resin appears naturally in the pine tree to protect and cover the wounds caused by the incisions, being collected and stored in plastic bags or vases [1]. The production potential of rosin in Portugal is immense, since the pine tree (*Pinus pinaster*) represented in 2010 about 22% of the forest area in the national territory, corresponding to 714,000 ha [11]. In Portugal, warm temperatures allow each pine to produce, on average, 3–4 kg of resin per year [3]. However, the annual production of resin in Portugal in 2005 was about 5000 tons, and there is currently installed capacity to increase considerably this value to 20,000 tons [3]. The applications of the pine resin derivatives are very broad and are mainly used by the processing industries as raw material present in industrial products, pharmaceuticals, and adhesives [3]. Some constituents of rosin, such as abietic acid, are being used to obtain new polymers based on renewable resources, and investigations have been made to use rosin as an additive in polymeric materials such as styrene-butadiene-styrene-SBS rubber; acrylonitrile butadiene styrene-ABS and ethylene vinyl acetate-EVA [8].

Rosin is not usually used as a matrix or binder in composites, given its brittle behaviour and the poor thermal properties compared to other resins on the market. However, experiments have been carried out with rosin and natural fibres, namely sisal fibres, in which tubes with some hardness and strength were produced [2]. Still in the resin field, studies were conducted to improve the flexibility of rosin [20]. In that study, it was concluded that the EVA and SBS polymers can improve the fragility problem and give more flexibility to the rosin [20].

Research

Due to the brittleness characteristic of this resin (rosin YT 321), experiments were carried out with other materials, so that they can be used as additives to increase rosin plasticity. Five different additives were considered: starch and cork powder as wastes of natural origin; recycled rubber (particle size <0.8 mm) and wax as wastes of synthetic origin; and EVA (ethylene vinyl acetate) in granules as a virgin material of synthetic origin. The investigation began with the preparation of several samples of rosin with each additive individually, varying the proportions between 10 and 50%. In all experiments, the rosin was deposited in a solid state in a pot and placed to heat in an electric stove. The temperature was controlled through the levels on the controls of the electric cooker and measured with a thermostat to determine the corresponding temperature. After being sufficiently fluid, additives were added and blended with the rosin, individually. All components were pre-weighed on a precision balance (resolution 0,1 g). In terms of toughness, it was possible to see that the combination of pine resin and cork powder produced a brittle mixture, since more voids were created that contribute to the brittleness of the samples. On the opposite side, the wax waste, when mixed with the pine resin, promoted an increase of the ductility of the samples. However, the use of this additive modifies the original visual appearance of the resin, making it greenish. The processing of rubber granules led to the emission of fumes and odours, proving to be a dangerous and flammable additive. The results with this material were not very positive since the rubber waste granulate did not completely melt with the resin and the original appearance of the rosin changes completely, making the samples black and opaque. By adding the powdered starch, it was possible to verify some difficulties in dissolution with the hot rosin, creating agglomerates of material throughout the sample. Despite this, some of the materials were able to melt and be mixed with the rosin. However, these materials did not produce relevant results in terms of rosin flexibility. The virgin EVA presented itself as the most promising additive, showing complete compatibility with the pine resin, giving more flexibility, especially in the proportion of 30%, without significantly altering their original aspect.

After improving the properties of the resin to be used as matrix, the biocomposites reinforced with flax were created. For the formulation of those biocomposites, bidirectional flax fabric was used as reinforcement, which was purchased from Easy Composites Ltd—UK, a leading distributor of composite materials in Europe. Several squares of flax fabric were cut with the measurements of the cavity of the silicone mould: 100 mm × 100 mm. In each experiment four layers of reinforcing tissue were considered. Then the matrix of the biocomposite was formulated. After processing the blend, the resin was cast into the mould where four layers of fabric fibres were already allocated. A silicone cap was placed on the sample to ensure that the thickness was as constant as possible and to allow a pressure of 3, 9 kPa, to compact the resin in the fibres, removing the excess from the mixture. The sample cure occurred at room temperature for 30 min. It is

Fig. 1 **a** Correctly formulated composite **b** Sliding of the tissue in the matrix

important to note that other experiments were carried out in which the sliding of the fibres was observed when a pressure of 14, 7 kPa was applied for compaction (Fig. 1b). In addition to the excess of pressure, sliding of the fibres can be explained by the high viscosity of this resin, derived from the amount of EVA additive.

Manufacture of Test Specimens

At this stage samples of both materials were obtained: non-fibre reinforced RE matrix [Rosin + 30% EVA] and the 10% flax fibre reinforced matrix [RE10F]. The RE10F test specimens, intended for bending tests, were extracted from the previously obtained composite plate. The process was started by cutting the edges of each laminate to measure the average thickness of each material. In this case, the length measurements are defined as a function of thickness, as indicated in ISO 14125, resulting in a ratio of twenty times the thickness of the composite [12]. Also according to the Standard, considering the biocomposites formulated as class II materials, all test pieces must have the same width, 15 mm [12]. Once the measurements were defined, the lines of the test pieces were marked, leaving margins of 1.5 mm between each cut (thickness of the blade) in order not to compromise the tolerance required by the Standard (Fig. 2a). The specimens were extracted by the cuts made through a vertical water-cooled electric saw (Fig. 2b), in the INEGI facilities composite laboratory. The extracted specimens were cleaned, the burrs were removed and the edges of each specimen were polished manually with P320 sandpaper (Fig. 2c). The non-fibre reinforced (RE) specimens for the rosin material and 30% EVA were obtained in individual moulds, since when cutting a plate of this material, it proved to be too brittle and eventually broken making these specimens unusable.

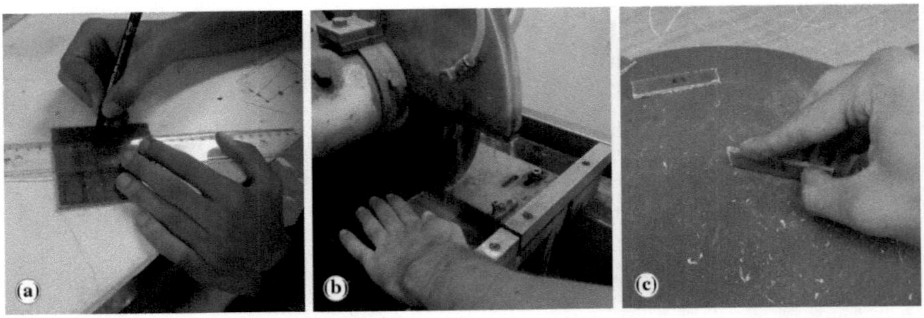

Fig. 2 **a** Marking of the measurements of the test sample **b** Cutting and extraction of test specimens **c** Sanding test specimens

Table 3 Test specifications according to each range of test specimens (ISO 14125)

Samples	Class	Span distance (mm)	Diameter of the rollers (mm)	Loading speed (mm/min)
RE	I	80	5	1
RE10F	II	46	5	1

Three-Point Bending Tests

Three-point bending tests were performed on both materials: RE and RE10F, according to ISO 14125. The materials were tested on a TIRAtest 2705—TIRA GmbH machine equipped with a load cell of 500 N. The samples were conditioned in a controlled environment at 23 °C and 50% relative humidity for 88 h until they were ready for testing. Table 3 lists the specifications of each test for each range of material. The tests were performed until the rupture of the specimen (Fig. 3c).

The results of the three-point bending tests are shown in Table 4. A 10% fibre content of flax (RE10F) increases the strength and stiffness of the non-fibre reinforced RE matrix [Rosin + 30% EVA] by almost 200%. In the case of the RE10F composite, it is found that the 10% fibre content leads to an increase in the deformation at break of 53% with respect to the non-reinforced matrix (RE).

Design Proposals

After the investigation with the material, a set of pieces was designed to demonstrate the applicability potential. The designed concept consists of semi-structural composite pieces with cylindrical-conical geometry, attached to other complementary pieces produced in cork. The concept intends to demonstrate the applicability of the material in different types of pieces of furniture and lighting (Fig. 4).

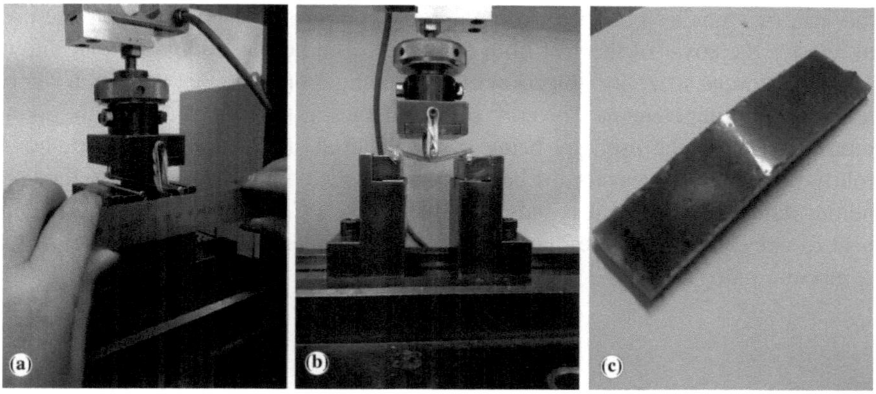

Fig. 3 **a** Measurement of the distance between supports **b** RE10F sample during the test **c** RE10F sample after breaking

Table 4 Three point flexural tests results

Samples	E [GPa]	σ [MPa]	ε [%]
RE	0.4 ± 0.1	7.6 ± 0.8	3.9 ± 0.5
RE10F	1.1 ± 0.2	23.0 ± 1.4	5.9 ± 0.8

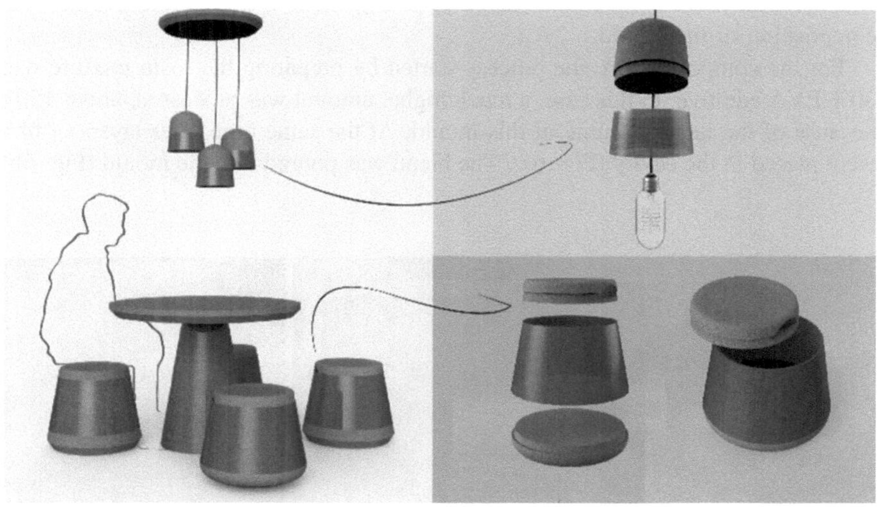

Fig. 4 Design proposal of the idealized concept

Two types of bench were designed, which differ in the geometry of the base, being able to be static or dynamic. The dynamic bench has this designation because it offers more freedom of movements to the user. To facilitate the transportation of the seat, two handles obtained by thinning the cork are allocated, allowing the user to remove the cover of the seat from the bench. The hollow composite part allows for some storage space for objects or even a spotlight on it, so that the bench can be used as a lighting element. The cork cover emerges as the seat of the bench, giving more comfort to the user, by being a soft material. The suspended lamp shown follows the same process for the bench, in which the composite is glued to the cork holder. For this product, it is only possible to use LED lamps which do not emit heat to not compromise the maximum working temperature of the material of the composite part.

Prototyping Process

For the creation of the prototype, a silicone mould was built. A small conical ceramic cup was acquired to serve as mould for the cavity of the part. On a metal base, a box with small MDF boards was constructed, considering the square between each block. Thereafter, the silicone was prepared as a function of the volume of the box, having been manually mixed in a plastic container with 10% catalyst. The mixture was then placed in a vacuum chamber to remove air bubbles. After this process, the silicone was poured into the mould until practically covering the entire height of the ceramic vessel, to obtain the base of the mould. The mould was cured at room temperature for 24 h. In Fig. 5c it is possible to see the final composition of the mould.

For the composite part, the process started by preparing the rosin mixture with 30% EVA additive. In this case, a much higher amount was processed, about 150 g because of the larger volume of this mould. At the same time, four layers of fibre were placed in the cavity (Fig. 6a). The blend was poured into the mould (Fig. 6b),

Fig. 5 **a** Ceramic cup with injection wax to reduce 40 mm of the size of the cavity **b** Casting and curing of the silicone mixture **c** Final composition of the mould

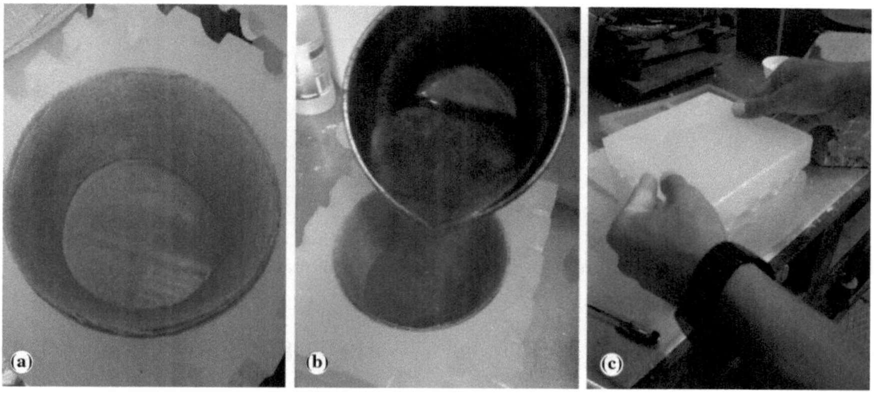

Fig. 6 **a** Positioning of the fibres in the cavity **b** Casting the mixture **c** Placing the compaction lid

with the lid helping the mixture to rise through the cavity and thus forming a substantially linear wall with the fibres (Fig. 6c). The cap also aided in fibre compaction. A pressure of 11, 8 kPa was applied through a block of stone placed on the lid. After 45 min curing at room temperature, the piece was demoulded. This process required caution so that the part would not be damaged on the surface.

For the complementary cork pieces, the process started by cutting small cork blocks, considering the volume of each piece, to save the maximum material (Fig. 7a). The blocks were later machined in the lathe, considering the diameters indicated in the technical drawings, since the projected parts present revolution geometry. Finishing operations were also carried out, including the manual sanding of each piece (Fig. 7b).

Following, on Fig. 8, are photographs of the prototypes obtained by the processes described above. The luminaire can be adapted to the ceiling, suspended, or placed on a desk or table, using an anchor base. The displayed seat (prototype) also

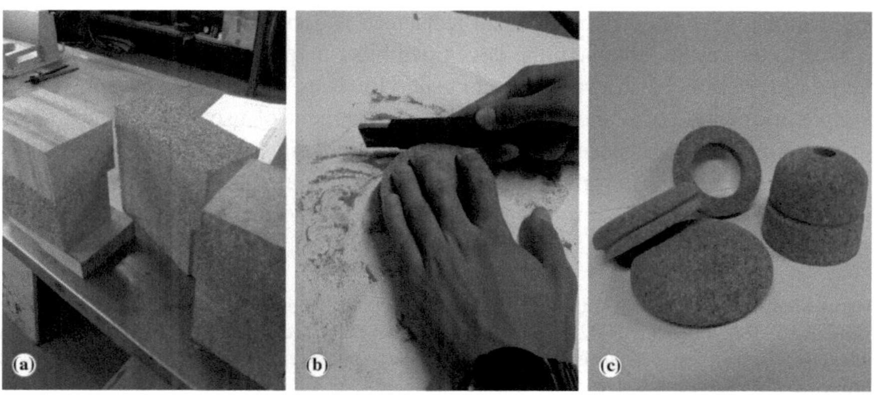

Fig. 7 **a** Cutting of cork blocks **b** Sanding of parts to improve finishing **c** Cork parts obtained

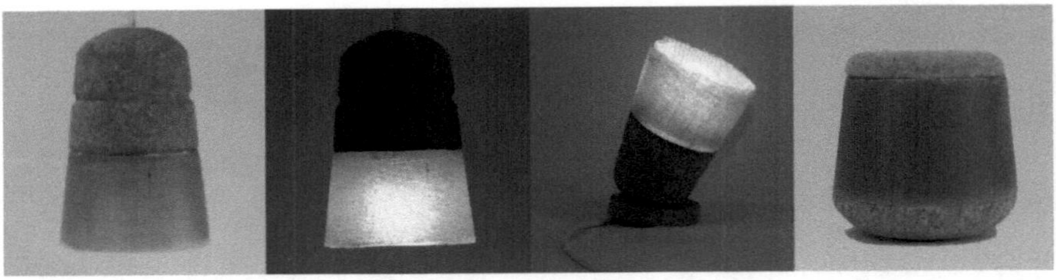

Fig. 8 Final prototypes obtained with the new biocomposite material

allows the installation of a portable light, transforming this object into a lighting element.

Conclusions

This research resulted in the development of sustainable products with the aim of contributing to the rehabilitation and extraction of pine resin in Portuguese forests and the relaunching of flax as an agricultural resource that is part of the historical and economic past of Portugal. The individual contact with these two raw materials allowed us to anticipate some drawbacks related to the strength of the materials, and the present research sought to optimize this aspect. In particular, it has been realized that the brittleness of the rosin can be reduced with the use of additives of natural and synthetic origin. In this case, the addition of the EVA polymer, in the proportion of 30%, allowed to confer some resistance and flexibility to the rosin. Despite this, the addition of this polymer raises the viscosity of the resin, which can promote slippage of the fibre fabric and thus hamper the processing method. Through the flexural tests, it was possible to realize that a content of 10% of flax fibre raises the strength and stiffness of the composite for about 200% when compared to the non-reinforced matrix, opening the opportunity to use it in applications where higher mechanical resistance is demanded.

Due to the results obtained, the developed biocomposite can be applied to some pieces of furniture, as shown in the concept. However, there are restrictions on the complexity of the parts, the size, and the environment of use. With regard to the proposed products, it is necessary to study their industrial viability, and efforts must be made to optimize design and production parameters. Further tests should be performed in a collaboration of an industrial partner to analyse all manufacturing stages.

Acknowledgements The Authors gratefully acknowledge the funding of Project NORTE-01-0145-FEDER-000022—SciTech—Science and Technology for Competitive and Sustainable Industries, cofinanced by Programa Operacional Regional do Norte (NORTE2020), through Fundo Europeu de Desenvolvimento Regional (FEDER).

References

1. Abad Viñas, R., Caudullo, G., & de Rigo, D. (2016). Pinus pinaster in Europe: Distribution, habitat, usage and threats. *European Atlas of Forest Tree Species*. Luxembourg: European Commission.
2. Afonso, A. T. (2013). *Desenvolvimento Sustentável em Engenharia Mecânica-Um caso de estudo: Projeto Autarkeia*. MS: University of Porto.
3. Anastácio, D., & Carvalho, J. (2008). *Sector dos Resinosos em Portugal*. DGRF, Lisbon: Evolução e Análise.
4. Bioplastics Magazine. (2015). *Newest research data show robust growth European wood and natural fiber composite market* [Online]. Available: http://www.bioplasticsmagazine.com/en/news/meldungen/20150621-New-WPC-study-nova-Institute.php. Accessed January 22, 2017.
5. Coppen, J. J. (1995). *Gums, resins and latexes of plant origin*. Canada: FAO.
6. Coppen, J. J., & Hone, G. (1995). *Non-wood forest products 2. Gum naval stores: Turpentine and rosin from pine resin*. Rome: FAO.
7. Dorgan, J. R. (2013). *Transforming the resource base: New bioplastics* [Online]. SPE. Available: http://www.4spepro.org/view.php?source=005111-2013-10-04. Accessed February 8, 2017.
8. Euro-Yser. (2013). Provisional Technical Data Sheet RD 27/01. In (AVEIRO), E. Y.-P. Q. S. A. (Ed.).
9. Faostat, F. (2014). Statistical database- Production/Crops of Flax fibre and tow.
10. Faruk, O., Bledzki, A. K., Fink, H.-P., & Sain, M. (2012). Biocomposites reinforced with natural fibers: 2000–2010. *Progress in Polymer Science, 37,* 1552–1596.
11. ICNF. (2013). *IFN6-Áreas dos usos do solo e das espécies florestais de Portugal continental*. Instituto da Conservação da Natureza e das Florestas: Resultados preliminares. Lisbon.
12. Iso, I. O. F. S. (1998). *Fibre-reinforced plastic composites—Determination of flexural properties ISO 14125: 1998*. Geneva: ISO.
13. Joshi, S. V., Drzal, L., Mohanty, A., & Arora, S. (2004). Are natural fiber composites environmentally superior to glass fiber reinforced composites? *Composites Part A Applied Science and Manufacturing, 35,* 371–376.
14. Kandachar, P., & Brouwer, R. (2001). Applications of bio-composites in industrial products. *MRS Online Proceedings Library Archive,* 702, null-null.
15. Marketsand Markets. (2016). *Natural fiber composites market by type (wood fiber and non-wood fiber), manufacturing process (compression molding, injection molding, and others), application (building & construction, automotive, and electrical & electronics), and region—Global forecasts to 2021* [Online]. Available: http://www.marketsandmarkets.com/Market-Reports/natural-fiber-composites-market-90779629.html. Accessed January 12, 2017.
16. Mohanty, A. K., Misra, M., & Drzal, L. T. (2005). Natural fibers, biopolymers, and biocomposites. CRC Press.
17. Pil, L., Bensadoun, F., Pariset, J., & Verpoest, I. (2016). Why are designers fascinated by flax and hemp fibre composites? *Composites Part A Applied Science and Manufacturing, 83,* 193–205.
18. Pilla, S. (2011). *Handbook of bioplastics and biocomposites engineering applications*. Wiley.
19. Quarshie, R., & Carruthers, J. (2014). Biocomposites: Technology overview. United Kingdom.
20. Sousa, C. (2015). *Resina e Design: Contributo para o aumento da extração da resina do pinheiro*. MS: University of Porto.
21. Taj, S., Munawar, M. A., & Khan, S. (2007). Natural fiber-reinforced polymer composites. *Proceedings-Pakistan Academy of Sciences, 44,* 129.
22. Tambyrajah, D. (2015). *Indulge & Explore Natural Fiber Composites*. Netherlands: NFCDesign Platform.

Development and Characterization of Microcrystalline Cellulose Based Novel Multi-scale Biocomposites

Sohel Rana, Shama Parveen, Subramani Pichandi and Raul Fangueiro

Abstract Recently, multi-scale composites have been developed by combining reinforcements from different length scales, in order to overcome the drawbacks of conventional composites as well to enhance their performances. Most of the research works on multi-scale composites have been performed using carbon based nanomaterials such as carbon nanotubes (CNTs) and nano fibres (CNFs). Looking at the outstanding characteristics of carbon based multi-scale composites, multi-scale composites based on plant based fibres and nano materials are also gaining research attention. This paper provides a brief overview of recent research studies on multi-scale composites, particularly focusing on plant based fibres and nanomaterials.

Keywords Multi-scale composites · Plant fibres · Nanocellulose
Fibre-matrix interface · Mechanical performance

Introduction

Hybrid and Multi-scale Composites

Hybrid composites are developed by combining two or more different types of reinforcing fibres. Hybridization is required to achieve some of the characteristics which cannot be achieved using only one set of reinforcing fibres. For example, composites based on carbon fibres and thermos-setting matrices provide very good stiffness and strength, but they suffer from low breaking strain and inferior ductility. In order to improve this behaviour, hybrid composites can be developed by mixing carbon and glass fibres at certain ratios to achieve higher breaking strain and ductility (this phenomenon is commonly known as "pseudo-ductility") [1, 2]. Also,

S. Rana (✉) · S. Parveen · S. Pichandi · R. Fangueiro
Centre for Textile Science and Technology (2C2T), University of Minho, Campus de Azurem, 4800-058 Guimaraes, Portugal
e-mail: soheliitd2005@gmail.com

R. Fangueiro and S. Rana (eds.), *Advances in Natural Fibre Composites*,
https://doi.org/10.1007/978-3-319-64641-1_15

hybrid composites developed using glass and carbon fibres or glass and carbon powder show better strain and damage sensing ability than carbon fibre composites by proving an alarm signal well before the composite breakage [1]. Figure 1 shows the schematic representation of carbon-glass hybrid composites with better self-sensing performance.

Multi-scale composites are a special class of hybrid composites where materials from different length scale such as macro, micro and nano materials are combined together. More frequently, multi-scale composites are developed by introducing nano reinforcement within the matrix or fibre surface of conventional composites [3–10]. The concept of multi-scale composites and its difference from conventional composite materials have been illustrated in Fig. 2.

As clear from Fig. 2, it can be understood that these composites can be developed by either introducing the nanomaterials on the fibre surface or dispersing them within the matrix of composites. In case of CNT based multi-scale composites, different approaches which have been tried to introduce CNTs on fibre surface are: (1) growing CNTs on fibre surface in CVD reactor, (2) spraying, (3) chemical grafting, (4) electro-phorectic deposition, (5) coating, and so on. Figure 3 shows the schematic representation of these various processes.

These different processes have their merits and demerits. For example, the process of CNT growth in chemical vapor deposition (CVD) reactor is

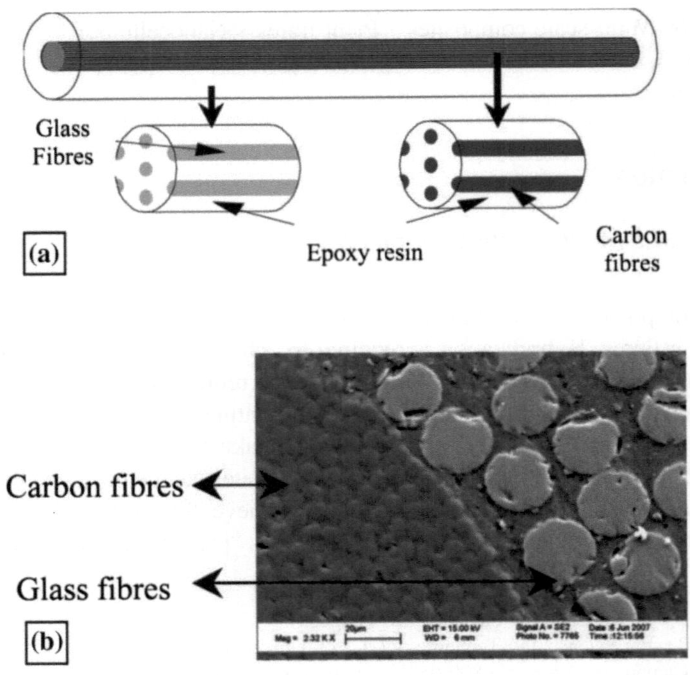

Fig. 1 Schematic diagram of **a** carbon fibre–glass fibre hybrid composites and **b** cross/sectional view of a sample observed by SEM [1]

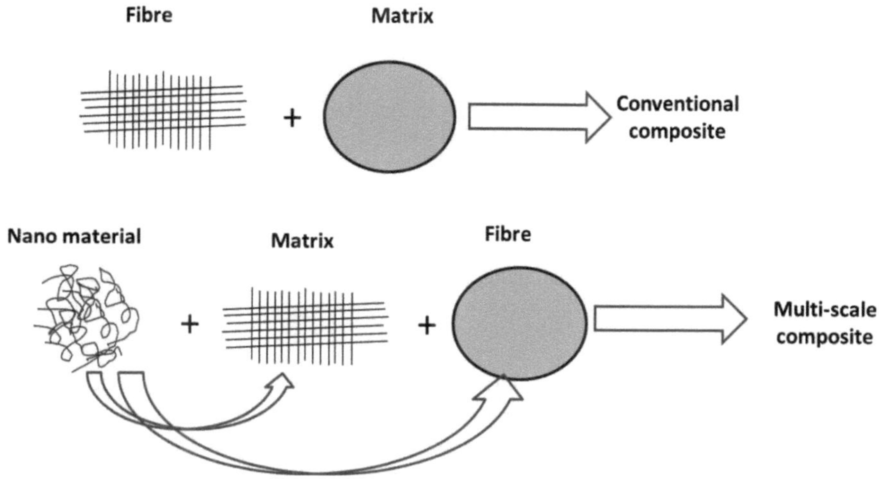

Fig. 2 Concept of multi-scale composites [10]

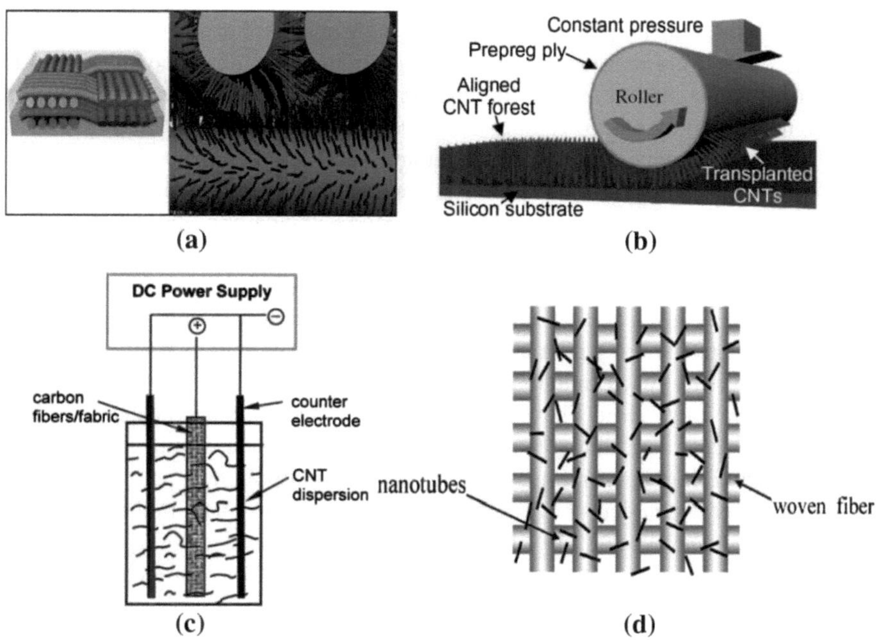

Fig. 3 Techniques of incorporating nanomaterials on fibre system: **a** nanomaterial growth, **b** transfer printing, **c** electrophoretic deposition **d** spraying process [10]

advantageous to control the alignment of grown CNTs. It is very interesting to note that CNTs can be grown vertically on the fibre surface, as shown in Fig. 3a. So, CNTs can be grown vertically on the fibre plies in laminated composites so that the

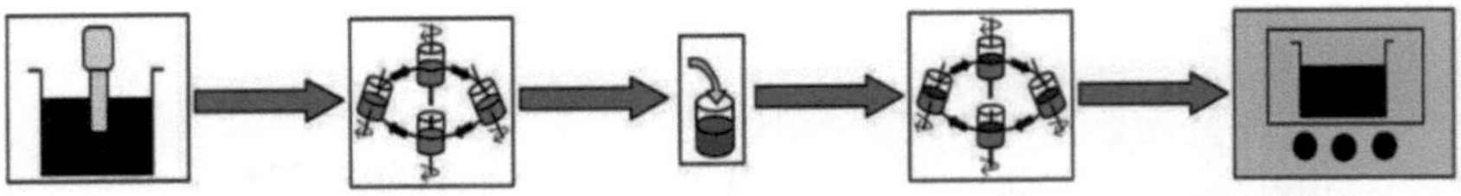

Fig. 4 Dispersion of CNT using ball milling technique for fabrication of multi-scale composites [10]

fibre layers will be stitched or interlocked by vertically grown CNTs. Laminated composites developed using this approach showed very good delamination resistance and inter-laminar fracture toughness [10]. In addition, CNTs grown on single fibres can lead to improved intra-laminar fracture toughness (within a single lamina of laminated composites). However, this process can only be used for high temperature resistant fibres due to very high processing temperature of CVD reactor and therefore, not suitable for plant based fibres. A variation of the growing process, "known as transfer printing process", is more suitable for plant based fibres, as it uses a temperature resistance substrate to grow the nanotubes and subsequently, transfer the nanotubes to the reinforcing fibres (Fig. 3b). Spraying CNTs on fibre surface using a spray gun or coating CNT suspension on fibre surface (Fig. 3d) are other simple techniques, but cannot guarantee uniform CNT distribution on the fibre surface due to chances of CNT agglomeration. Electrophoretic deposition of CNTs on carbon fibre surface has been performed using carbon fibre as the cathode of the electrophoresis chamber containing solution of negatively charged functionalized CNTs, which are attracted to the oppositely charged electrode, i.e., carbon fibres. It is clear that this process can be used for only functionalized nanotubes. Dispersion of nanomaterials within the matrix of composite materials is another frequently used method of fabricating multi-scale composites [3–9]. Different techniques which have been used for dispersing nanomaterials within composite's matrix are ultrasonication, high speed mechanical stirring and shear mixing, calendaring, ball milling, using surfactants and polymers, and so on [3–9]. These different techniques have been extensively studied and multi-scale composites have been successfully developed. Figure 4 shows the CNT dispersion within the matrix of multi-scale composites using ball milling process.

Properties of Carbon Based Multi-scale Composites

The properties of multi-scale composites fabricated using the above two approaches are usually different. When the nanomaterials are dispersed all over the matrix of composites, the matrix dominated properties of composites, mainly the fracture toughness, bulk electrical and thermal conductivity improved to a greater extent as compared to the composites, in which nanomaterials are dispersed only on the fibres. However, it has been noticed that some nanomaterials get usually detached from the fibre surface and enter into the inter-fibre matrix regions, influencing the matrix dominated properties of composites. In general, multi-scale composites results in better in-plane mechanical properties, fracture toughness, delamination and impact resistance, electrical and thermal conductivity, thermal stability, etc. as compared to the conventional composites. The improvement of mechanical properties in different CNT based multi-scale composites is listed in Table 1. In addition, multi-scale composites can also be designed to exhibit some special characteristics such as sensing capability of strain and damage, EMI shielding, gas barrier properties and so on.

Plant Fibre Based Multi-scale Composites

Plant fibres and composites reinforced with plant fibres are being used by different industrial sectors, including mainly automobiles and construction. The driving forces towards the growing use of plant fibre based products are increasing environmental concerns as well as low cost, light weight and lower health hazards while using plant fibre based materials [11]. However, lower mechanical properties, inferior fire, temperature and chemical resistance, higher moisture absorption and degradation of plant fibres make them unsuitable for various structural applications demanding high strength and stability. An extensive research is going on currently on plant fibre based composites to overcome the existing problems. Most of the research and investigations have recommended to modify the surface of natural fibres using various chemical and physical treatments. The chemical treatments, including treatments with alkali, peroxide, silane, ozone, etc., usually modify the surface chemistry of the fibre, reducing hydrophilicity and improving compatibility with different matrices [12]. However, chemical treatments are not environment friendly and also, some treatments may reduce the mechanical properties of fibres. Physical treatments, on the other hand, such as plasma and corona discharge techniques, are clean and dry and modify the surface at nano level without changing the bulk properties [12]. Some plant fibres have been already treated with atmospheric plasma with significant enhancements in tenacity, surface properties and interface with polymeric matrices (Fig. 5).

Novel Approaches for Plant Fibre Surface Modification

Recently, plant fibre surface has been modified with nanomaterials to improve their surface characteristics and interface with polymers as well as to introduce sensing characteristics. In a recent study, multi-walled CNTs were deposited on the surface

Table 1 Mechanical properties of CNT based multi-scale composites developed using different types of CNT and techniques [10]

Type of CNT and concentration	Fibre/matrix and composite fabrication method	Property improvement
Amino functionalized DWCNT, 0.1 and 0.3 wt%, dispersed in resin	Glass fabric/epoxy, Resin transfer molding	Improvement of 20% in interlaminar shear strength
MWCNT, thin-MWCNT, amine functionalized double walled CNT 0.5 wt%, dispersed in resin	Carbon fibre/epoxy, Preparation of prepreg in drum winder and laminates in vacuum bag	Fracture toughness improved by 80% for MWCNTs and modifying epoxy by compatibilizer
MWCNT functionalized and non-functionalized 1 wt%, dispersed in resin	Glass fibre/epoxy, vacuum assisted resin transfer molding	Improvement of 14% in tensile strength, 20% in Young's modulus, 5% in shear strength
Silane functionalized MWCNTs, 1 wt%, Dispersed in resin	Basalt/epoxy, VARTM	Flexural modulus and strength increased by approximately 54% and 34%, respectively
Silane and acid functionalized MWCNTs, 1 wt%	Basalt/epoxy, autoclave processing	Flexural modulus, strength and fracture toughness of silane-treated CNT based composites were 10, 14 and 40% greater than those of acid-treated CNT based composites
DWCNT-NH_2, 0.025-0.1 wt %, dispersed in resin	Carbon fibre/epoxy, Vacuum Infusion Technique	Enhancement in flexural modulus by up to 35%, 5% improvement in flexural strength, 6% improvement in absorbed impact energy, and 23% decrease in the mode I interlaminar toughness
Amine functionalized MWCNTs, 1 wt%, dispersed in resin	Carbon fibre/epoxy, Hand Layup-Vacuum bag processing	Increase in Young's modulus, inter-laminar shear strength, and flexural modulus by 51.46%, 39.62%, and 38.04%, respectively
SWNT, 0.1 wt%, dispersed in matrix	Carbon fabric/epoxy, Hand layup- compression moulding	Improvements of 95% in Young's modulus, 31% in tensile strength, 76% in compressive modulus and 41% in compressive strength
SWNT, 0.1 wt%, sprayed on to fibres in midplane ply	Glass/vinyl ester, VARTM	Up to 45% increase in shear strength over control samples
Vertically aligned CNT on prepreg surface for ply stitching, 1 vol.%, transfer printing method	Carbon fibre/epoxy, Autoclave processing	Increase in fracture toughness by 1.5–2.5 times in Mode I, and 3 times in Mode II

(continued)

Table 1 (continued)

Type of CNT and concentration	Fibre/matrix and composite fabrication method	Property improvement
MWCNT, 1–2 vol.% in composite, grown on surface of the fibres	Alumina fibre/epoxy, Vacuum bag processing	Improvement of steady-state toughness by 76%, in-plane tension-bearing stiffness by 19%, critical strength by 9%, and ultimate strength by 5%
CNT, 1–3 vol.% in composite, grown on fibre surface	Alumina fibre/epoxy, Vacuum bag processing	Enhancement of 69% in interlaminar shear Strength
MWCNTs, 0.25 wt% on fibre surface, electrophoretic deposition	Carbon fibre/epoxy, VARTM technique	Enhancement of the interlaminar shear strength by 27%

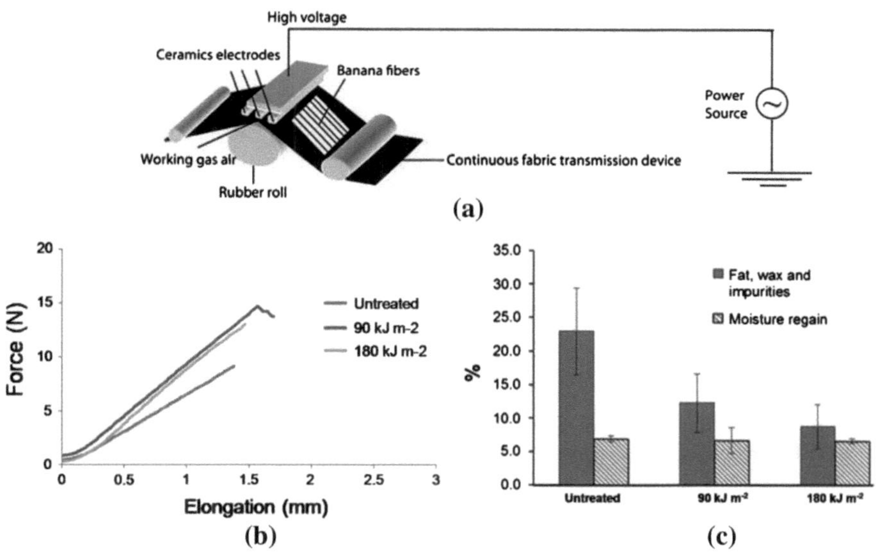

Fig. 5 **a** Schematic diagram of atmospheric plasma treatment machine, **b** tensile force-elongation curves of untreated and plasma treated Quiscal fibre and **c** effect of plasma treatment on impurities and moisture content of Quiscal fibres [12]

of jute fabrics using a scalable dip coating process and subsequently, multi-scale composites were fabricated by impregnating the CNT deposited jute fabrics with epoxy resin [13]. The schematic of this process is shown in Fig. 6.

It can be seen that MWCNT networks were formed on the fibre surface and some CNTs were also present with the matrix near to the fibre surface. The developed multi-scale composites showed improvement in dielectric properties as compared to neat jute/epoxy composites and also sensitivity to temperature, moisture and stress/strain.

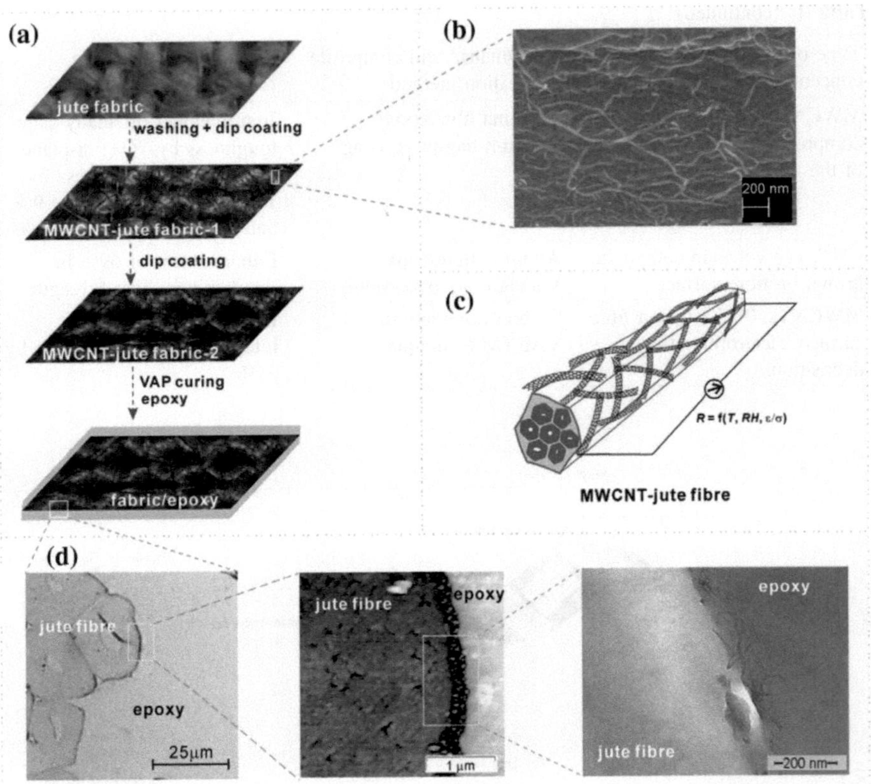

Fig. 6 **a** Optical photographs of as-obtained, once and twice MWCNT dispersion treated jute fabrics, **b** FE-SEM image of MWCNT-jute fibre, **c** schematic presentation of single MWCNT-jute fibre, **d** From *left* to *right* are the SEM, EFM phase and TEM images of the cross-section of the MWCNT-jute fibre/epoxy composite, VAP: vacuum-assisted processing [13]

Although CNTs are being widely used for industrial applications, the cost and toxicity of CNTs are restricting their use in specific applications. Recently, there is a growing interest on nanomaterials based on plant fibres such as nano cellulose. Different forms of nano and micro cellulosic materials have been extracted from plants (Fig. 7) such as bacterial cellulose (BC), nano fibrillated cellulose (NFC), nano crystalline cellulose (NCC), micro-fibrillated cellulose (MFC) and micro crystalline cellulose (MCC) [14–16]. These cellulosic materials have interesting mechanical and optical properties and are already finding widespread applications in different industrial sectors such as packaging, biomedical, textiles, automobiles, construction, composite reinforcements, etc. The mechanical properties of nanocellulose have been listed in Table 2 and their different application areas are provided in Table 3.

Cellulose nano and micro materials have been studied to a considerable extent as reinforcement of composite materials. Similar to CNT, these nano/micro

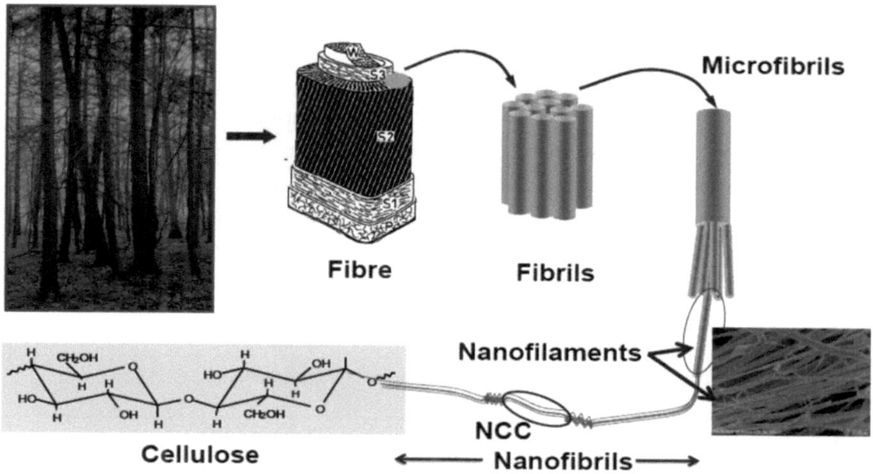

Fig. 7 Schematic representation of defibrillation of plant cellulose into cellulose nano materials [14–16]

Table 2 Comparison of mechanical properties of nanocellulose with other nanomaterials [14–16]

Material	Density (g/cm^3)	CTE (10^{-6}/K) Axial	Tensile Strength (GPa) Axial	Elastic Modulus (GPa)	
				Axial	Transverse
Crystalline cellulose	1.6	0.1	7.5	120–220	11–57
Kevlar-49 Fiber	1.4	2	3.5	124–130	2.5
Clay nanoplatelets	–	–	–	170	–
Carbon nanotubes	–	–	11–63	270–950	0.8–30
Boron nanowhiskers	–	6	2–8	250-360	–

reinforcement have been either introduced within the fibre surface, as mentioned before, or dispersed within the matrix. Growing BC on plant fibre surface has been reported and multi-scale composites have been fabricated using the BC grown natural fibres. In a recent study, BC was grown on sisal fibres by adding them in the fermentation medium of *Acetobacter xylinum.* Sisal fibres were seen to be covered by BC after 4–7 days of incubation at 30 °C [17, 18]. The schematic diagram of the fermenter vessel and BC grown on sisal fibres can be seen from Fig. 8.

The growth of BC on to plant fibres such as sisal and hemp resulted in significant improvement in the interface with polymeric matrices such as poly(L-lactic acid) (PLA) and cellulose acetate butyrate (CAB). Interfacial shear strength (IFSS), as determined from the single fibre pull-out tests, was considerably improved after the growth of BC, as can be noticed from Fig. 9.

The improved interface with polymeric matrices, in turn, resulted in significant increase in the mechanical properties of composite materials. Improvements of

Table 3 Application areas of nanocellulose [14–16]

High volume applications	Low volume applications	Novel and emerging applications
Cement	Wallboard facing	Sensors—medical, environmental, industrial
Automotive body	Insulation	Reinforcement fiber-construction
Automotive interior	Aerospace structure	Water filtration
Packaging coatings	Aerospace interiors	Air filtration
Paper coatings	Aerogels for the oil and gas industry	Viscosity modifiers
Paper filler	Paint-architectural	Purification
Packaging filler	Paint-special purpose	Cosmetics
Replacement-plastic packaging	Paint-OEM applications	Excipients
Plastic film replacement		Organic LED
Hygiene and absorbent products		Flexible electronics
Textiles for clothing		Photovoltaics
		Recyclable electronics
		3D printing
		Photonic films

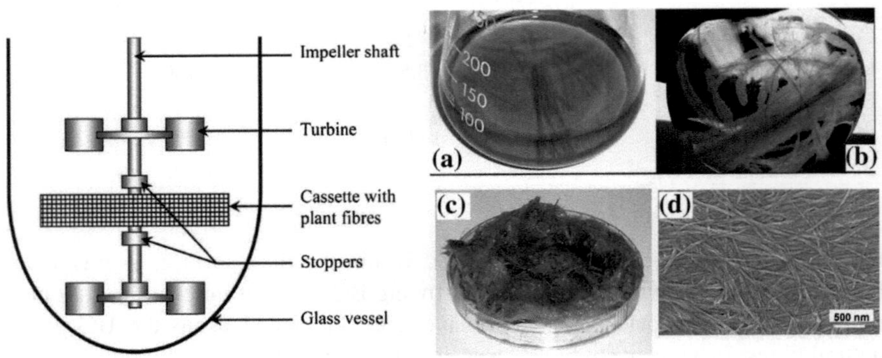

Fig. 8 Schematic drawing of the fermentor vessel used (*left*) and photographs of sisal fibers before and after 2 days of bacterial culture (*right*) [17, 18]

strength and stiffness of PLA matrix based composites in the range of 40–70% were achieved, as can be noticed from the results presented in Table 4.

Besides growth of nano cellulose on plant fibres, nano cellulose was also coated by researchers to fabricate multi-scale composites. Nano cellulose was prepared from waste jute fibres by chemical treatments and subsequently, nano cellulose suspensions were coated to jute fibres and dried [19]. The nano cellulose coated jute fibres were then used to fabricate jute/epoxy multi-scale composites. It was

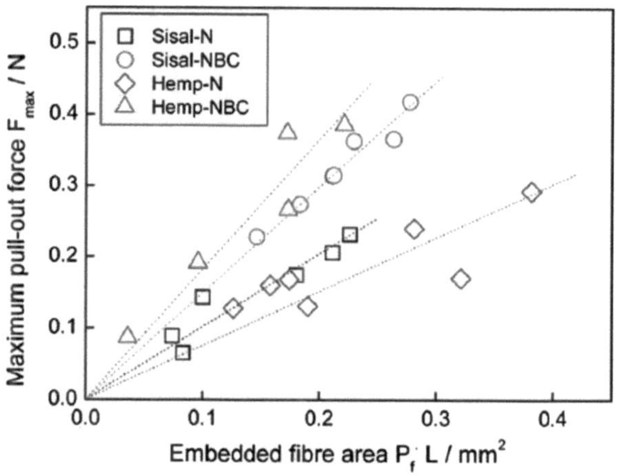

Fig. 9 Single fiber pullout results for hemp and sisal fibers in CAB matrix; (*square*) natural sisal fiber (Sisal-N); (*circle*) sisal fiber modified with bacterial cellulose (Sisal-NBC); (*diamond*) natural hemp fiber (Hemp-N); (*triangle*) hemp fiber modified with bacterial cellulose (Hemp-NBC) [17]

Table 4 Tensile properties of untreated and BC-treated sisal/PLLA composites [18]

Sisal Fibre	Properties at 0° fibre orientation		Properties at 90° fibre orientation	
	Modulus (GPa)	Strength (MPa)	Modulus (GPa)	Strength (MPa)
Untreated	7.91 ± 0.77	78.9 ± 8.5	2.09 ± 0.05	10.0 ± 1.3
BC-treated	11.21 ± 0.69	113.8 ± 8.1	3.12 ± 0.10	16.8 ± 2.4

observed that through nanocellulose coating tensile modulus, flexural strength, flexural modulus, fracture toughness and storage modulus increased by 21, 47, 48, 32 and 56%, respectively. The improvement of mechanical properties was attributed to the improved interfacial interactions between fibre and matrix due to nanocellulose present on fibre surface. The fracture toughness improvement resulted from the increased fibre pull-out during fracture of the nanocellulose coated composites [19].

Generating cellulose nano fibrils on the fibre surface through mechanical fibrillation is another approach to improve the interface of natural fibres with polymeric matrices. Interfacial shear strength of surface fibrillated regenerated cellulosic fibres with low density polyethylene (LDPE) matrix was improved significantly using this approach [20].

Improvement of interface of plant fibres with cementitious matrix through surface modification using BC has also been achieved [21]. BC powder has been coated on to bagasse fibres (Fig. 10) and subsequently, multi-scale cementitious composites have been fabricated.

It was observed that coating with BC resulted in more hydration products on the fibre surface as compared to the uncoated fibres. Moreover, BC coating protected the fibres and led to less penetration of alkaline ions into the lumens of fibres. This

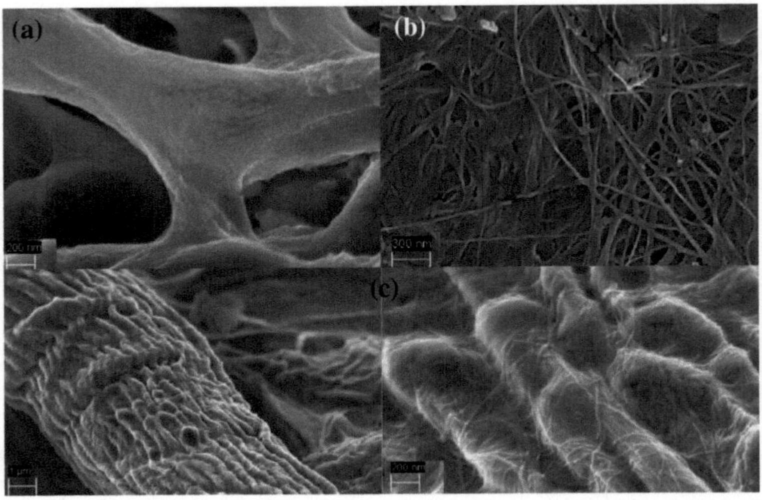

Fig. 10 FE-SEM micrographs of **a** bagasse fibers **b** BC and **c** BC-coated bagasse fibers

resulted in reduced fibre mineralization in case of BC coated fibres. Additionally, presence of BC on fibre surface increased the surface OH groups and improved the interaction with cement matrix. These positive effects were responsible for significant enhancement in flexural modulus, strength and fracture toughness of BC coated fibre composites as compared to the controlled samples, as shown in Fig. 11.

Very recently, nano cellulose based multi-scale composites have been developed by the second approach, i.e., introducing nano cellulose within the matrix. In this research, a combination of micro and nano cellulose fibres was introduced within cementitious composites and it was observed that the addition of multi-scale reinforcements (3%) resulted in 50% improvement of fracture energy of unreinforced cementitious composites [22].

Multi-scale biocomposites has also been developed by dispersing MFC within the matrix of bamboo fibre/PLA composites [23]. A calendaring process using three-roll-mill was used to homogeneously disperse MFC within PLA matrix. It was observed that MFC addition did not have a significant influence on strength and stiffness, but the fracture energy of bamboo/PLA composites enhanced to considerable extent by developing multi-scale composites. As shown in Fig. 12, the strain energy until fracture improved by 200% by dispersing 1 wt% MFC using 5 μm gap in the three-roll mill. The well dispersed MFC network around the bamboo fibres could prevent the sudden crack path through the reinforcing bamboo fibres leading to strong increase in the fracture energy.

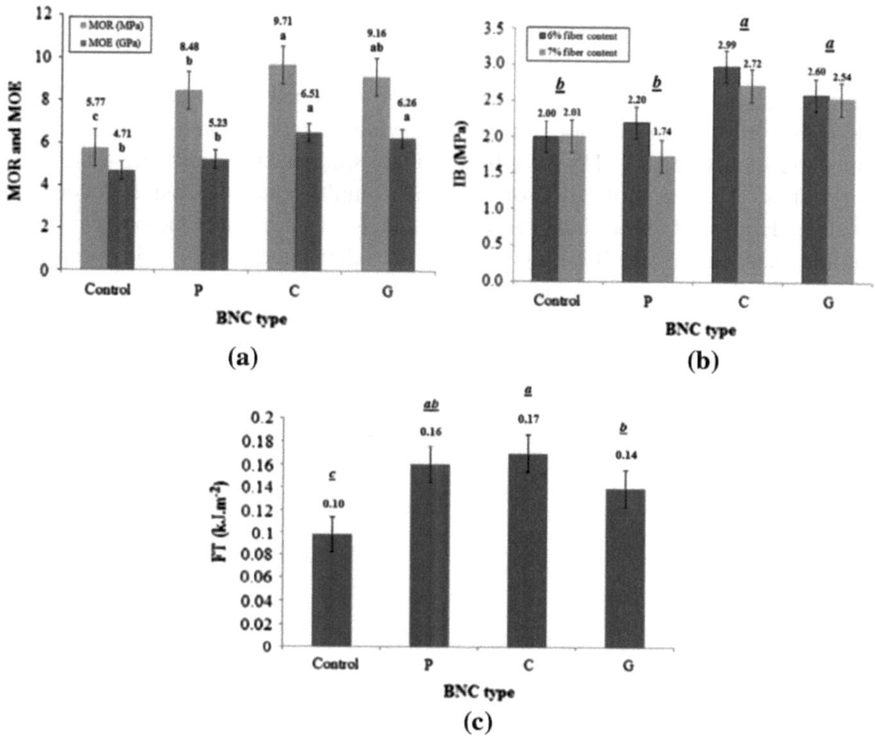

Fig. 11 Influence of BC on the mechanical properties of bagasse composites: **a** Modulus of elasticity (MOE) and modulus of rupture (MOR), **b** Interfacial bonding (IB) strength and **c** Fracture toughness (FT) [21]

Fig. 12 Influence of nano cellulose conc. and processing conditions on strain energy of bamboo/PLA composites

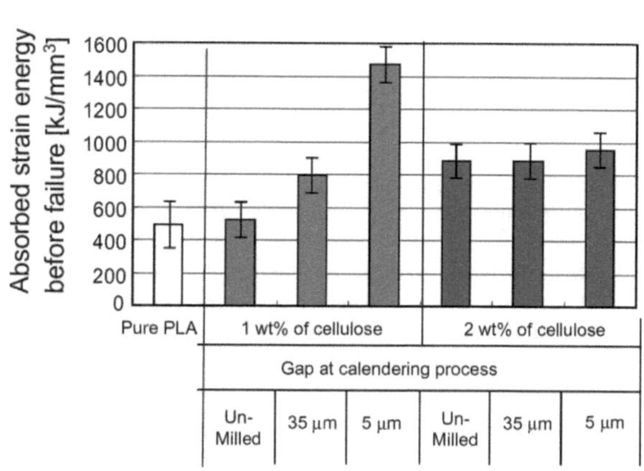

Conclusions

Multi-scale composites have great potential for advanced industrial applications as they can exhibit multi-functional properties, which can be tailored according to the applications. Although CNTs have been widely studied for developing multi-scale composites, practical application of CNT based multi-scale composites is very rare till date. High cost, processing problems and toxicity are the factors which are limiting their use in specific purposes. On the other hand, multi-scale composites based on bio-based materials such as micro and nano cellulose have been studied to a lesser extent. However, the initial results obtained with these composites are highly promising and also, advantageous looking at the sustainability and environmental issues. In order to produce these composites at industrial scale, further research and developments are required to overcome the practical problems such as processing problems, low production, necessity of special equipments, etc.

References

1. Rana, S., Subramani, P., Fangueiro, R., & Correia, A. G. (2016). A review on smart self-sensing composite materials for civil engineering applications. *AIMS Materials Science, 3* (2), 357–379.
2. Rana, S., Zdraveva, E., Pereira, C., Fangueiro, R., & Correia, A. G. (2014). Development of hybrid braided composite rods for reinforcement and health monitoring of structures. *The Scientific World Journal, 2014*(2014), 1–9.
3. Rosado, K. P., Rana, S., Pereira, C., & Fangueiro, R. (2013). Self-sensing hybrid composite rod with braided reinforcement for structural health monitoring. *Materials Science Forum, 730–732,* 379–384.
4. Rana, S., Alagirusamy, R., & Joshi, M. (2009). A review on carbon epoxy nanocomposites. *Journal of Reinforced Plastics Composites, 28,* 461–487.
5. Rana, S., Alagirusamy, R., & Joshi, M. (2011). Single-walled carbon nanotube incorporated novel three phase carbon/epoxy composite with enhanced properties. *Journal of Nanoscience and Nanotechnology, 11*(8), 7033–7036.
6. Rana, S., Alagirusamy, R., & Joshi, M. (2011). Development of carbon nanofibre incorporated three phase carbon/epoxy composites with enhanced mechanical, electrical and thermal properties. *Composites Part A Applied Science and Manufacturing, 42*(5), 439–445.
7. Rana, S., Bhattacharyya, A., Parveen, S., Fangueiro, R., Alagirusamy, R., & Joshi, M. (2013). Processing and performance of carbon/epoxy multi-scale composites containing carbon nanofibres and single walled carbon nanotubes. *Journal of Polymer Research, 20*(12), 1–11.
8. Rana, S., Parveen, S., & Fangueiro, R. (2015). Advanced carbon nanotube reinforced multi-scale composites. In Ehsan Bakerpur (Ed.), Advanced composite materials: Manufacturing, properties, and applications, De Gruyter Open.
9. Rana, S., Alagirusamy, R., Fangueiro, R., & Joshi, M. (2012). Effect of carbon nanofiber functionalization on the in-plane mechanical properties of carbon/epoxy multiscale composites. *Journal of Applied Polymer Science, 125*(3), 1951–1958.
10. Rana, S., & Fangueiro, R. (Eds.) (2016) *Advanced composite materials for aerospace engineering: Processing, properties and applications*. Woodhead Publishing.

11. Rana, S., Pichandi, S., Parveen, S. & Fangueiro, R. (2014). Natural plant fibers: Production, processing, properties and their sustainability parameters. In *Roadmap to Sustainable Textiles and Clothing* (pp. 1–35). Singapore: Springer.
12. Relvas, C., Castro, G., Rana, S., & Fangueiro, R. (2015). Characterization of physical, mechanical and chemical properties of quiscal fibres: The influence of atmospheric DBD plasma treatment. *Plasma Chemistry and Plasma Processing, 35*(5), 863–878.
13. Zhuang, R.-C., Doan, T. T. L., Liu, J.-W., Zhang, J., Gao, S.-L., & Mäder, E. (2011). Multi-functional multi-walled carbon nanotube-jute fibres and composites. *Carbon, 49*(8), 2683–2692.
14. Parveen S, Rana S, Fangueiro R. (2017). Macro and nano dimensional plant fibre reinforcements for cementitious composites. In *Sustainable & non-conventional construction materials using inorganic bonded fiber composites*. Woodhead Publishing.
15. Moon, R. J., Martini, A., Nairn, J., Simonsen, J., & Youngblood, J. (2011). Cellulose nanomaterials review: Structure, properties and nanocomposites. *Chemical Society Reviews, 40*(7), 3941–3994.
16. Peng, B. L., Dhar, N., Liu, H. L., & Tam, K. C. (2011). Chemistry and applications of nanocrystalline cellulose and its derivatives: A nanotechnology perspective. *The Canadian Journal of Chemical Engineering, 89*(5), 1191–1206.
17. Pommet, M., Juntaro, J., Heng, J. Y. Y., Mantalaris,A., Lee, A. F., Wilson, K., et al. (2008). Surface modification of natural fibers using bacteria: Depositing bacterial cellulose onto natural fibers to create hierarchical fiber reinforced nanocomposites. *Biomacromolecules* 9, No. 6: 1643–1651.
18. Juntaro, J., Pommet, M., Kalinka, G., Mantalaris, A., Shaffer, M. S. P., & Bismarck, A. (2008). Creating hierarchical structures in renewable composites by attaching bacterial cellulose onto sisal fibers. *Advanced materials* 20, No. 16: 3122–3126.
19. Jabbar, A., Militký, J., Wiener, J., Kale, B. M., Ali, U., & Rwawiire, S. (2017). Nanocellulose coated woven jute/green epoxy composites: Characterization of mechanical and dynamic mechanical behavior. *Composite Structures* 161: 340–349.
20. Karlsson, J. O., Gatenholm, P., Blachot, J.-F., & Peguy, A. (1996). Improvement of adhesion between polyethylene and regenerated cellulose fibers by surface fibrillation. *Polymer composites* 17, No. 2: 300–304.
21. Mohammadkazemi, F., Doosthoseini, K., Ganjian, E., & Azin, M. (2015). Manufacturing of bacterial nano-cellulose reinforced fiber—cement composites. *Construction and Building Materials, 101,* 958–964.
22. Peters, S., Rushing, T., Landis, E., & Cummins, T. (2010). Nanocellulose and microcellulose fibers for concrete. *Transportation Research Record: Journal of the Transportation Research Board, 2142,* 25–28.
23. Okubo, K., Fujii, T., & Thostenson, E. T. (2009). Multi-scale hybrid biocomposite: Processing and mechanical characterization of bamboo fiber reinforced PLA with microfibrillated cellulose. *Composites Part A Applied Science and Manufacturing, 40*(4), 469–475.

Investigation of Mechanical and Thermomechanical Properties of Nanocellulose Coated Jute/Green Epoxy Composites

Abdul Jabbar, Jiří Militký, Azam Ali and Muhammad Usman Javed

Abstract The present study was aimed to investigate the effect of nanocellulose coating on the mechanical and thermomechanical properties of jute/green epoxy composites. Cellulose was purified from waste jute fibers, converted to nanocellulose by acid hydrolysis and subsequently 3, 5 and 10 wt% of nanocellulose suspensions were coated over woven jute reinforcement. The composites were prepared by hand layup and compression molding technique. The surface topologies of treated jute fibers, jute cellulose nanofibrils (CNF), nanocellulose coated jute fabrics and fractured surfaces of composites were characterized by scanning electron microscopy (SEM). The prepared composites were evaluated for tensile, flexural, fatigue, fracture toughness and dynamic mechanical properties. The results revealed the improvement in tensile modulus, flexural strength, flexural modulus, fatigue life and fracture toughness of composites with the increase in concentration of nanocellulose coating over jute reinforcement except the decrease in tensile strength. Dynamic mechanical analysis (DMA) results also showed the increase in storage modulus and reduction in tangent delta peak height of nanocellulose coated jute composites.

Keywords Jute fibers · Nanocellulose coating · Mechanical properties Fracture toughness · Dynamic mechanical analysis

---

A. Jabbar (✉) · J. Militký · A. Ali · M.U. Javed
Department of Materials Engineering, Faculty of Textile Engineering, Technical University of Liberec, Studentská 1402/2, 461 17 Liberec, Czech Republic
e-mail: abduljabbarntu@gmail.com

A. Jabbar · M.U. Javed
Faculty of Engineering and Technology, National Textile University, Faisalabad 37610, Pakistan

R. Fangueiro and S. Rana (eds.), *Advances in Natural Fibre Composites*,
https://doi.org/10.1007/978-3-319-64641-1_16

Introduction

Natural fiber reinforced polymer composites (NFPC) have gained considerable attention in the recent years due to their environment and economic benefits and low energy consumption [13, 21]. Natural fibers offer many advantages over their synthetic counterparts (e.g. glass and carbon) such as cost effectiveness, easy to process, renewable, recyclable, available in huge quantities, low fossil-fuel energy requirements and the most importantly their high specific strength to weight ratio [13, 18]. Thus natural fibers are considered promising candidates for replacing conventional synthetic reinforcing fibers in composites for semi-structural and structural applications [10].

Cellulose is an abundant, renewable and biodegradable naturally occurring material on earth that can be obtained from numerous resources [5]. It is an infinite source of raw material for environment friendly and biocompatible products. Lignocellulosic fibers such as jute, hemp and flax etc. are rich in cellulose, abundantly available and easy to handle and process. Cellulose micro/nano fillers in polymers have already attracted considerable interest by improving the strength and stiffness of resulting composites [5, 7, 19]. Various types of cellulosic resources are used as precursors to extract and purify cellulose fibrils from lignocellulosic fibers. In the previous literature, cellulose micro/nano fibrils obtained from different precursors such as hemp fibers [9], pineapple [8], isora fibers [5] and jute fibers [11] are used as reinforcing filler in polymer matrices and resulted in the improvement of composite properties. However, in the present study, waste jute fibers are used as precursor to extract and purify nanocellulose which is subsequently coated over the woven jute fabric instead of using it as filler in matrix. Afterwards, the nanocellulose coated jute fabric is used as reinforcement in green epoxy polymer to form a composite material. Therefore, the objective of the present study is to characterize the mechanical and dynamic mechanical properties of composites reinforced with nanocellulose coated jute fabric. To the author's best knowledge, there is no study available in open literature yet that fulfills the above mentioned objective using nanocellulose coated jute fabric as reinforcement and green epoxy resin as matrix system.

Materials and Methods

Material

Woven jute fabric produced from tossa jute (*C. olitorius*) fibers having an areal density of 600 g m^{-2} with 5-end satin weave design was produced on a shuttle loom with 386 tex yarn. Warp and weft densities of the fabric were 6.3 threads per cm and 7.9 threads per cm respectively. Waste jute fibers, sourced from a jute mill, were used as precursor for purification and extraction of nanocellulose. Green

Table 1 Resin system characteristics

Characteristics	Resin + Hardener (CHS-Epoxy G520 + TELALIT 0600)
Viscosity (Pa s, 25 °C)	3.8
Mixing ratio (pbw)	100:32
Minimal curing temperature (°C)	20
Minimal potlife (23 °C, hours)	6
T_g (°C)	200
Flexural strength (MPa)	115
Tensile strength (MPa)	65
Elongation (%)	4
Impact strength (kJ/m^2)	20

epoxy resin CHS-Epoxy G520 and hardener TELALIT 0600 were supplied by Spolchemie, Czech Republic. The main characteristics of the resin system are reported in Table 1 as specified by the manufacturer. Sulphuric acid (H_2SO_4) and sodium hydroxide (NaOH) were supplied by Lach-Ner, Czech Republic. Sodium sulfate (Na_2SO_4) and sodium hypochlorite (NaOCl) were supplied by Sigma-Aldrich, Czech Republic.

Purification and Extraction of Nanocellulose from Waste Jute Fibers

The main steps followed to purify and extract nanocellulose from waste jute fibers are depicted in Fig. 1. Waste jute fibers were chopped to approximate length of 5–10 mm and immersed in 2% sodium hydroxide (NaOH) solution for 2 h at 80 °C temperature maintaining a liquor ratio of 50:1. The process was repeated three times. The fibers were then washed with tap water several times The jute fibers were then bleached with 7.0 g/l sodium hypochlorite (NaOCl) solution at room temperature for 2 h under pH 10–11 and subsequently antichlored with 0.1% sodium sulphate (Na_2SO_4) at 50 °C for 20 min. Finally, the fibers were washed with tap water several times until the final pH was maintained at 7.0 and then allowed to dry at room temperature for 48 h and at 100 °C in an oven for 2 h. Alkali and bleaching treatments were used to purify cellulose and to remove maximum amount of hemicellulose and lignin from the fibers. The bleached jute fibers were milled using a high-energy planetary ball mill of Fritsch pulverisette 7. The sintered corundum container of 80 ml capacity and zirconium balls of 10 mm diameter were chosen for 20 min of milling. The ball to material ratio (BMR) was kept at 10:1 and the speed was kept at 850 rpm. Acid hydrolysis of milled jute fibers was conducted for 1 h at 45 °C under mechanical stirring using 65% (w/w) H_2SO_4. The fiber content during acid hydrolysis was 5% (w/w). The whole process flow chart is shown in Fig. 1. After hydrolysis, the suspension was diluted with cold water (4 °C) to stop

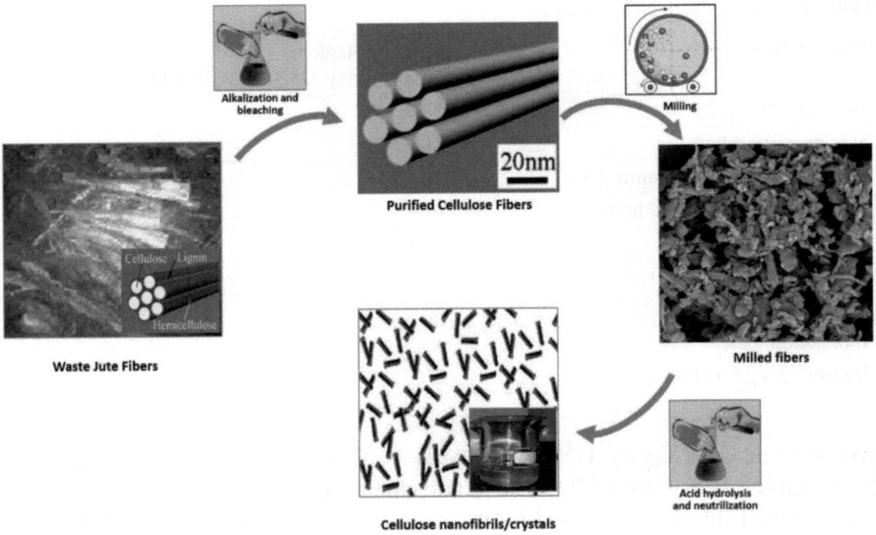

Fig. 1 The process flow chart for the purification and extraction of nanocellulose

the reaction, neutralized with NaOH solution and discolored by NaOCl solution. The supernatant was removed from the sediment and replaced by new distilled water several times. Finally, 3, 5 and 10% (w/w) nanocellulose suspensions were prepared by increasing cellulose concentration and decreasing water concentration through vacuum filtration using Buchner funnel.

Coating of Jute Fabric with Nanocellulose

The prepared nanocellulose suspensions (3, 5, and 10 wt%) were ultrasonicated for 5 min with Bandelin ultrasonic probe and then applied on the surface of jute fabric by roller padding at room temperature. Finally, the coated fabrics were dried at 70 °C for 60 min.

Preparation of Composites

The composites were prepared by hand layup method. The resin and hardener were mixed in a ratio of 100:32 (by weight) according to manufacturer recommendations, before hand-layup. The prepared resin mixture was poured on fabric layers and spread out by a hand roller. The gentle rolling action of hand roller confirmed the wetting of jute fabrics and the excess resin was squeezed out of the panel layup by the roller. The composite layup along with Teflon sheets were sandwiched between a pair of steel plates and cured at 120 °C for 1.0 h in mechanical convection oven with predetermined weight on it to maintain uniform pressure of about 50 kPa. The fiber volume fraction (V_f) of all composites was in the range of 0.25–0.27. The prepared composite samples were designated as CF0 (uncoated), CF3 (3 wt% nanocellulose coated), CF5 (5 wt% nanocellulose coated) and CF10 (10 wt% nanocellulose coated).

Characterization and Testing

Scanning Electron Microscopy (SEM)

The topologies of alkali treated and bleached jute fibers, cellulose nanofibrils and nanocellulose coated jute fabrics were observed with Vega-Tescan TS5130 Scanning Electron Microscope at 30 kV accelerating voltage. The surface of fibers was gold coated prior to SEM inspection. Additionally, in order to measure the size distribution using SEM image, the topology of cellulose nanofibrils was also observed on Zeiss Ultra Plus field emission scanning electron microscope (FE-SEM) at low accelerating voltage (1.0 kV) and low probe current ($\approx$10 pA) to eliminate charging effect and sample damage due to interaction with primary electrons. The software used for image analysis was NIS Elements BR 3.22.

Mechanical Characterization of Composites

Tensile properties of composites were characterized on an MTS series 370 servo-hydraulic load frame equipped with 647 hydraulic wedge grip of 100 kN load capacity at a cross head speed of 2 mm/min and gauge length of 100 mm in accordance with ASTM D3039-00 using rectangular specimens of dimension $200 \times 20 \times$ h mm^3, where "h" is the actual thickness of specimen in the range of 4.0–4.5 mm. Flexural test was performed in three point bending mode on a universal mechanical testing machine, Shimadzu AGS-J with 5 kN load cell, following ASTM D790 standard. The specimens of dimension $160 \times 12.7 \times$ h mm^3 were used maintaining a span to thickness ratio of 32:1 according to standard recommendations.

The tension-tension fatigue performance of composites was experimentally evaluated at two different stress levels (80 and 70% of the ultimate tensile stress σ_u) assuming the same geometry of specimens as the one adopted in quasi-static tensile tests. Tests were performed at gage length of 100 mm under constant stress amplitude, stress ratio (R) of 0.1 (ratio of maximum to minimum stress during a loading cycle) and frequency of 5 Hz. Three specimens were tested for each stress level up to final failure of specimens. Tests were realized on the same MTS series 370 servo-hydraulic loading machine equipped with 647 hydraulic wedge grip of 100 kN load capacity.

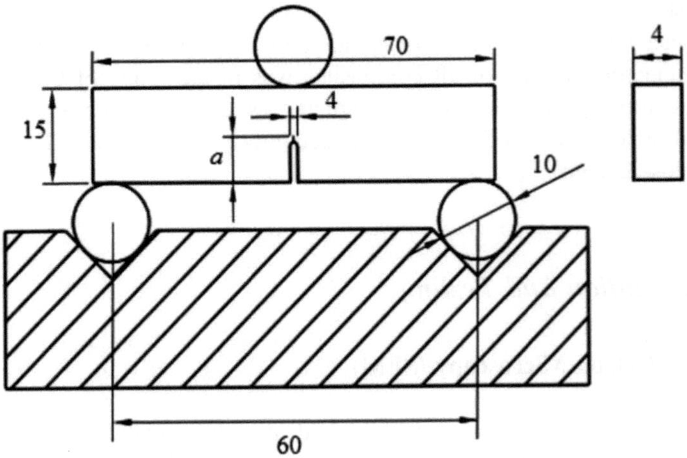

Fig. 2 SENB (single edge notch bending) specimen and fixture dimensions for fracture toughness test (dimensions in mm)

Fracture toughness, K_{Ic}, of composites was determined by three point bending method using the single edge notch bend (SENB) specimens in accordance with the standard test method ASTM D5045-99. A sharp crack of length "*a*" between 0.45 W and 0.55 W was introduced by using a notch maker CEAST NOTCHVIS and a fresh razor blade at the notch tip ('W' is the width of specimen). The specifications of fixture and specimens are shown in Fig. 2.

The tests were performed using universal mechanical testing machine, Shimadzu AGS-J with 5 kN load cell, at a cross head speed of 10 mm/min. Five specimens were tested for each condition to get an average value. Statistical analysis of tensile, flexural and fracture toughness properties was done by one-way analysis of variance (ANOVA) and probability value $p \leq 0.05$ was considered as an indicative of statistical significance compared to the control sample (CF0 composite).

Dynamic Mechanical Characterization of Composites

The dynamic mechanical properties of composites were measured in 3-point bending mode using Q800 Dynamic Mechanical Analysis (DMA) instrument of TA instruments (New Castle DL, USA). The testing conditions were controlled in the temperature range of 30–180 °C, with a heating rate of 3 °C/min, fixed frequency of 1 Hz, preload of 0.1 N, amplitude of 20 µm, and force track of 125%. The samples having a thickness of 4–4.5 mm, width of 12 mm and span length of 50 mm were used for DMA. Two replicate samples were tested for each condition and average values were reported.

Results and Discussion

SEM Study of Chemically Treated Jute Fibers and Jute Cellulose Nanofibrils

Surface topologies of jute fibers after alkali treatment, bleaching and milling are presented in Fig. 3. Figure 3a shows the jute fibers bound together in the form of fiber bundles by cementing materials e.g. hemicellulose and lignin but after repeated alkali treatment, splitting of fibers is observed due to destruction of mesh structure with a little rough and clean surface, may be due to majority of the removal of non-cellulosic materials [24, 28] as shown in Fig. 3b.

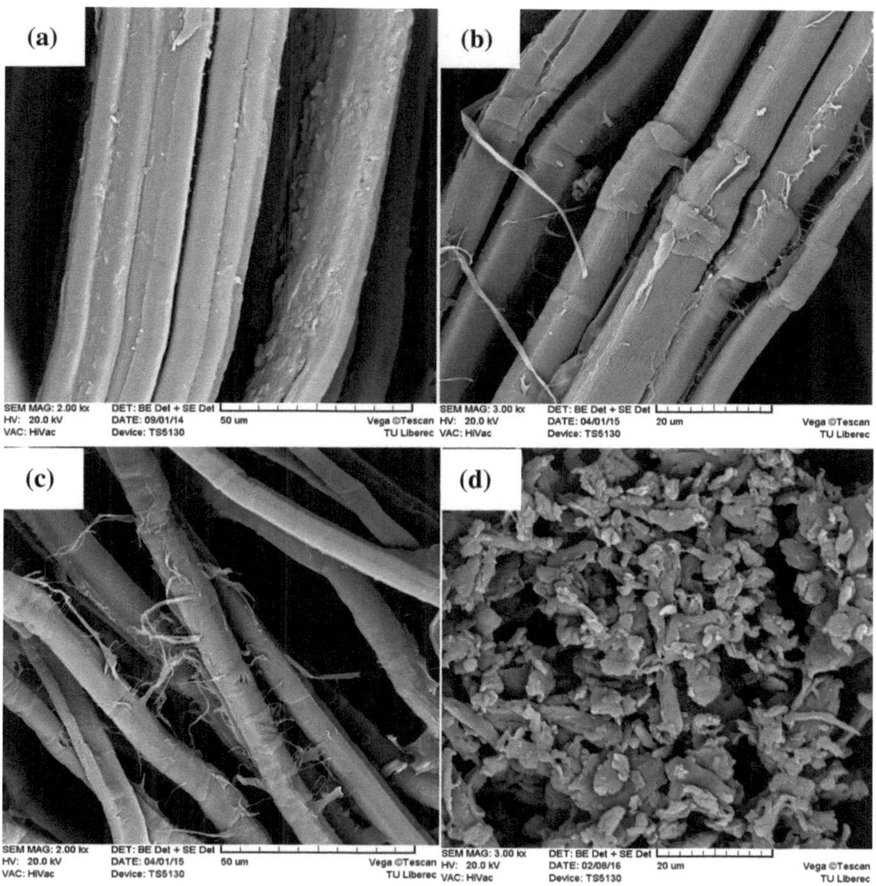

Fig. 3 SEM images of jute fibers **a** untreated, **b** alkali treated, **c** bleached and **d** pulverized (milled)

The alkali treated fibers are further separated to individual fibers with more clean surface and fibrillation on the surface, may be due to more delignification after bleaching treatment [2], as shown in Fig. 3c. Figure 3d displays the milled jute fibers with size distribution of fibers in the micron range.

A high resolution FE-SEM image at nanoscale level is precisely analyzed to measure the size (width) of jute cellulose nanofibrils (CNF), after acid hydrolysis, as shown in Fig. 4a. The histogram of size distribution of 50 measurements is

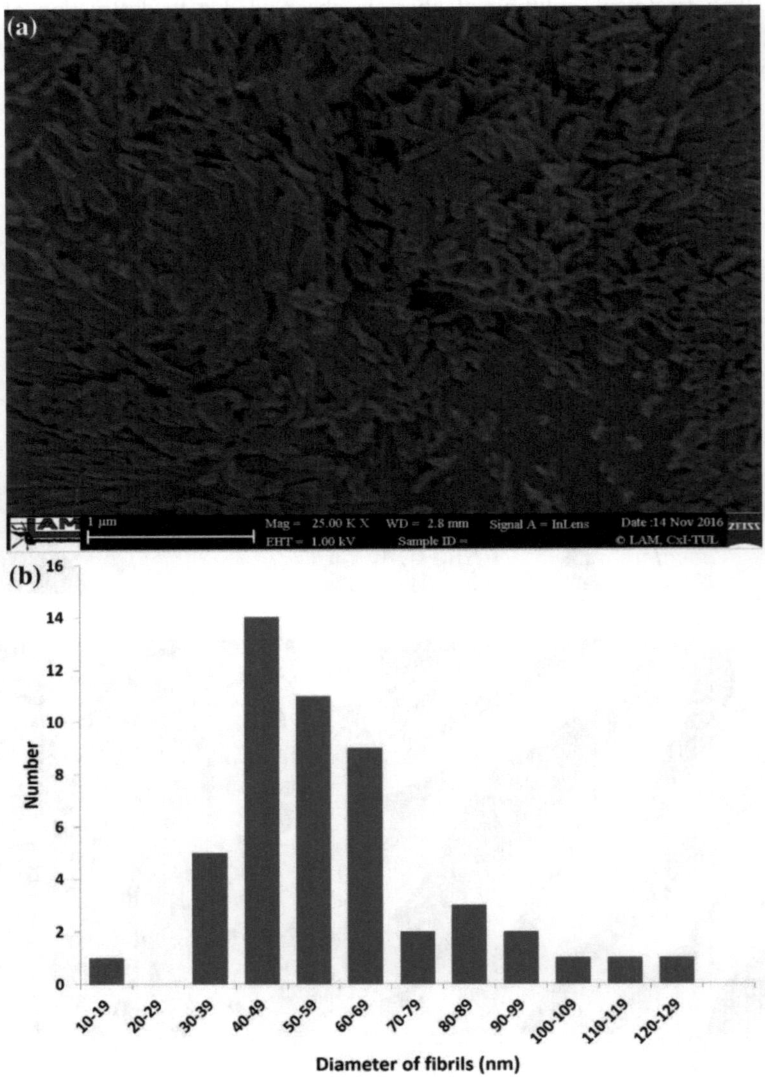

Fig. 4 **a** FE-SEM image of jute cellulose nanofibrils. **b** Histogram of width distribution of cellulose nanofibrils

shown in Fig. 4b. The calculated average diameter (width) of CNF was found to be 57.40 ± 20.61 nm.

Surface Topology of Nanocellulose Coated Fabrics

The topological changes that occur on the surface of jute fabric after nanocellulose coating are shown in Fig. 5. It is apparent that there is depositing of nanocellulose on the fabric surface forming a layer. The layer thickness increases gradually with the increase in nanocellulose concentration as clear in Fig. 5b–d and the fabric surface coated with 10 wt% nanocellulose suspension is covered almost completely with cellulose as shown in Fig. 5d.

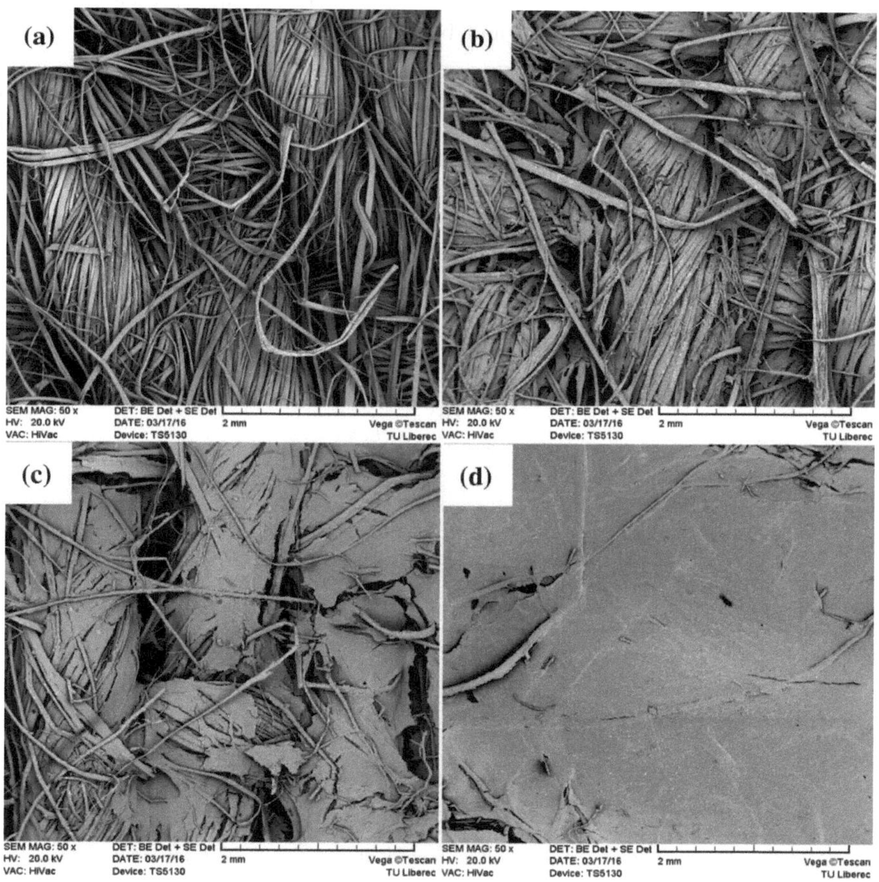

Fig. 5 Surface topology of jute fabric coated with; **a** 0 wt%, **b** 3 wt%, **c** 5 wt%, **d** 10 wt% of nanocellulose suspensions

Tensile Properties

Figure 6a shows the typical stress-strain curves of uncoated and coated jute composites with different nanocellulose content. The stress-strain curve for uncoated (CF0) composite shows almost linear behavior until failure whereas the curves for composites coated with different nanocellulose content show linear behavior followed by change in slope showing nonlinear behavior, thus presenting plastic deformation and gradual debonding of fibers from the matrix just before failure. The tensile failure behavior also reveals more brittle nature of CF0 composite as compared to nanocellulose coated composites (Fig. 6a).

Figure 6b presents the average values and standard deviation of ultimate tensile strength and tensile modulus of composites. The tensile properties indicate the decrease in strength and increase in modulus (stiffness) of composites with the increase in nanocellulose concentration. For CF10 composite, the tensile modulus increases from 4.6 to 5.58 GPa showing 21% increase, thus presenting better fiber/matrix interfacial interaction and bonding which would be effective at the early stages of loading.

However, the decrease in tensile strength of composites with the increase in nanocellulose concentration may be explained by the differences in failure strains of nanocellulose coated jute fabric reinforcement and the matrix. In other words, the reinforcement does not come into effect when the failure strain of matrix is much greater than that of reinforcement. So the composite shows a failure before the stress is transferred from matrix to reinforcement [7]. The ANOVA for the tensile strength (p = 0.000) and tensile modulus (p = 0.008) showed the statistically significant difference between the means at the 95.0% confidence level.

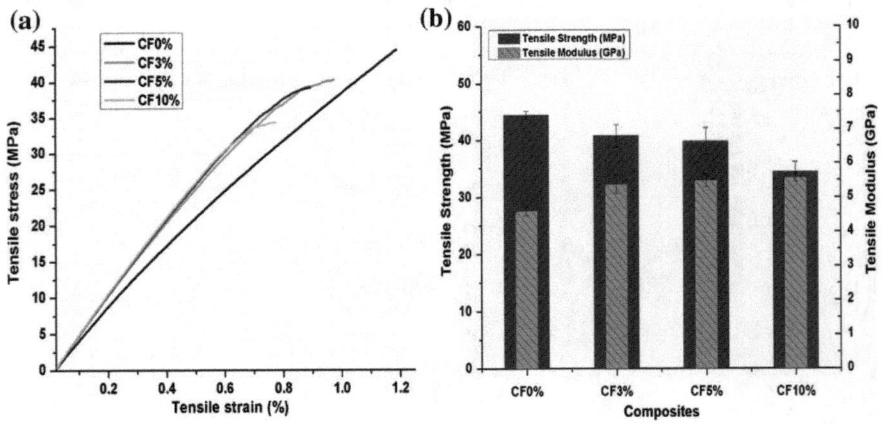

Fig. 6 **a** Tensile stress-strain curves and **b** tensile strength and modulus of uncoated and nanocellulose coated jute composites

Flexural Properties

The flexural strength and modulus increase with the increase in nanocellulose concentration in composites. The flexural strength increases from 32.94 MPa for CF0 composite to 32.94, 43.53 and 48.66 MPa for CF3, CF5 and CF10 composites, respectively thus allowing 26, 32 and 47% increase on average as shown in Fig. 7. Similarly, flexural modulus increases from 3.83 GPa for CF0 composite to 4.81, 4.73 and 5.67 GPa for CF3, CF5 and CF10 composites, respectively thus permitting 25, 23.5 and 48 increase on average. These findings may suggest the strong interaction between matrix and reinforcement after nanocellulose coating which increases with the increase in nanocellulose content. The other possibility of the enhancement of flexural properties may be the increase in bending stiffness/rigidity of jute reinforcement after coating with nanocellulose which also increases with the increase nanocellulose concentration [15, 16]. The ANOVA for the flexural strength ($p = 0.000$) and flexural modulus ($p = 0.000$) also showed the statistically significant difference between the means at the 95.0% confidence level.

Fatigue Life

The S-N (fatigue life) curves of all composites in the considered stress range are shown in Fig. 8. The experimental fatigue data was fitted by semi-logarithmic

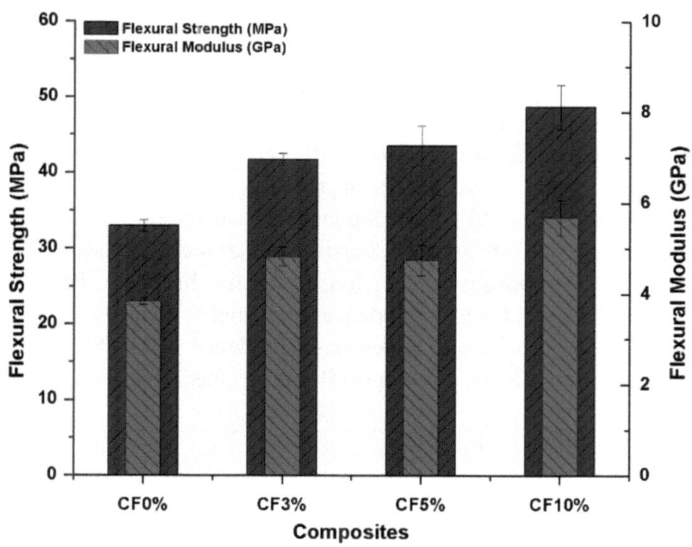

Fig. 7 Flexural strength and modulus of uncoated and nanocellulose coated jute/green epoxy composites

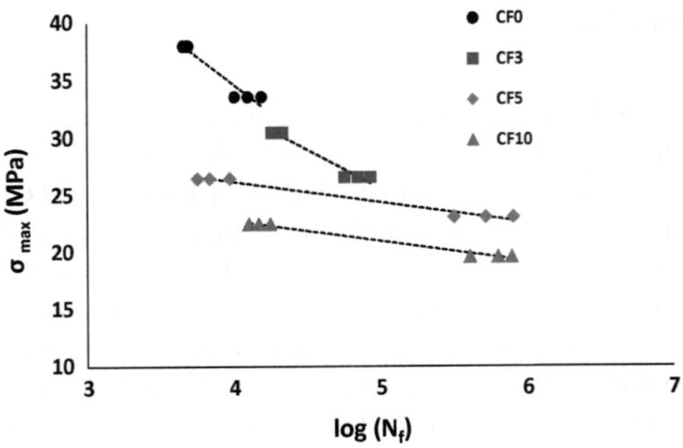

Fig. 8 Maximum stress (σ_{max}) versus logarithm of number of cycles to failure log(N_f) and semi-logarithm fitting for each composite

function, $\sigma_{\max} = k\log N_f + a$ [4] (where k and a are the parameters to be defined by the least square method as given in Table 2) to have a reliable predicting of the fatigue life corresponding to other stress levels in the considered stress range, which are not directly determined by testing. The quality of the fitting is related to the coefficient of determination R^2 [23]. Values of R^2 close to 1 confirm the better goodness-to-fit of experimental and predicted data. The curves in Fig. 8 have R^2 values in the range 0.92–0.98 (Table 2). The negative slope k values of the linear fitting curves, listed in Table 2, show a decreasing trend with the increase in nanocellulose content of the composites. This predicts an increase in fatigue life for composites containing higher contents of nanocellulose coated over woven jute reinforcement than that of uncoated jute composite. This is clearer in Fig. 9 when comparing the average number of cycles to failure for the two applied stress levels. CF3 composite has higher fatigue life at 80% stress level (low cycles regime) whereas decreasing the stress level to 70% (high cycles regime) causes a significant increase in the fatigue life of CF5 and CF10 composites.

The reliability of the above results on the fatigue life can be assessed by the confidence level index, based on the Student's t-distribution [23]. The confidence levels in Table 3 show that, in strict statistics terms: the hypothesis "the nanocellulose coated jute composites have a longer fatigue life than the uncoated jute composite (CF0)" is valid with confidence level higher than 99% for both stress levels except for CF5 composite which has confidence level >88% at 80% stress level. The better fatigue life of nanocellulose coated jute composites can be

Table 2 Parameters of the linear fitting of S-N curves

Composites	k	a	*Adj.* R^2
CF0	−9.73 (3.06)	73.63 (11.90)	0.92287
CF3	−6.82 (1.69)	59.74 (7.73)	0.95104
CF5	−1.80 (0.32)	33.43 (1.55)	0.97462
CF10	−1.80 (0.26)	30.03 (1.30)	0.98312

Note Standard deviation in parentheses

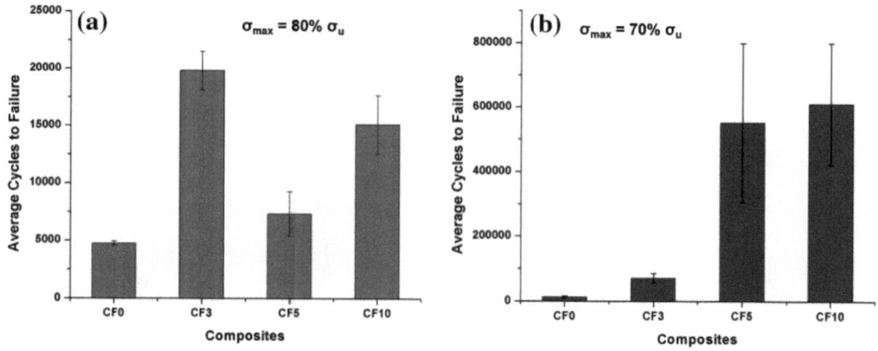

Fig. 9 Comparison of average fatigue life of composites: **a** 80% and **b** 70% of σ_u

Table 3 Confidence levels for three hypotheses ('>' means longer fatigue life)

Stress level (%)	Confidence level for hypotheses		
	CF3 > CF0 (%)	CF5 > CF0 (%)	CF10 > CF0 (%)
70	99.8	99.4	99.9
80	99.9	88.8	99.3

interpreted as a better damage tolerance of these materials mainly due to increase in intermolecular and physical interactions thus forming a rigid and stiff network due to nanocellulose coating over woven jute reinforcement.

Fracture Toughness

Figure 10a shows the typical K_Q versus displacement curves presenting the crack growth behavior of composites and Fig. 10b presents the fracture toughness (K_{1c}) with respect to nanocellulose content in composites. The fracture mode was brittle for CF0 composite showing slip-stick behavior but as the nanocellulose content over jute reinforcement increased, the fracture mode changed from brittle to a little ductile and K_{1c} values increased from 2.64 MPa $m^{1/2}$ for CF0 composite to 3.20, 3.21 and 3.49 MPa $m^{1/2}$ for CF3, CF5 and CF10 composites, respectively resulting in 21, 21.5 and 32% increase on average (Fig. 10b). Thus, fracture toughness increases linearly with the increase in concentration of nanocellulose coating over reinforcement. The ANOVA for the fracture toughness, K_{1c} ($p = 0.024$) showed the statistically significant difference between the means at the 95.0% confidence level.

Crack deflection, plastic deformation, voids, crack pinning/bridging, fiber pull-out and debonding are the known toughening mechanisms in epoxy matrices, found in literature [30]. Figure 11a–d exhibits SEM images for all nanocellulose contents which clearly shows fiber fracture, fiber pullout and some voids for all composites.

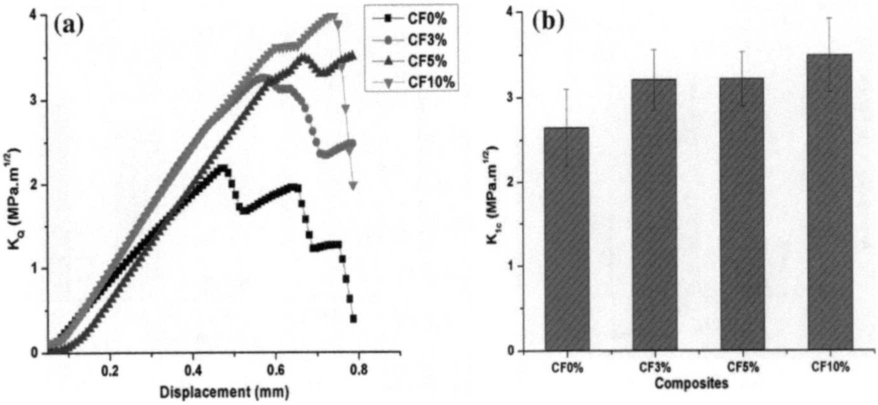

Fig. 10 **a** Typical K_Q versus displacement curves and **b** fracture toughness (K_{1c}) of uncoated and cellulose coated jute/green epoxy composites

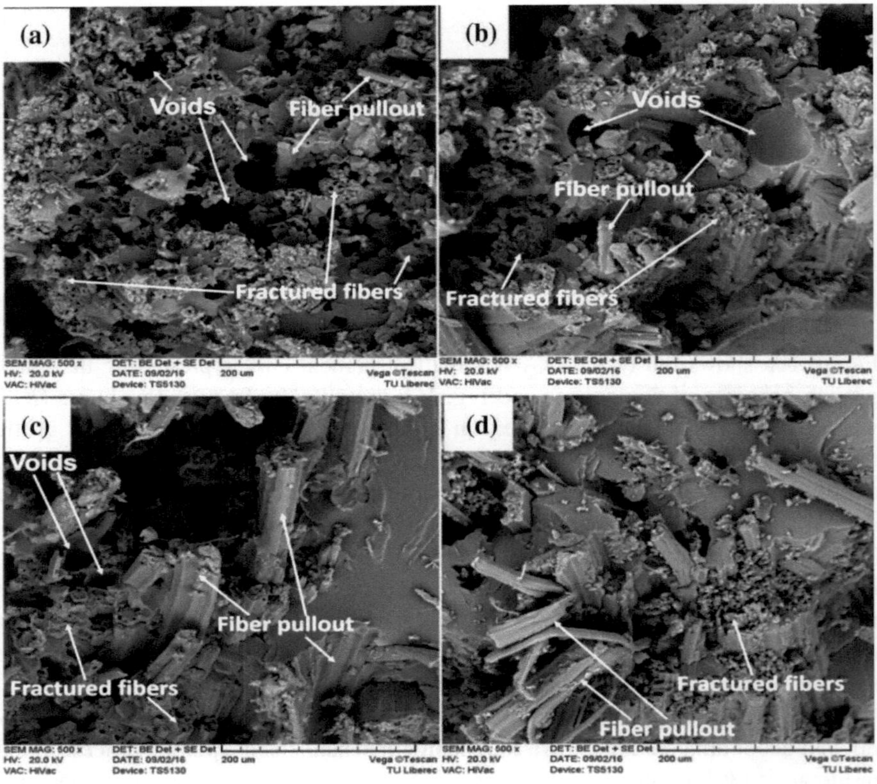

Fig. 11 Fracture surface topology of nanocellulose coated jute/green epoxy composites; **a** CF0, **b** CF3, **c** CF5 and **d** CF10

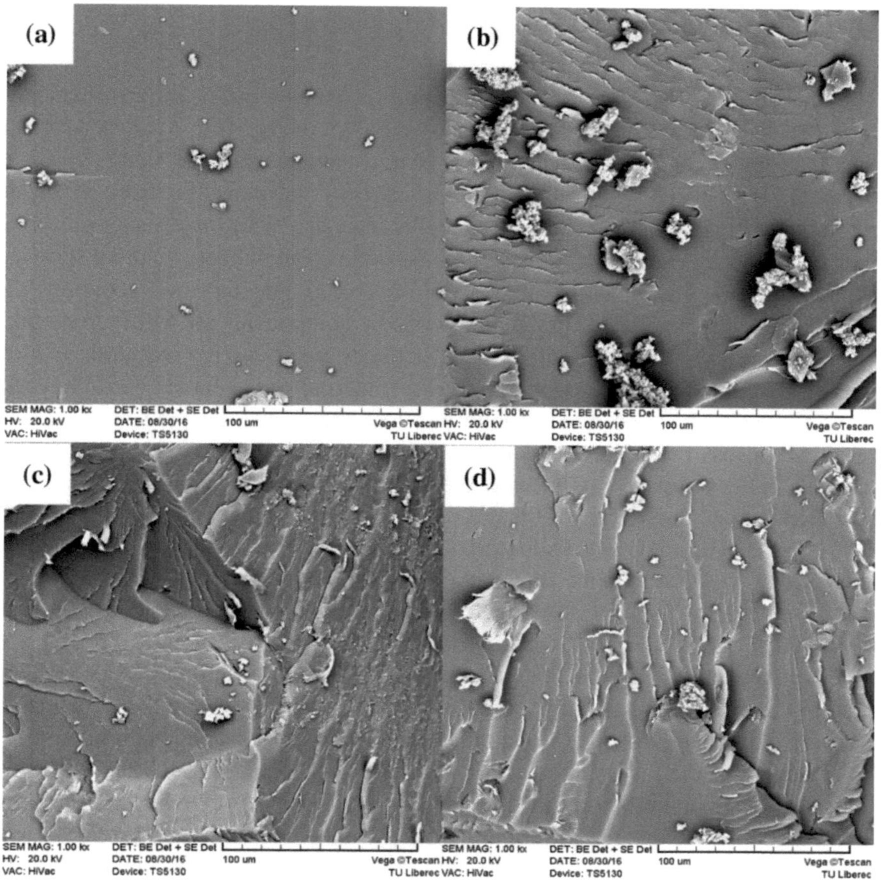

Fig. 12 Fracture surfaces in the matrix region of jute/green composites; **a** CF0, **b** CF3, **c** CF5 and **d** CF10

However, fiber pullout is a little more prominent mechanism for nanocellulose coated composites, may be due to increased fiber debonding during fracture resulting in increased crack propagation length during deformation and hence fracture toughness.

Figure 12 shows the fracture surface topologies in the matrix region. It can be seen from Fig. 12a that the fracture surface of green epoxy matrix of CF0 composite is very smooth and featureless, which indicates typical brittle fracture behavior with lack of significant toughness mechanism [1]. However, Fig. 12b–d show rougher fracture surfaces and river patterns in the matrix regions of nanocellulose coated jute composites. An increase in fracture surface roughness can be used as an indicator to the presence of plastic deformation and crack deflection mechanisms, which increase fracture toughness by increasing crack propagation length during deformation [31].

Dynamic Mechanical Analysis

The change in storage modulus as a function of temperature of different nanocellulose coated and uncoated jute composites is shown in Fig. 13a. It can be seen that the shape of storage modulus curves is almost same for all the samples and E' decreases with increase in temperature because of the transition from glassy to rubbery state. However, the storage modulus increases with increasing concentration of nanocellulose in composites both in glassy and rubbery regions showing superior reinforcing effects of nanocellulose coated jute fabrics throughout the specified temperature range. In the glassy region, components are highly immobile, close and tightly packed and intermolecular forces are strong [25] resulting in high storage modulus but as temperature increases intermolecular forces become weak, the components become more mobile and lose their close packing arrangement, resulting in loss of stiffness and hence storage modulus. Figure 13a reveals that uncoated jute composite (CF0) has the lowest storage modulus throughout the specified temperature range. CF0 has 4.05 GPa of E' (measured at 30 °C), however, CF3, CF5, CF10 composites show maximum values of 4.72, 5.27 and 6.35 GPa, respectively resulting in an increase of 16, 30 and 56% in E'. Moreover, E' (measured at 150 °C) increased from 0.28 GPa for CF0 composite to 0.48, 0.71 and 0.94 GPa for CF3, CF5, CF10 composites, respectively representing 71, 153 and 235% increase. This shows that when nanocellulose concentration over the jute fabric is increased, the stiffness effect of reinforcement is progressively increased not only in the glassy region but also in the rubbery region. The above findings may be attributed to the fact that, as cellulose nanofibrils coated on jute fabric possess large surface area, it promotes the interfacial interactions between the reinforcement and matrix thus reducing the mobility of polymer chains [6] and better stress transfer at the interface [27]. The other fact is the increase in the stiffness of reinforcement with increasing nanocellulose concentration. Furthermore, the formation of rigid and stiff network interconnected by hydrogen bonds is the accepted theory to explain the excellent mechanical properties of composites incorporated with nanocellulose [17, 20, 26, 29].

Figure 13b presents the loss modulus (E'') versus temperature of different nanocellulose coated and uncoated jute composites. The rapid rise in loss modulus in a system indicates an increase in the polymer chains free movements at higher temperatures due to a relaxation process that allows greater amounts of motion along the chains that is not possible below the glass transition temperature [22]. Figure 13b also revealed that the value of loss modulus is increased with the increase in concentration of nanocellulose coating as compared to uncoated composite (CF0) representing a higher amount of energy dissipation associated with an increase in internal friction. It is interesting to note that T_g is decreased from 95 °C for CF0 composite to 88 and 91 °C for CF3 and CF5 composites, respectively but T_g of CF0 and CF10 composites are almost same. Simultaneously, a positive shift in T_g values to higher temperatures is noted with the increase in nanocellulose

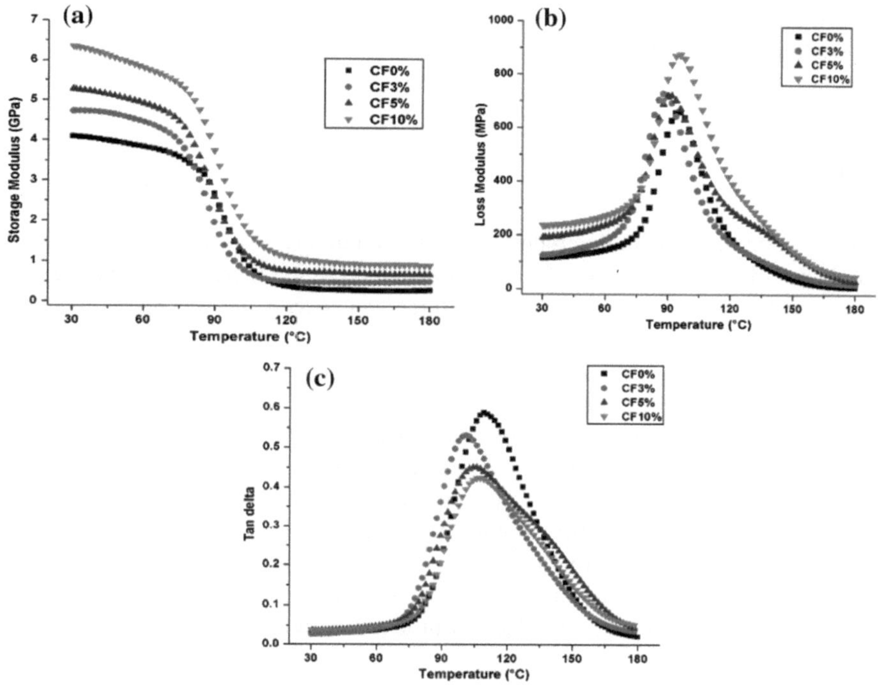

Fig. 13 Dynamic mechanical properties of nanocellulose coated and uncoated jute composites; **a** storage modulus, **b** loss modulus and **c** tan delta

concentration in the system (Fig. 13b). A possible reason may be the change in crosslinking density of the network [3].

The damping factor (tanδ) as a function of temperature of composites reinforced with different concentration of nanocellulose coated jute fabric is shown in Fig. 13c. The highest value of tanδ peak is observed for CF0 composite resulting in more energy dissipation whereas a reduction is tanδ peak height is observed in composites with the increase in concentration of nanocellulose coating. The peak height of 0.59 for CF0 composite is reduced maximum to 0.42 for CF10 composite representing a 40% decrease (Fig. 13c) thus indicating that there are both strong intermolecular and physical interactions contributing to greater molecular restrictions at the interface and less energy dissipation and that the storage modulus is influenced more by the increase in nanocellulose concentration than loss modulus in the composites. The width of tanδ peak becomes broader for nanocellulose coated jute composites especially for CF5 and CF10. The damping in materials generally depends on the molecular motions at the interfacial region [12]. Therefore, the increase in width of tanδ peak of nanocellulose coated composites is suggestive of increased volume of interface [25] and an increase in the inhibition of relaxation processes in composites thus decreasing the mobility of polymer chains within the system and a higher number of chain segments upon increase in concentration of nanocellulose in composites [14].

Conclusion

Tensile modulus, flexural strength, flexural modulus, fatigue life and fracture toughness of composites were found to improve with the increase in concentration of nanocellulose coating over jute reinforcement except the decrease in tensile strength. Dynamic mechanical test also revealed the increase in storage modulus and reduction in tangent delta peak height of composites with the increase in nanocellulose concentration over jute reinforcement. Based on the analysis of results, the improvement in mechanical and dynamic properties were likely attributed to the increase in interfacial interaction between reinforcement and matrix due to large surface area exposed by nanocellulose coated over jute reinforcement, the formation of rigid and stiff network interconnected by hydrogen bonds and the increase in the stiffness of reinforcement with increasing nanocellulose concentration. Whereas, the differences in failure strains of coated jute reinforcement and the matrix might be the reason of reduction in tensile strength of these composites.

Acknowledgements The authors acknowledge the financial support from Technical University of Liberec under student grant scheme (SGS-21158) to carry out this work.

References

1. Alamri, H., & Low, I. M. (2012). Characterization of epoxy hybrid composites filled with cellulose fibers and nano-SiC. *Journal of Applied Polymer Science,* 126.
2. Beg, M. D. H., & Pickering, K. L. (2008). Accelerated weathering of unbleached and bleached Kraft wood fibre reinforced polypropylene composites. *Polymer Degradation and Stability, 93,* 1939–1946.
3. Brunner, A., Necola, A., Rees, M., Gasser, P., Kornmann, X., Thomann, R., et al. (2006). The influence of silicate-based nano-filler on the fracture toughness of epoxy resin. *Engineering Fracture Mechanics, 73,* 2336–2345.
4. Carvelli, V., Betti, A., & Fujii, T. (2016). Fatigue and Izod impact performance of carbon plain weave textile reinforced epoxy modified with cellulose microfibrils and rubber nanoparticles. *Composites Part A Applied Science and Manufacturing, 84,* 26–35.
5. Chirayil, C. J., Joy, J., Mathew, L., Koetz, J., & Thomas, S. (2014). Nanofibril reinforced unsaturated polyester nanocomposites: Morphology, mechanical and barrier properties, viscoelastic behavior and polymer chain confinement. *Industrial Crops and Products, 56,* 246–254.
6. Chirayil, C. J., Mathew, L., Hassan, P., Mozetic, M., & Thomas, S. (2014). Rheological behaviour of nanocellulose reinforced unsaturated polyester nanocomposites. *International Journal of Biological Macromolecules, 69,* 274–281.

7. Cho, M.-J., & Park, B.-D. (2011). Tensile and thermal properties of nanocellulose-reinforced poly (vinyl alcohol) nanocomposites. *Journal of Industrial and Engineering Chemistry, 17,* 36–40.
8. Costa, L. M. M., de Olyveira, G. M., Cherian, B. M., Leão, A. L., de Souza, S. F., & Ferreira, M. (2013). Bionanocomposites from electrospun PVA/pineapple nanofibers/*Stryphnodendron adstringens* bark extract for medical applications. *Industrial Crops and Products, 41,* 198–202.
9. Dai, D., Fan, M., & Collins, P. (2013). Fabrication of nanocelluloses from hemp fibers and their application for the reinforcement of hemp fibers. *Industrial Crops and Products, 44,* 192–199.
10. Dányádi, L., Móczó, J., & Pukánszky, B. (2010). Effect of various surface modifications of wood flour on the properties of PP/wood composites. *Composites Part A Applied Science and Manufacturing, 41,* 199–206.
11. Das, K., Ray, D., Banerjee, C., Bandyopadhyay, N., Sahoo, S., Mohanty, A. K., et al. (2010). Physicomechanical and thermal properties of jute-nanofiber-reinforced biocopolyester composites. *Industrial and Engineering Chemistry Research, 49,* 2775–2782.
12. Dong, S., & Gauvin, R. (1993). Application of dynamic mechanical analysis for the study of the interfacial region in carbon fiber/epoxy composite materials. *Polymer Composites, 14,* 414–420.
13. Jabbar, A., Militký, J., Kale, B. M., Rwawiire, S., Nawab, Y., & Baheti, V. (2016). Modeling and analysis of the creep behavior of jute/green epoxy composites incorporated with chemically treated pulverized nano/micro jute fibers. *Industrial Crops and Products, 84,* 230–240.
14. Júnior, J. H. S. A., Júnior, H. L. O., Amico, S. C., & Amado, F. D. R. (2012). Study of hybrid intralaminate curaua/glass composites. *Materials and Design, 42,* 111–117.
15. Kale, B. M., Wiener, J., Militky, J., Rwawiire, S., Mishra, R., & Jabbar, A. (2016). Dyeing and stiffness characteristics of cellulose-coated cotton fabric. *Cellulose, 23,* 981–992.
16. Kale, B. M., Wiener, J., Militky, J., Rwawiire, S., Mishra, R., Jacob, K. I., et al. (2016). Coating of cellulose-TiO_2 nanoparticles on cotton fabric for durable photocatalytic self-cleaning and stiffness. *Carbohydrate Polymers, 150,* 107–113.
17. Kargarzadeh, H., Sheltami, R. M., Ahmad, I., Abdullah, I., & Dufresne, A. (2015). Cellulose nanocrystal: A promising toughening agent for unsaturated polyester nanocomposite. *Polymer, 56,* 346–357.
18. Ku, H., Wang, H., Pattarachaiyakoop, N., & Trada, M. (2011). A review on the tensile properties of natural fiber reinforced polymer composites. *Composites Part B Engineering, 42,* 856–873.
19. Lee, S.-Y., Mohan, D. J., Kang, I.-A., Doh, G.-H., Lee, S., & Han, S. O. (2009). Nanocellulose reinforced PVA composite films: Effects of acid treatment and filler loading. *Fibers and Polymers, 10,* 77–82.
20. Lu, J., Wang, T., & Drzal, L. T. (2008). Preparation and properties of microfibrillated cellulose polyvinyl alcohol composite materials. *Composites Part A Applied Science and Manufacturing, 39,* 738–746.
21. Lu, T., Jiang, M., Jiang, Z., Hui, D., Wang, Z., & Zhou, Z. (2013). Effect of surface modification of bamboo cellulose fibers on mechanical properties of cellulose/epoxy composites. *Composites Part B Engineering, 51,* 28–34.
22. Martínez-Hernández, A., Velasco-Santos, C., De-Icaza, M., & Castano, V. M. (2007). Dynamical–mechanical and thermal analysis of polymeric composites reinforced with keratin biofibers from chicken feathers. *Composites Part B Engineering, 38,* 405–410.
23. Meloun, M., & Militky, J. (2011). *Statistical data analysis: A practical guide*. Limited: Woodhead Publishing.
24. Mukherjee, A., Ganguly, P., & Sur, D. (1993). Structural mechanics of jute: The effects of hemicellulose or lignin removal. *Journal of the Textile Institute, 84,* 348–353.
25. Pothan, L. A., Oommen, Z., & Thomas, S. (2003). Dynamic mechanical analysis of banana fiber reinforced polyester composites. *Composites Science and Technology, 63,* 283–293.

26. Raquez, J.-M., Murena, Y., Goffin, A.-L., Habibi, Y., Ruelle, B., Debuyl, F., et al. (2012). Surface-modification of cellulose nanowhiskers and their use as nanoreinforcers into polylactide: A sustainably-integrated approach. *Composites Science and Technology, 72,* 544–549.
27. Ray, D., Sarkar, B., Das, S., & Rana, A. (2002). Dynamic mechanical and thermal analysis of vinylester-resin-matrix composites reinforced with untreated and alkali-treated jute fibres. *Composites science and technology, 62,* 911–917.
28. Ray, D., Sarkar, B. K., Rana, A., & Bose, N. R. (2001). Effect of alkali treated jute fibres on composite properties. *Bulletin of Materials Science, 24,* 129–135.
29. Samir, M. A. S. A., Alloin, F., Sanchez, J.-Y., & Dufresne, A. (2004). Cellulose nanocrystals reinforced poly (oxyethylene). *Polymer, 45,* 4149–4157.
30. Wetzel, B., Rosso, P., Haupert, F., & Friedrich, K. (2006). Epoxy nanocomposites–fracture and toughening mechanisms. *Engineering Fracture Mechanics, 73,* 2375–2398.
31. Zhao, S., Schadler, L. S., Hillborg, H., & Auletta, T. (2008). Improvements and mechanisms of fracture and fatigue properties of well-dispersed alumina/epoxy nanocomposites. *Composites Science and Technology, 68,* 2976–2982.

The Use of Sedimentation for the Estimation of Aspect Ratios of Charged Cellulose Nanofibrils

Amaka Joy Onyianta and Rhodri Williams

Abstract In this study, the aspect ratios of carboxymethylated and TEMPO-oxidised cellulose nanofibrils (CNF) were estimated by gel point analysis using a sedimentation approach. The flocculation and subsequent sedimentation of the CNF aqueous suspensions, which were stabilised by negative repulsive forces, were made possible after screening in concentrated salt media. The aspect ratios of the CNFs were then calculated from the linear fit (gel point) of the plot of concentration of the CNFs against the relative sediment height, using the crowding number theory.

Keywords Cellulose nanofibrils · Carboxymethylation · TEMPO-oxidation
Aspect ratio · Sedimentation

Introduction

Cellulose nanofibrils extracted from various sources of cellulose (e.g. wood, plants, tunicates, bacteria and algae) are topical materials which are widely researched and used in academia and in industry [8, 10]. This wide usage arises because of their renewability, biocompatibility and biodegradation potentials coupled with improved mechanical strength, lightweight properties, optical properties, barrier properties and structuring capabilities. Various cellulose fibre pre-treatments such as mechanical, chemical and biological pre-treatments have been widely adopted as a means of reducing the energy requirements associated with the mechanical fibrillation process to yield cellulose nanofibrils with acceptable qualities [10, 12]. These pre-treatment processes can lead to reduction in fibre dimensions, intro-

A.J. Onyianta (✉) · R. Williams
Material Chemistry and Processing, School of Engineering and Built Environment, Edinburgh Napier University, Edinburgh EH10 5DT, United Kingdom
e-mail: a.onyianta@napier.ac.uk

R. Williams
e-mail: r.williams@napier.ac.uk

R. Fangueiro and S. Rana (eds.), *Advances in Natural Fibre Composites*,
https://doi.org/10.1007/978-3-319-64641-1_17

duction of repulsive forces or fibre swelling, and partial hydrolysis of disordered regions of cellulose chains by enzymes. Moreover, these different pre-treatment routes alongside the different cellulose starting materials, different mechanical treatments (homogenisation, microfluidisation, grinding, and ultrasound sonication) and duration of treatment [8], have the ability to significantly affect the morphology, surface chemistry and rheological behaviours of the resulting CNF. The use of diverse techniques and instruments to characterise CNF material [8] has enabled a better understanding of the material properties. This has led to the production of materials that are fit for specific purposes. However, the characterisation of CNF is continuously being developed.

The aspect ratio (A), which is the ratio of length to width of CNF plays an important role in its ability to function as efficient reinforcements and as rheology modifiers. The measurement of widths are usually carried out by analysing images from field emission scanning electron microscopy (FE-SEM), transmission electron microscopy (TEM) and atomic force microscopy (AFM) [9]. Depending on the microscopic method used, the widths can vary for the same sample. CNFs with low fibril width (3–4 nm) have been reported after pre-treatment by 2,2,6,6-tetramethylpiperidine-1-oxyl (TEMPO) mediated oxidation followed by "mild" mechanical treatment [7, 17]. Other pre-treatment methods such as carboxymethylation [22] and use of enzymes [13] produce CNFs with 5–15 nm and 5–20 nm widths respectively. Mechanical pre-treatments such as beating, refining, sonication, and milling, followed by any of the mechanical treatments mentioned earlier yield fibrils within 15 and 100 nm [19].

The measurement of lengths have been made on individualised TEMPO-oxidised CNF from TEM images, denoted as number average length (L_{n}) and weighted average length (L_{w}) [20]. Determination of average lengths of low aqueous concentrations of CNF have also been carried out from viscoelastic [6] and shear viscosity [20] measurements. The lengths obtained from these rheological measurements appear to be up to 2 times higher than the lengths measured from TEM images. It should be noted that the rheological measurements in this study were made from aqueous dispersions, therefore the influence of the anionic groups on the rheological response cannot be ignored. This method of length determination was stated to be most suitable for CNFs that are well diluted and individualised without networks.

However, the presence of fibril entanglements and networks on most CNF materials make it difficult to measure the length of fibrils from microscopic images, in order to determine their aspect ratios. The aspect ratios of uncharged CNFs were estimated by gel point analysis using sedimentation and yield stress approaches [15, 21, 23]. The theory involved in the use of sedimentation for the determination of aspect ratios is based on the fact that particles in suspension can either adopt a dilute, semi-dilute or concentrated states. The threshold between the dilute and the semi-dilute regions is known as the connectivity threshold or the gel point ($\boldsymbol{\Phi}_c$). This is defined as "the boundary between the dilute and semi-dilute regions, the lowest volume fraction where the particles first form a continuous network" [21]. The gel point can be obtained from sedimentation experiments, which is then inputted into the crowding number equation to calculate the aspect ratio [21]. This method of aspect ratio determination has not yet been applied on charged CNFs because of the presence of repulsive forces which stabilise the fibrils in suspension. Therefore, taking the advantage of the screening effect of inorganic salts on charged CNF materials, we estimated the aspect ratio of carboxymethylated CNF and TEMPO-oxidised CNF by gel point analysis using the sedimentation approach.

Materials and Methods

A commercial grade of never-dried hardwood bleached sulphite cellulose pulp (32% solid content) was used for all the pre-treatments. Monochloroacetic acid, sodium hydroxide (NaOH), Isopropanol, ethanol, methanol, acetic acid, sodium bicarbonate ($NaHCO_3$), 2,2,6,6-tetramethylpiperidine-1-oxyl (TEMPO), sodium bromide (NaBr), sodium hypochlorite (NaClO, 10–15%), sodium chloride (NaCl), were all reagent grade and used as received from Sigma Aldrich (United Kingdom). Ultrapure water (Purelab Option-Q ELGA DV 25) was used throughout the experiment. The total surface charge on the cellulose was obtained from conductometric titrations. The fibrils width distribution were measured from FE-SEM images using ImageJ software. Prior to the sedimentation experiments, the CNFs were first screened in varying concentrations of NaCl to determine the concentration at which to carry out the sedimentation experiments.

Pre-treatments and Sample Preparation

Carboxymethylation: The carboxymethylation pre-treatment was carried out on the never-dried cellulose pulp using the procedures described by Wågberg et al. [22]. In summary, the never-dried pulp was dispersed in ultrapure water at approximately 0.5 wt% before being solvent exchanged to ethanol by filtration. The ethanol-exchanged cellulose was added to a solution of monochloroacetic acid in isopropanol and allowed to stand for 30 min. After which the mixture was gradually added to a solution of NaOH in methanol. The entire mixture was added to isopropanol that had been heated to 65 °C in a 2 L jacketed reactor, and allowed to react for 1 h under continuous stirring. The reaction mixture was filtered and washed with ultrapure water and 0.1 M acetic acid. The carboxymethylated cellulose was converted to its sodium form using 4 wt% $NaHCO_3$ before a final wash with ultrapure water was carried out. Aqueous dispersions (1 wt%) were prepared from the insoluble cellulose obtained after the pre-treatment and passed 5 times through the 200 μm and 100 μm Z-shaped interaction chamber of the M-110EH-30 microfluidiser (Microfluidics, USA) at 172.4 MPa. The resulting CNF was stored in the refrigerator.

TEMPO-Oxidation: The second sample was prepared by TEMPO-oxidation of cellulose using the general procedure described by Saito and Isogai [16]. The cellulose pulp was dispersed in ultrapure water at 1 wt% concentration and transferred to a glass reactor. Appropriate amounts of TEMPO and NaBr were added and stirred until completely dissolved. The reaction was started by adding NaClO (5000 μmol/g) in dropwise manner. The pH was maintained at 10 ± 0.2 during the reaction by adding 0.5 M NaOH. After 1 h, the reaction was quenched with 20 mL of ethanol. Aqueous dispersions (1 wt%) were also prepared from the insoluble cellulose obtained after oxidation and passed 5 times through the microfluidiser using the same conditions as with the carboxymethylation pre-treatment. The resulting CNF was also stored in the refrigerator.

Characterisation

Surface charge analysis: The total surface charge of the never-dried cellulose and carboxymethylated cellulose were determined by conductometric titration according to the SCAN-CM 65:02 method [1]. In summary, cellulose samples were each dispersed in 0.1 M HCl and allowed to stand for 15 min before being washed until the conductivity was less than 5 μS/cm. Then 490 mL of ultrapure water and 10 mL of 0.05 M NaCl were added to 0.5–1 g dry weight of cellulose pulp in a beaker. This was stirred thoroughly using a magnetic stirrer and titrated against 0.05 M NaOH. The surface charge on the TEMPO-oxidised cellulose was determined using the method described by Saito and Isogai [16]. The total surface charge was determined from the analysis of the conductivity curve using OriginPro 8 (OriginLab Corporation, USA).

Field emission-scanning electron microscopy (FE-SEM): Micrographs at 45,000× magnifications were acquired from the pre-treated samples using FE-SEM S-4800 (Hitachi, Japan). Prior to analysis, 20 μL of 0.0005 wt% slurry was deposited on the mica surface on the SEM stub and left overnight to air dry. This was sputter coated with a layer of gold for 85 s using EMITECH K550X (Quorumtech, UK) gold sputter coater. The fibril widths were measured from approximately 150 fibrils using ImageJ software.

Sedimentation: The sedimentation method used by Varanasi et al. [21] was adopted with minor variations. Prior to the sedimentation experiments, the carboxymethylated CNF was dispersed in a range of 0M to 3M NaCl while the TEMPO oxidised CNF was dispersed in 0M to 4M NaCl, at 0.07 w/v% cellulose content. This was done to examine the charge screening potentials of different concentrations of salt on the CNFs. Subsequently, 2M and 3M NaCl solution was respectively chosen as the medium for the sedimentation of the carboxymethylated and TEMPO-oxidised samples. These salt concentrations represent an average salt concentration where the levelling off of the relative sediment heights were attained. 0.02–0.08 wt% suspensions were prepared from 1 wt% original aqueous suspensions of the CNF samples and allowed to sediment in 60 mL glass vials for 96 h before photographs were captured. The sediment height (h_s) and the initial suspension height (h_0) were measured using ImageJ software. The plot of concentration (wt. %) against the relative sediment height (h_s/h_0) was fitted with a quadratic equation ($ax^2 + bx$), where the linear fit is equal to the gel point/connectivity threshold [23]. Using the equivalent densities of 1.83 M NaCl (1.07 g/cm^3), 3.06 M NaCl (1.11 g/cm^3) and cellulose (1.5 g/cm^3), weight fractions were converted to volume fraction using Eq. 1 since $\Phi \ll 1$.

$$\Phi = C(\rho_l/\rho_f) \tag{1}$$

Here, Φ is the volume fraction, ρ_l is the density of liquid, ρ_f is density of fibre, C is weight fraction (g of fibre/g of suspension).

The aspect ratio was then estimated from the crowding number theory using Eq. 2 [21].

$$\mathrm{A} = 4.90\Phi_{\mathrm{C}}^{-0.5} \tag{2}$$

Results and Discussion

Total Surface Charge

Results from the surface charge analysis show that the surface of the cellulose was modified with carboxymethyl groups (550 ± 4 µmol/g of CH_2COO^-) and carboxyl groups (1063 ± 16 µmol/g of COO^-) from the respective carboxymethylation and TEMPO-oxidation pre-treatments. This in comparison with the 30 µmol/g surface charge of the starting material. The total surface charge obtained after the pre-treatments are within the literature reported values of 515–610 µmol/g for carboxymethylated cellulose and 500–1500 µmol/g for TEMPO-oxidised cellulose [2, 4, 11, 18, 22]. It should be noted that various factors such as pulp type, amount of reagent, time, temperature and pH can greatly contribute to the variations in total charged groups reported after pre-treatments [20]. This makes comparison of results across literature difficult.

Fibril Morphology and Width Distribution

It can be seen from the FE-SEM images of the carboxymethylated CNF and the TEMPO-oxidised CNF (Fig. 1) that both CNFs possess an interconnected and network structure. An average width of 11 ± 3 and 9 ± 2 nm was obtained for the carboxymethylated CNF and the TEMPO-oxidised CNF respectively. Images from FE-SEM are known to be of a lower resolution compared to those from TEM [3].

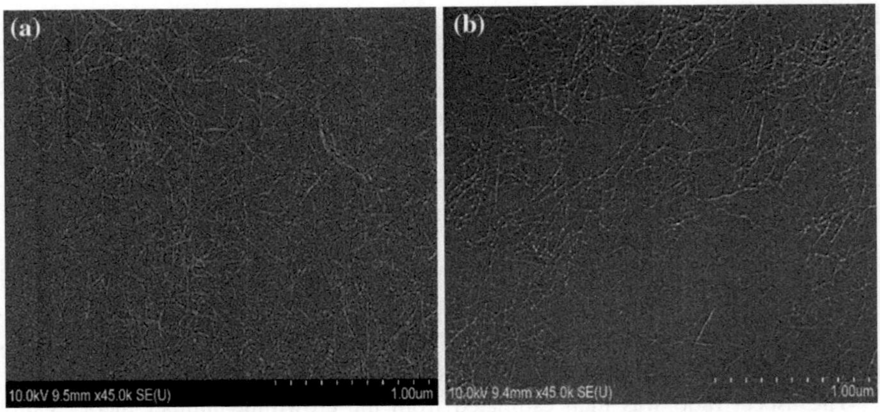

Fig. 1 FE-SEM image of the carboxymethylated CNF (**a**) and the TEMPO-oxidised CNF (**b**)

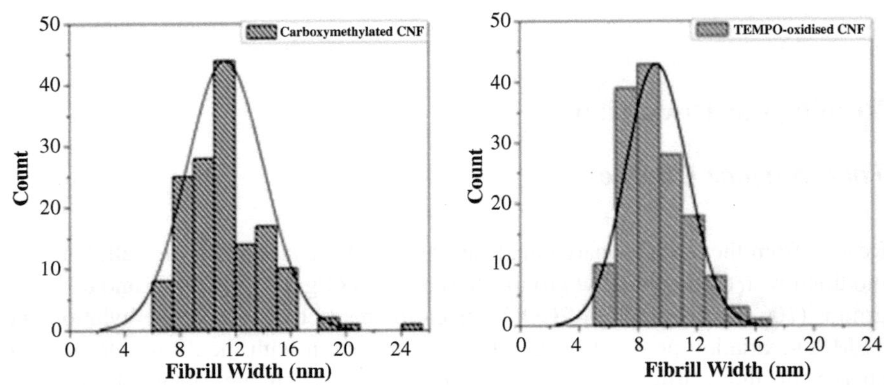

Fig. 2 Fibril width distribution for carboxymethylated CNF and the TEMPO-oxidised CNF

Hence, the measured fibril widths reported in this study appear larger than those reported in other studies where TEM images were used. The fibril width distribution displayed in Fig. 2 shows a more polydisperse fibril width for the carboxymethylated CNF than for the TEMPO-oxidised CNF.

Estimation of Aspect Ratio by Gel Point Analysis

According to the well-known Derjaguin-Landau-Verwey-Overbeek (DVLO) theory, increasing the concentration of electrolytes in dispersions of charged particles would lead to the screening of the surface charges, resulting in the aggregation of the particles. This theory has been used to study the aggregation behaviours of

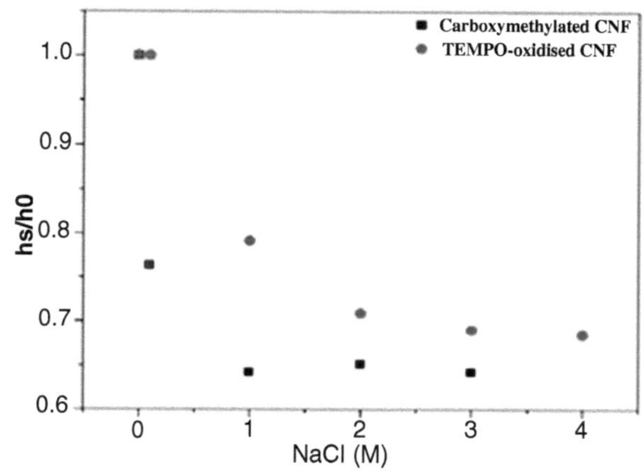

Fig. 3 Effect of increased salt concentration on the sedimentation of 0.7 w/v% charged CNF materials

CNFs and cellulose nanocrystals [5, 14]. Figure 3 shows the effect of the increased salt concentration on the charged CNFs. It can be seen that the relative sediment height decreased with increase in salt concentration, indicating that the charges on the CNFs are being screened by the salt. A levelling off of the relative sediment height at higher salt concentration can be clearly seen for all the CNF materials. The carboxymethylated CNF having a lower surface charge levelled off at a lower salt concentration. On the other hand, the TEMPO-oxidised CNF having a higher surface charge required a higher salt concentration (3M) before a levelling off can be attained.

The aggregation and subsequent sedimentation of the charged CNFs in high salt media was used to estimate their aspect ratios. Figure 4 shows the quadratic fit of the plot of concentration against the relative sediment height. The results from the analysis and the estimated aspect ratios are displayed in Table 1. It can be seen that the aspect ratio of the TEMPO-oxidised CNF is higher than that of the carboxymethylated CNF. FE-SEM images and width distribution also show that the

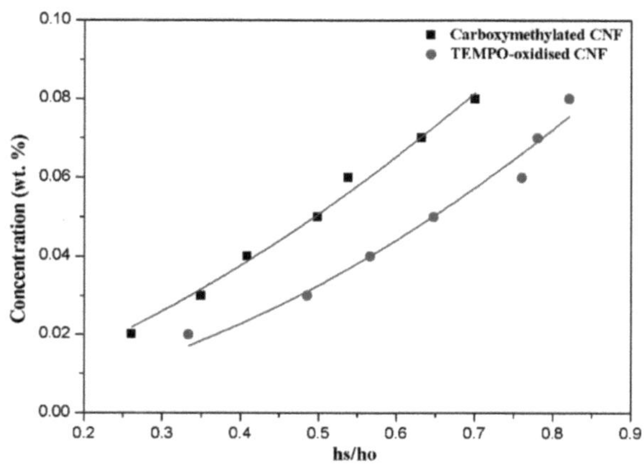

Fig. 4 Quadratic fit of initial concentration against relative sediment height

Table 1 Sedimentation results obtained from quadratic regression fits

Sample	Quadratic regression	Gel point (wt. fraction)	Gel point (vol. fraction)	Aspect ratio
Carboxymethylated CNF	$y = 0.07x^2 + 0.065x$, $R^2 = 0.98$	0.00065	0.00046	229 ± 18
TEMPO-oxidised CNF	$y = 0.08x^2 + 0.02x$, $R^2 = 0.97$	0.0002	0.00015	400 ± 30

fibril width of the carboxymethylated CNF are thicker than those of the TEMPO-oxidised CNF. Therefore the higher aspect ratio of TEMPO-oxidised CNF can be attributed to its narrower fibril width. The estimated aspect ratio of the TEMPO-oxidised CNF reported in this study falls within the range (310–623) of those reported from shear viscosity length measurements [20].

Conclusions

The estimation of aspect ratios of carboxymethylated and TEMPO-oxidised CNFs were carried out using gel point analysis by sedimentation. The aggregation and subsequent sedimentation of the charged CNFs were made possible by dispersing them in high concentrations of NaCl solution. An aspect ratio of 229 was estimated for the carboxymethylated CNF while an aspect ratio of 400 was estimated for the TEMPO-oxidised CNF. The higher estimated aspect ratio of the TEMPO-oxidised CNF was attributed to its narrower fibril widths compared to those of carboxymethylated CNF.

Acknowledgements The authors would like to thank the Edinburgh Napier University for the provision of the 25th Anniversary studentship grant.

References

1. Scandinavian Pulp, Paper and Board Testing Committee. (2002). Total acidic group content.
2. Benhamou, K., Dufresne, A., Magnin, A., Mortha, G., & Kaddami, H. (2014). Control of size and viscoelastic properties of nanofibrillated cellulose from palm tree by varying the TEMPO-mediated oxidation time.*Carbohydrate Polymers*, *99*, 74–83.
3. Dufresne, A. (2012). *Nanocellulose: from nature to high performance tailored materials.* Berlin; Boston: Walter de Gruyter GmbH.
4. Fukuzumi, H., Saito, T., Iwata, T., Kumamoto, Y., & Isogai, A. (2009). Transparent and high gas barrier films of cellulose nanofibers prepared by TEMPO-mediated oxidation. *Biomacromolecules, 10,* 162–165.
5. Fukuzumi, H., Tanaka, R., Saito, T., & Isogai, A. (2014). Dispersion stability and aggregation behavior of TEMPO-oxidized cellulose nanofibrils in water as a function of salt addition. *Cellulose, 21,* 1553–1559.

6. Ishii, D., Saito, T., & Isogai, A. (2011). Viscoelastic evaluation of average length of cellulose nanofibers prepared by TEMPO-mediated oxidation. *Biomacromolecules, 12,* 548–550.
7. Isogai, A., Saito, T., & Fukuzumi, H. (2011). TEMPO-oxidized cellulose nanofibers. *Nanoscale, 3,* 71–85.
8. Jonoobi, M., Oladi, R., Davoudpour, Y., Oksman, K., Dufresne, A., Hamzeh, Y., et al. (2015). Different preparation methods and properties of nanostructured cellulose from various natural resources and residues: a review. *Cellulose, 22,* 935–969.
9. Kangas, H., Lahtinen, P., Sneck, A., Saariaho, A.-M., Laintinen, O., & Hellén, E. (2014). Characterization of fibrillated celluloses. A short review and evaluation of characteristics with a combination of methods. *Nordic Pulp & Paper Research Journal, 29,* 129–143.
10. Klemm, D., Kramer, F., Moritz, S., Lindström, T., Ankerfors, M., Gray, D., et al. (2011). Nanocelluloses: A new family of nature-based materials. *Angewandte Chemie International Edition, 50,* 5438–5466.
11. Naderi, A., Lindström, T., Sundström, J., Fiberteknologi, Centrum för Biofibermaterial, B., Kth, Skolan För, K., Centra & Fiber- OCH, P. 2014. Carboxymethylated nanofibrillated cellulose: Rheological studies. *Cellulose, 21,* 1561–1571.
12. Nechyporchuk, O., Belgacem, M. N., & Bras, J. (2016). Production of cellulose nanofibrils: A review of recent advances. *Industrial Crops and Products.*
13. Pääkkö, M., Ankerfors, M., Kosonen, H., Nykänen, A., Ahola, S., Österberg, M., et al. (2007). Enzymatic hydrolysis combined with mechanical shearing and high-pressure homogenization for nanoscale cellulose fibrils and strong gels. *Biomacromolecules, 8,* 1934–1941.
14. Phan-Xuan, T., Thuresson, A., Skepö, M., Labrador, A., Bordes, R., & Matic, A. (2016). Aggregation behavior of aqueous cellulose nanocrystals: The effect of inorganic salts. *Cellulose, 23,* 3653–3663.
15. Raj, P., Mayahi, A., Lahtinen, P., Varanasi, S., Garnier, G., Martin, D., et al. (2016). Gel point as a measure of cellulose nanofibre quality and feedstock development with mechanical energy. *Cellulose, 23,* 3051–3064.
16. Saito, T., & Isogai, A. (2004). TEMPO-mediated oxidation of native cellulose. The effect of oxidation conditions on chemical and crystal structures of the water-insoluble fractions. *Biomacromolecules, 5,* 1983–1989.
17. Saito, T., Kimura, S., Nishiyama, Y., & Isogai, A. (2007). Cellulose nanofibers prepared by TEMPO-mediated oxidation of native cellulose. *Biomacromolecules, 8,* 2485–2491.
18. Siró, I., Plackett, D., Hedenqvist, M., Ankerfors, M., & Lindström, T. (2011). Highly transparent films from carboxymethylated microfibrillated cellulose: The effect of multiple homogenization steps on key properties. *Journal of Applied Polymer Science, 119,* 2652–2660.
19. Taheri, H., & Samyn, P. (2016). Effect of homogenization (microfluidization) process parameters in mechanical production of micro- and nanofibrillated cellulose on its rheological and morphological properties. *Cellulose,* 1–18.
20. Tanaka, R., Saito, T., Ishii, D., & Isogai, A. (2014). Determination of nanocellulose fibril length by shear viscosity measurement. *Cellulose, 21,* 1581–1589.
21. Varanasi, S., He, R., & Batchelor, W. (2013). Estimation of cellulose nanofibre aspect ratio from measurements of fibre suspension gel point. *Cellulose, 20,* 1885–1896.
22. Wågberg, L., Decher, G., Norgren, M., Lindström, T., Ankerfors, M., & Axnäs, K. (2008). The build-up of polyelectrolyte multilayers of microfibrillated cellulose and cationic polyelectrolytes. *Langmuir, 24,* 784–795.
23. Zhang, L., Batchelor, W., Varanasi, S., Tsuzuki, T., & Wang, X. (2012). Effect of cellulose nanofiber dimensions on sheet forming through filtration. *Cellulose, 19,* 561–574.

Electrical Conductivity of PLA Films Reinforced With Carbon Nano Particles from Waste Acrylic Fibers

Salman Naeem, Syed Qummer Zia Gilani, Vijay Baheti, Jakub Wiener, Jiri Militky, Saima Javed, Azam Ali, Zafar Javed and Syed Zameer ul Hassan

Abstract The acrylic fibrous waste was effectively transformed into activated carbon through physical activation in single stage carbonization under charcoal using at different carbonization temperatures 800, 1000 and 1200 °C. The characterization of carbonized fibers was performed with the help of XRD and EDX in order to find out the degree of crystallinity and proportion of different elements in carbonized fibres prepared at different temperatures. Later on, the carbonized fibers were pulverized in dry conditions by high energy planetary ball milling to get activated carbon micro/nano particles. In addition to refinement of size, the electrical conductivity of pulverized carbon particles was found to increase with increase in milling time. Subsequently, the particles were incorporated into poly lactic acid (PLA) in different loadings from 1 to 10% to develop conductive green composite films. The composite films were then characterized for electrical, thermo-mechanical and thermal properties.

Keywords Acrylic · Activated carbon · Stabilization · Carbonization Graphitization

S. Naeem (✉) · S.Q.Z. Gilani · V. Baheti · J. Wiener · J. Militky · A. Ali
Faculty of Textile Engineering, Department of Material Engineering,
Technical University of Liberec, Liberec, Czech Republic
e-mail: salman.ntu@gmail.com

S. Javed
Department of Microbiology and Molecular Genetics,
Punjab University Lahore, Lahore, Pakistan

Z. Javed
Department of Garments Manufacturing, National Textile University,
Faisalabad, Pakistan

S.Z. ul Hassan
Department of Textile Engineering, Balochistan University
of Information Technology, Engineering and Management Sciences, Quetta, Pakistan

R. Fangueiro and S. Rana (eds.), *Advances in Natural Fibre Composites*,
https://doi.org/10.1007/978-3-319-64641-1_18

Introduction

Millions of tons of textile waste is being landfilled every year. Last year Turkey landfilled around two hundred eighty thousand tons of textile waste which is only 2.62% of municipal solid waste. Environmental protection agency states that on average 31 kg of textile and clothing waste is generated by one person. The results are not too much difference in Europe where on average one UK citizen discards 30 kg of textile waste every year with a population of more than 500 million for the 27 countries of EU, this equates to approximately more than 18.3 million tons of clothing and textile waste generated every year. Textile waste is considered as one of the fastest growing sectors in terms of household waste and the amount of waste is forecast to increase as sales of new textiles and clothing continue to increase [1]. The main categories of clothing and textile waste are composed of synthetic materials such as acrylic, nylon and polyester and natural fibres such as wool, flax and cotton [2]. Due to strict rules and regulations for conserving nature, it is preferred and encouraged to utilize the fibrous wastes in value added applications.

Theoretically any source material rich in carbon content can be used for the formation of carbon. Practically wood, coconut shells, peat and fruit stones and a number of other materials are also used, in laboratories municipal wastes [3], synthetic polymers [4], tyres and acrylic fibres [5] are also used to produce carbon fibres. However on industrial scale different kinds of synthetic and natural materials are in use for precursor materials like Polyacrylonitrile, phenolic resins, polyamides, pitch and cellulose. Literature shows that PAN based precursors are better than other precursors like rayon and pitch etc. due to greater carbon yield and higher melting point. The polyacrylonitrile (PAN) based precursors are dominant as compared with other precursors for the formation of carbon fibers. The PAN polymer is high melting polymer, relatively hard and insoluble in which carbon chains are connected to one another as shown in Fig. 1.

PAN based carbon fibers have dominant edge over pitch and viscose rayon due to the following reasons.

- Because of its structure as shown in Fig. 5, it permits pyrolysis at faster rate without much disturbance in its basic structure as well as preferred orientation in molecular chains along the fiber axis as present in original fiber.
- The polymer decomposes before melting.
- When pyrolyzed at 1000 °C and above, high yield of carbon becomes possible.

Fig. 1 Molecular structure of polyacrylonitrile [6]

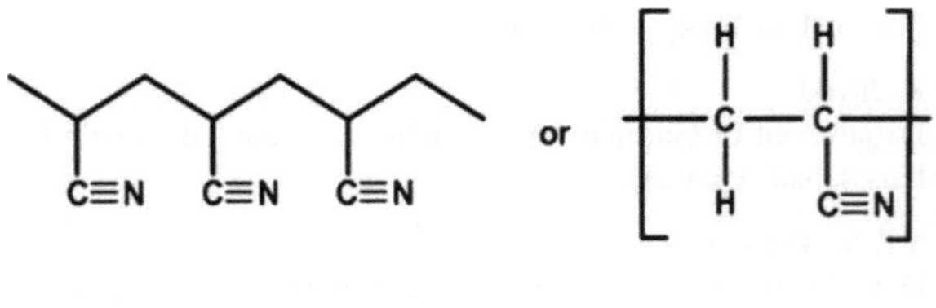

- Improvement in molecular chains is possible during thermal stabilization as it becomes plastic in nature around 180 °C and by different post spinning modifications [7].

The researchers are now trying to reduce the cost of carbon materials for using different cheap and alternate materials. In this context the idea of using acrylic fibrous waste is a favorable approach not only to reduce the cost of carbon but also for helping eco-friendly environment.

Recently, the development of low cost, high performance and functional composites has gained much importance. In this context, biopolymers are attractive materials because they can be easily biodegraded by microorganism even in local environment to their component elements [8, 9]. Biopolymer is a type of polymer that is produced by living organisms like RNA, DNA, cellulose, starch and protein. Amongst these cellulose is the most common organic compound and biopolymer found on the earth. From structural point of view, biopolymers are not same as conventional polymers. Biopolymers have a well-defined primary structure that folds to form compact shapes, however polymers have simple and random structure. The biopolymers and bioplastic market is rapidly increasing and is witnessed a current annual growth rate of 12.0% from 2016 to achieve a market size around 5.08 US billion dollars by the year 2021 [10].

Polylactic acid is biodegradable thermoplastic aliphatic polyester produced from renewable resources like tapioca roots, corn starch or sugarcane. Since 2010, its consumption was second highest in terms of volume as compared with other biopolymers [11]. PLA biopolymer films accounted for roughly around 25% of the global industry share in 2015. It has become popular over other biopolymers due to its easy process ability, high mechanical strength and faster rate of decomposition. As a result, its use is increased in applications of food packaging, milk and water bottles, biodegradable plastics [12, 13]. Besides its advantages, it has some drawbacks as well like its thermal and mechanical properties are not stable at high temperature [14]. To improve performance and to incorporate functional characteristics, the addition of nano-fillers seems to be a favorable approach [15].

Keeping in view the above described advantages different groups of researchers are in search to explore and improve properties by incorporating different particles. For instance, the addition of cellulosic nano particles helped to improve the mechanical and thermo-mechanical properties of Polylactic acid significantly. Likewise addition of conductive particles in insulating polymers generate a new kind of intelligent materials called as conductive polymer nano-composites [16–19].

Carbon black particles and nano-tubes incorporation causes substantial impact on electrical and thermal properties. The well-known extra-ordinary properties like high electrical conductivity, large surface area and low percolation threshold makes them superior candidate over other conventional conductive fillers.

In this work, waste of acrylic fibers was converted into activated carbon by using high temperature furnace. The waste acrylic fibers were first stabilized at 250 °C under the application of tension. The stabilized fibers were then carbonized under the layer of charcoal at different temperatures for getting higher values of electrical conductivity. Later theses fibers were converted to carbon nano-particles through the action of ball milling process. After making the particles these were incorporated into PLA films with different proportions in order to develop green conductive composites.

Experimental

Material and Method

The acrylic waste in fibrous form was taken from Grund Industries, Czech Republic. Every year huge amount of acrylic fibrous waste is generated by Czech Republic in the form of bath mats. Earlier these type of wastes were discarded or recycled by industries. However in recent years due to implementation of new rules and regulations the cost of discarding or recycling has increased. Hence in this changed scenario it is recommended to explore alternate methods for effective utilization of textile waste. Hence selection of acrylic fibrous waste seems to be attractive for the formation o activated carbon. The acrylic fibers were removed from bath mats by using mechanical cutting method. The physical characteristics of acrylic fibers as shown in Table 1.

Polylactic acid was taken from NatureWorks LLC, USA, it has average molecular weight of 200,000 and a density of 1.26 g cm^{-3}. chloroform was used as a solvent for dissolving PLA. It was purchased from Sigma Aldrich, Czech Republic.

Stabilization and Carbonization of Acrylic Fibrous Wastes

The acrylic fibers after removing from bath mats were stabilized at 250 °C (heating rate of 50 °C h^{-1}). The stabilized fibers were then pyrolysis under the layer of charcoal in order to have gradual reaction of atmospheric air with carbonized acrylic fibrous waste. The carbonization behavior was studied under two variables:

- Final pyrolysis temperature (FPT)
- Heating rate per hour (HRPH)

Table 1 Physical properties of acrylic fibers

Fineness [tex]	117
Tenacity [cN/tex]	23.84
Elongation [%]	45
Wet shrinkage [%]	2.5

Three levels of final temperature were selected that is 800, 1000 and 1200 °C. Heating rate was varied from 150 to 300 °C h^{-1} and finally to 450 °C h^{-1}. These variables were optimized to get higher electrical conductivity.

Effect of Carbonization Temperature on the Properties of Activated Carbon Fibers

The stabilized fibers were carbonized at high temperature under charcoal for single stage carbonization and activation. The final carbon fibers were prepared at heating rate of 300 °C h^{-1} having final temperatures (800, 1000 and 1200 °C) with no holding time. The process of stabilization at 250 °C in case of acrylic fibers results in cyclization, dehydrogenation, and oxidation of polyacrylonitrile structure [20]. During stabilization nitrile groups form non-meltable ladder structure, which not only improves mechanical properties but yield of resulting carbon as well. During subsequent carbonization of stabilized fibers, the ladder polymer further cross links to form turbostratic carbon structure and the orientation of basal planes leads to graphite like structure. After carbonization physical and analytical characterization was done on the prepared activated carbon fibers. Physical characterization included yield %, shrinkage, flexibility and dusting properties.

The shrinkage of activated carbon fibers was measured as per ASTM D 2259 standard available for testing textile fibers. The shrinkage was evaluated from change in length of fibres before and after carbonization. Similarly like shrinkage, yield of activated carbon before carbonization and after carbonization at different temperatures was calculated by using Eq. 1.

$$Yield = \frac{Final\,weight\,of\,activated\,web}{Initial\,weight\,of\,acrylic\,web} \times 100 \quad (1)$$

The flexibility or stiffness was evaluated from bending length by employing the principle of cantilever bending of the web under its own weight as per ASTM D 1388 standard.

Effect of Carbonization Temperature on Electrical Conductivity

Electrical resistance of carbonized fibers prepared at 800, 1000 and 1200 °C was measured with the help of Hewlett Packard 4339 B high resistance meter. The electrical resistance was measured at 100 voltage, the temperature and relative humidity were respectively 22 °C and 29.5% at the time of measurement of electrical resistance. Because of high electrical conductivity of resulting carbon fibres, another method of measuring electrical resistance by using multimeter was also implied.

Preparation of Nanoparticles from Carbonized Acrylic Fibrous Wastes

After getting the optimum parameter for pyrolysis, dry pulverization of carbonized fibers at predetermined settings was performed by using high energy planetary ball milling of Fritsch pulverisette 7, Germany. The sintered corundum container for ball milling having a capacity of 80 ml along with zirconium balls with a diameter of 10 mm chosen for 1–3 h of dry milling. The ball to material ratio (BMR) in this study was maintained at 10:1 however the speed was retained at 850 rpm.

Characterization of Activated Carbon Nanoparticles

Morphology of carbon Particles. After every hour of dry ball milling the Particle size distribution was analyzed with the help of Malvern zetasizer nano series which is based on the principle of dynamic light scattering from the brownian motion of particles. Before characterization carbon particles were disperse in deionized water after they were ultra-sonicated for five minutes with the help of bandelin ultrasonic probe. Further scanning electron microscope (SEM) at 30 kV and field emission scanning electron microscope (FESEM) at 5 kV accelerated voltage was used for analyzing morphology of milled activated carbon particles. Carbon particles having weight 0.01 g were dispersed in acetone solution of 100 ml and a drop of dispersion was put on aluminum foil and then coated with gold after drying.

Electrical Conductivity. After every 30 min of dry ball milling the milled carbon particles were dispersed by using Bandelin SONOPLUS ultrasonic probe in distilled water. Later the dispersion was used for measuring electrical conductivity with the help of conductometer having different concentration of activated carbon particles (0.5% by weight to 4.0%).

EDX analysis. For determining proportion of different elements in activated carbon prepared at different temperatures (800, 1000 and 1200 °C) energy dispersive X-ray spectroscopy (EDX) analysis was performed.

XRD analysis. In order to find out the change of crystallinity by increasing temperature, the X-ray diffraction analysis was performed with the help of PAN analytical X pert MPD diffraction system. XRD is a technique which is used for the determination of crystalline content in the material and investigation of unit cell dimensions. Degree of crystallinity in carbonized fibers was calculated by using Eq. 2 [21].

$$I_c = 1 - \frac{I_1}{I_2} \tag{2}$$

I_1 is intensity at minimum peak
I_2 is intensity at maximum peak

Preparation of PLA Films

Firstly, 5% by weight Polylactic acid was dissolved into chloroform to make a solution of 100 g L^{-1}. The pre-weighted amounts of carbon nano-particles derived from acrylic fibrous waste (1–10 wt% related to mass of films) were mixed in 100 ml solution of PLA and chloroform. The mixture was first stirred by the help of magnetic stirrer for 3 h at room temperature then ultra-sonicated for 15 min on BANDELIN Ultra Sonic Probe Mixer with 50-horn power. The prepared homogenous solution was poured in mould (6 cm × 4 cm) on a Teflon sheet. The films were kept at room temperature and after 20 h the prepared film were removed from moulds, folded and hot pressed at 140 °C. This helps to improve uniform dispersion of particles in the polymer. The process of hot press was repeated three times to ensure uniform distribution of particles in the polymer.

Characterization of PLA Films

Dynamic mechanical analysis (DMA). Dynamic mechanical analysis of neat PLA films and carbon reinforced PLA films was performed on DMA DX04T RMI instrument. The measurements were carried out at strain amplitude of 0.05%, temperature range of 35–100 °C, heating rate of 5 °C/min, a constant frequency of 1 Hz and jaw distance of 30 mm.

Differential scanning calorimetry (DSC). DSC 6 Perkin Elmer instrument was used for determining crystallization and melting behavior of neat PLA film and PLA reinforced with carbon particles manufactured at different temperatures. The PLA films with and without carbon particles having weight of 7 mg were used in the presence of nitrogen. The samples were heated to 200 °C with a heating rate of 5 °C/min.

Thermo gravimetric analysis (TGA). For determining thermal stability of neat PLA and PLA reinforced with carbon particles prepared at different temperatures thermo gravimetric analysis (TGA) TGA/SDTA 851, Metler Toledo was used. The PLA films having weight of 9 mg were used at 900 °C with a heating rate of 10 °C min^{-1} in inert atmosphere with the help of nitrogen.

Electrical properties. The resistivity of PLA carbon films prepared was measured according to ASTM D 257-14 at temperature 22 °C and relative humidity 40%. The specific voltage potential of 1 V using direct current was applied across opposite ends of activated carbon web and resultant current flowing across the sample was measured after 15 ± 1 s. Electrical conductivity of activated carbon webs was measured by two different techniques.

- Parallel electrode method
- Concentric electrode method

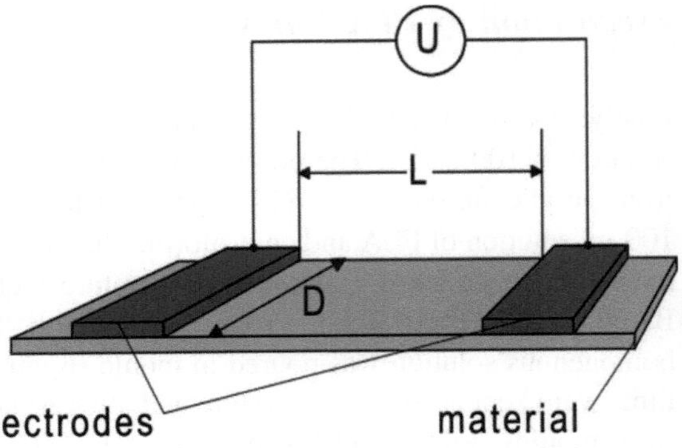

Fig. 2 Surface resistivity measurement set up for parallel electrode

Electrical conductivity or specific conductance is the reciprocal of electrical resistivity and measures the ability of a material to conduct electric current. Electrical conductivity is commonly represented by Greek letter σ (sigma). S.I unit of conductivity is Siemens per meter (*S/m*). The parallel electrode instrument is shown in Fig. 2. By using this apparatus surface resistance was determined.

Surface resistivity is calculated by using Eq. 3.

$$\rho = \mathrm{R} \times \frac{D}{L} \tag{3}$$

ρ Surface resistivity (Ω)
D Width of electrode (5 cm)
R Electrical resistance (Ω)
L Distance between electrode (2 cm)

The second method used for measuring conductivity is shown in Fig. 3. In this method conductivity was measured at 22 °C temperature with relative humidity of 62%. The specific voltage potential of 1 V was applied across the activated carbon fibers and readings were measured at 15 s. The conductivity was calculated by using Eq. 4.

$$\text{Surface resistivity} = \rho = R \times 9.99 \tag{4}$$

Results and Discussions

Effect of Carbonization Temperature on the Properties of Activated Carbon Fibers

With increasing temperature to 1200 °C from 800 °C, the available oxygen in high temperature furnace reacted with the newly formed carbon that caused a decrease in the yield of final carbon due to which more shrinkage and rigid structure was achieved.

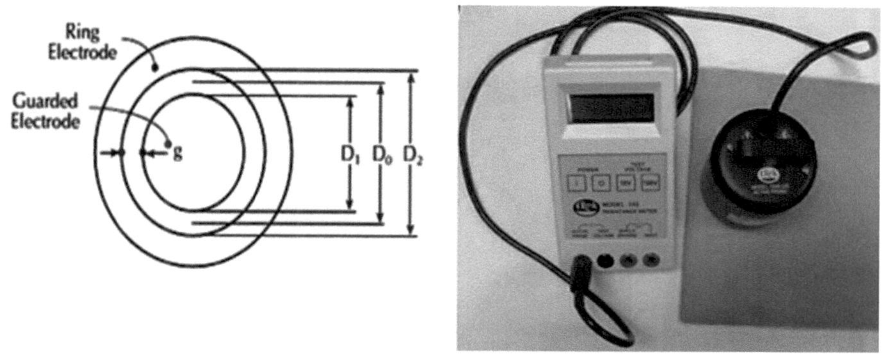

Fig. 3 Surface resistivity measurement set up for concentric ring electrode

Hence activated carbon prepared by using high temperature (1200 °C) exhibited poor dusting behavior and flexibility as shown in Table 2. The yield of finally prepared carbon slightly increased by increasing heating rate to 450 °C h^{-1} from 150 C h^{-1} as at slow heating rate prepared carbon remained in contact with high temperature for prolonged time which adversely affect not only on the yield but on flexibility as well.

Migration and hoping are two quite possible mechanisms for the transportation of electrons as shown in Fig. 4 in carbon fibres produced at different carbonization temperatures.

The property of electrical conductivity is because of two possible reasons which are:

- Jumping of electrons through interfaces/defects among closely disordered layers of graphite.
- Electron migration among different layers of graphite [22–24].

The activated carbon fibers prepares at 1200 °C showed more electrical conductivity as compared to activated carbon fibres prepared at 1000 and 800 °C

Table 2 Effect of carbonization temperature on physical properties of activated carbon fibers

Run	Final pyrolysis temperature [°C]	Heating rate [°C h^{-1}]	Yield (%)	Resistivity [Ω mm]	Flexibility
1	800	150	60.33 ± 3.13	169.65	Good
2	800	300	61.27 ± 3.63	1174.50	Good
3	800	450	63.33 ± 2.55	323.11	Average
4	1000	150	55.88 ± 2.33	4.69	Average
5	1000	300	57.12 ± 1.83	3.21	Poor
6	1000	450	58.55 ± 1.98	2.08	Poor
7	1200	150	46.45 ± 1.89	0.67	Poor
8	1200	300	45.11 ± 1.60	0.52	Poor
9	1200	450	49.11 ± 2.12	0.65	Poor

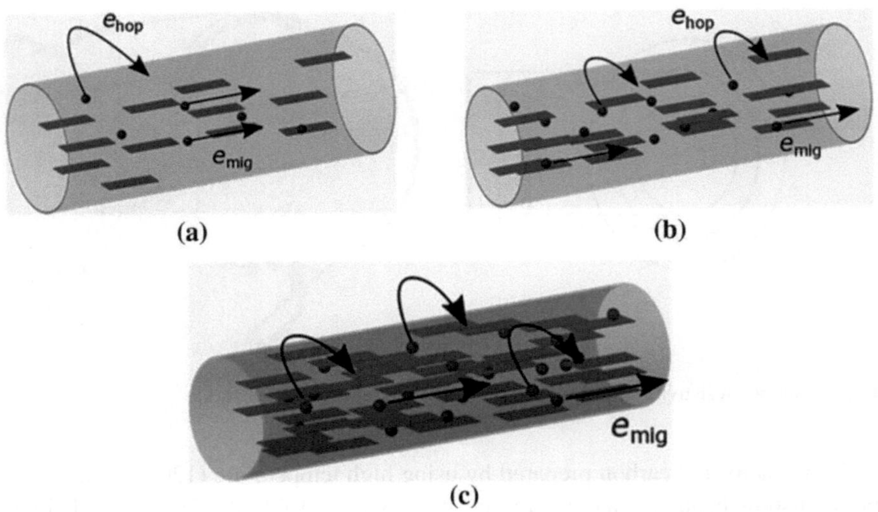

Fig. 4 Charge transport mechanism in **a** 800 °C activated carbon fiber **b** 1000 °C activated carbon fiber **c** 1200 °C activated carbon fiber

because of ease in migration through different graphite layers. Further the rise in electrical conductivity is because of more parallel orientation of carbon layers and higher degree of crystallinity at higher temperature as shown in figure.

Characterization of Carbon Nano Particles

Morphology of carbon *Particles*. Activated carbon fibers are dry milled from one to three hours not only to reduce size of carbon particles but also to increase surface area of carbon particles. In the initial one hour of dry ball milling the reduction in particle size is higher and on average particle size diameter was reduced to 1563 mm. However as the time of milling was increased the size of particles kept on decreasing and after three hours of dry milling the size of particles reached to 521 nm. When milling time was increased for prolonged time not only reduction in size was observed but also particle size distribution converted to almost unimodal distribution from multimodal distribution as temperature increased inside chamber due to continuous collision of balls [16, 17]. This increase in temperature inside the chamber caused cold welding of carbon nano particles and also a layer of particles also deposited on the walls of milling chamber. After three hours of milling the morphology of activated carbon nano particles was investigated by using scanning electron microscopy (SEM) as can be seen from Fig. 5a–c. It was observed that after milling the mixture comprised on both nano-segments and nano particles.

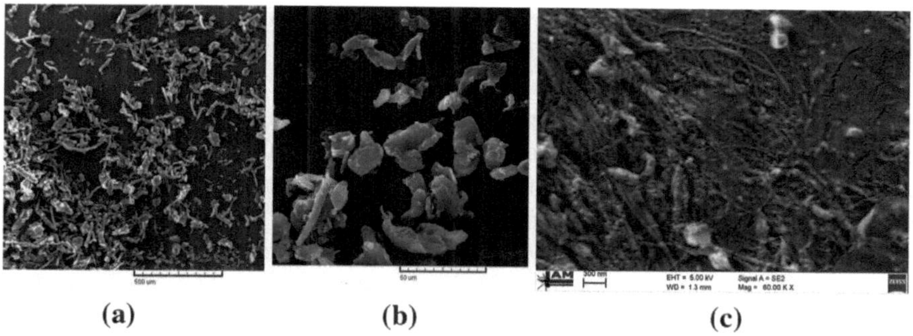

Fig. 5 Microstructure of activated carbon nano particle particles after milling **a** one hour, **b** two hour, **c** three hour

Electrical Conductivity. The electrical conductivity of activated carbon nano particles in aqueous dispersion with different concentrations (0.5–4.0 wt%) was measured as shown in Fig. 6.

The impact of milling time on electrical conductivity of activated carbon particles having low concentration (1% by weight and below) of activated carbon particles and it was found that electrical conductivity of activated carbon nano particles significantly increased by increasing milling time.

This behavior can be described in terms of reduction in size of carbon particles and increasing the surface area with the increase of milling time. However as the concentration of carbon nano particles was increased from 2% by weight to 4%, there is steady rise of electrical conductivity by increasing ball milling time. This rise in electrical conductivity is due to accomplishment of percolated network of carbon nano particles in aqueous dispersion because of high loading of particles.

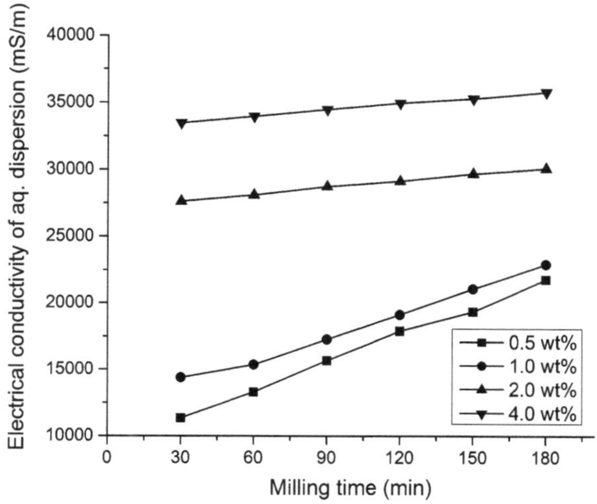

Fig. 6 Effect of dry milling time on electrical conductivity of activated carbon particles

EDX analysis. In order to determine relative proportion of different elements in activated carbon particles energy dispersive x-ray analysis was carried out. It can be seen from Table 3 that as the carbonization temperature was increased from 800 to 1000 °C and finally to 1200 °C the oxygen content kept on decreasing while carbon content increased. The activated carbon formed at 1200 °C showed 6.61% oxygen and 93.39% carbon content due to removal of different atoms at high temperature which caused increase in proportion of carbon [25].

XRD analysis. The XRD analysis helped to determine the crystallinity in activated carbon particles prepared at different carbonization temperatures. From Fig. 7a–c the XRD pattern of activated carbon particles prepared at different temperatures (800, 1000 and 1200 °C) can be seen.

The crystalline percentage of activated carbon particles was found to be 82.21, 86.7 and 92.41% respectively at 800, 1000 and 1200 °C. The rise in degree of crystallinity with increase of temperature is because of more content of carbon and simultaneously due to more parallel orientation of carbon basal planes.

Table 3 Effect of carbonization temperature on elemental composition of activated carbon particles

Element	App conc.	Intensity	Weight [%]	Atomic [%]
800 °C				
C K	0.26	2.12	0.13	91.76
O K	0.01	0.76	0.01	8.24
1000 °C				
C K	0.37	2.12	0.18	91.87
O K	0.02	0.76	0.02	8.13
1200 °C				
C K	0.18	2.10	0.09	93.39
O K	0.01	0.74	0.01	6.61

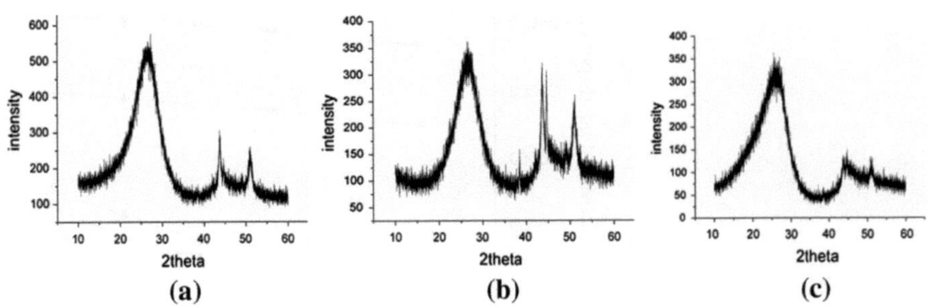

Fig. 7 Effect of carbonization temperature on crystallinity of activated carbon particles **a** at 800 ° C, **b** at 1000 °C, **c** at 1200 °C

Characterization of PLA Films

Electrical conductivity. Poly(lactic acid)/carbon films were produced by adding different amounts of acrylic-derived carbon particles prepared in this study into the solution of PLA with chloroform and later evaporating the solvent. The obtained composite films of PLA carbon showed conductivity at the loading of 10% by weight of carbon particles prepared at 1200 °C after 3 h of dry milling. In another study it was revealed that by the addition of 10% carbon particles, PLA films showed conductivity lower than 1 Ω m [26]. However by the addition of 3% carbon black particles, electrical conductivity was achieved in high density polyethylene (HDPE) [27]. The electrical conductivity of carbon particles prepared from acrylic fibrous waste can be seen from table. The contact resistance between carbon particles decreases when the surface area between them is increased due to which particles prepared at 1200 °C with three hours of milling was selected for making the films. The origin of electrical-conduction process is explained because of conductivity at high concentration of carbon particles which enables transport of electrical charges across micro contacts between the carbon particles dispersed in a homogenous way in PLA films. The path of electrical conduction is established because of particle to particle network which creates a network of channeling extending throughout the film. Both the instruments gave similar values of conductivity as show. The films prepared at 5% showed very less conductivity but the film prepared at 10% by weight showed increase in conductivity due to more establishing of electrical network as can be seen from Table 4.

Dynamic mechanical analysis. From Fig. 8a and Table 5 the load behavior of activated carbon particles (ACP)/PLA composite films and neat PLA films can be observed. When storage modulus of neat PLA with activated carbon particles ACP/PLA was analyzed, the improvement was observed over entire temperature span. The maximum improvement in modulus was observed in case of 5 wt% composite where storage modulus was increased from 2.04 to 3.84 GPa at 30 °C. The rise of storage modulus is because of higher stiffness characteristics of activated carbon particles while transfer of stress from matrix to carbon particles. However further increasing of loadings of particles of carbon 10 wt%, the value of storage modulus decreased to 2.46 GPa due to agglomeration and poor dispersion of particles that increased the stress concentration points.

By increasing temperature from 30 to 60 °C, the storage modulus of neat PLA decreased faster than PLA composite films. The storage modulus was dropped by 14 times in neat PLA, whereas the drops of 5 and 9 times were observed in case of 1 and 10 wt% composites films, respectively. The significant drop in storage modulus of neat PLA at 60 °C is due to the softening of matrix and easier movement of PLA chains. The relatively smaller drop in case of composite films is

Table 4 Electrical resistivity of PLA-carbon films

1	PLA-5% carbon	230 ± 2.33 Ωm
2	PLA-10% carbon	1.8 ± 1.90 Ωm

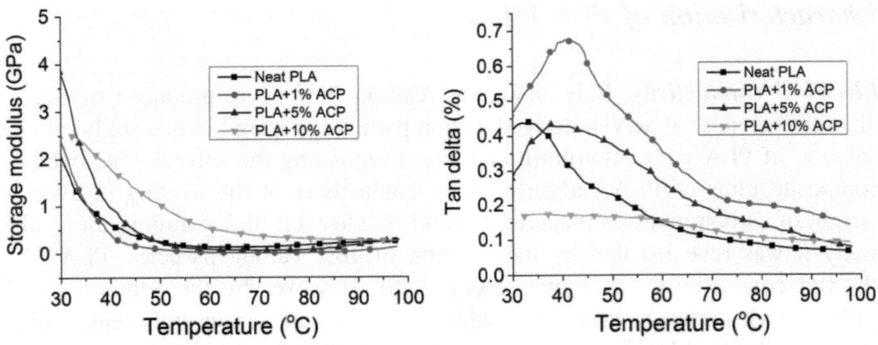

(a) Storage modulus (b) Damping factor

Fig. 8 Dynamic mechanical properties of PLA composites

Table 5 Storage modulus of ACP/PLA composites films at different temperature

Sample name	E′ (30 °C) (GPa)	E′ (60 °C) (GPa)	E′ (90 °C) (GPa)
Neat PLA	2.04 ± 0.15	0.14 ± 0.20	0.22 ± 0.02
PLA+1% ACP	2.49 ± 0.50	0.52 ± 0.80	0.27 ± 0.21
PLA+5% ACP	3.84 ± 0.35	0.09 ± 0.46	0.18 ± 0.04
PLA+10% ACP	2.46 ± 0.24	0.26 ± 0.55	0.04 ± 0.26

attributed to the presence of ACP, which restricted the motion of PLA chains. At 60 °C, PLA composite films of 1 and 10 wt% ACP showed 271%, and 85.00% respective increase in storage modulus as compared to neat PLA films. The higher storage modulus values of PLA composite films compared to neat PLA above 60 °C are attributed to the nucleating behavior of ACP, which improved crystallinity of PLA through transcrystallization.

Tan delta or mechanical loss factor can be explained as loss in modulus to storage modulus. When the content of carbon particles increased in PLA films the peak of tan delta was shifted positively as can be seen from Fig. 8. In case of 1 and 5 wt% activated carbon particles/PLA films, the shift is 13 and 11 °C respectively. The positive shift in tan delta is because of more surface interaction between ACP and matrix which restricted or reduced mobility of matrix chains around them. However when 10 wt% particles of activated carbon were added there was negligible peak because of inability for damping characteristic which is due to compact filling of particles.

Differential scanning calorimetry analysis. The behavior of PLA films with activated carbon particles and neat PLA is analyzed by using DSC thermogram as can be seen from Fig. 9. The glass transition temperature (T_g), followed by cold crystallization (T_{cc}) temperature and melting (T_m) temperature of polymer can be analyzed on the thermograms of all the samples. From Table 6 it is clear that value

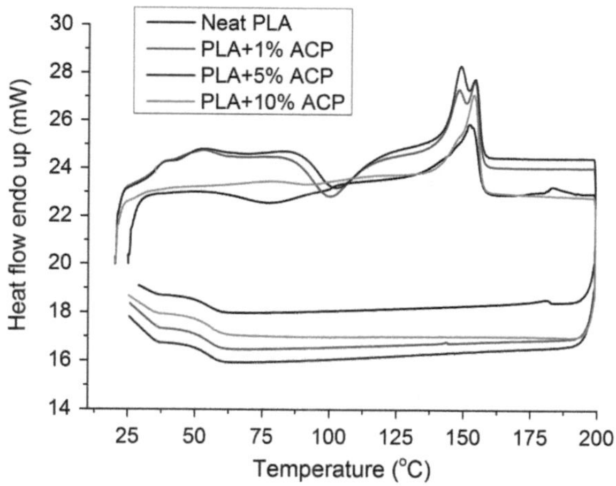

Fig. 9 Differential scanning calorimetry of PLA composites

Table 6 Behavior of neat and ACP/PLA composite films on application of heat

Sample	Tg (°C)	Tcc (°C)	Tm (°C)
Neat PLA	42 ± 0.3	80 ± 1.1	147 ± 0.1
PLA+1% ACP	43 ± 0.5	107 ± 1.2	151 ± 0.1
PLA+5% ACP	45 ± 0.6	111 ± 1.4	151 ± 0.2
PLA+10% ACP	47 ± 1.2	96 ± 1.5	153 ± 0.5

of glass transition temperature increased by increasing concentration of activated carbon particles. The maximum increase in glass transition temperature from 42 to 47 °C from neat PLA to PLA with carbon particles at a loading of 10 wt%. The glass transition temperature depends on steric effects, flexibility of chains, molecular weight of polymer, branching of chains, cross linking density and intermolecular interactions among the chains. The steady increase in T_g by 2, 7 and 12% as compared with neat PLA film can be explained because of reduced PLA chain flexibility after adding 1, 5 and 10 wt% of activated carbon particles. The peak for cold crystallization became broader and moved to high temperature after adding activated carbon particles as compared with the peak of cold crystallization for neat PLA. The higher T_{cc} witnessed during heating may be due to slower crystallization induced by activated carbon particles. From Table 6 it is clear that value of T_m improved at lower concentration of ACP. The addition of activated carbon particles (1–5 wt%) caused increase in the value of T_m from 147 to 151 because of adhesion between PLA matrix and carbon particles. A careful analysis of thermograms of composites revealed small endotherms before main melting peaks. At high temperature (T_{m2}) the melting peak indicates towards more crystalline structure and shoulder peak at low temperature (T_{m1}) is due to less crystalline structure.

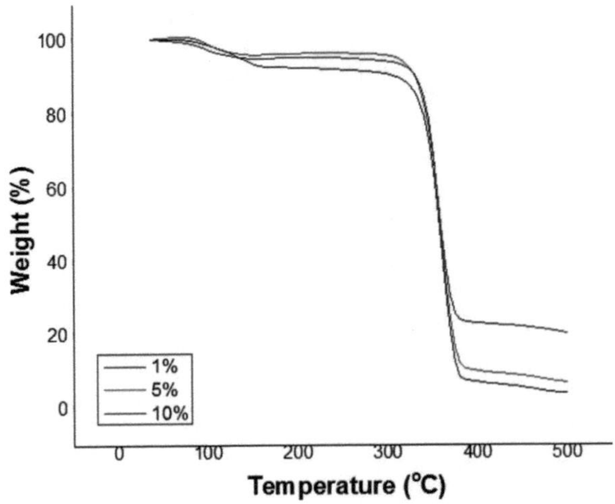

Fig. 10 Thermal stability of PLA composite films

This clearly indicates that poly lactic acid develops heterogeneous crystalline morphology by adding activated carbon particles.

Thermogravimetric analysis. Thermogravimetric analysis was performed for determining thermal stability of neat PLA films and PLA reinforced with carbon particles with different percentage of loading. Thermogravimetric analysis PLA reinforced with carbon particles exhibited oxidative and thermal stability, decomposition temperature and mass loss while heat treated as shown in Fig. 10. The initiation of mass loss experienced in all the films around 80–90.

The neat poly lactic acid film degraded with negligible residue however PLA films reinforced with carbon particles left with residue related to carbon. The amount of residue kept on rising as carbon content was raised from 1 to 5% and ultimately to 10% which clearly indicates that thermal stability of films increased after adding carbon particles. The 20.19% increase in mass was achieved by adding 10% loading of carbon particles. The increase in thermal stability is because of more homogenous distribution of carbon in PLA matrix which caused positive impact on the volatile products of PLA. The results are very much similar with previously reported thermal stability of PLA with halloysite and silica nano-particles [28, 29].

Conclusions

The objective of this study was to successfully utilize acrylic fibrous waste for the formation of activated carbon and later on incorporated with biopolymer to get functional characteristics of PLA composites. As the temperature of carbonization was increased from 800 to 1000 °C and finally to 1200 °C, electrical conductivity also kept on increasing due to more parallel orientation of carbon basal planes.

The higher electrical conductivity of activated carbon sample prepared at 1200 °C was attributed to more graphitization, which was confirmed from presence of sharp diffraction peak observed in XRD spectra and higher degree of crystallinity. As the temperature during carbonization was increased, the degree of crystallinity along with carbon content in carbonized fibres also increased, which is the main cause for decrease in resistivity. Hence carbon fibres produced by carbonization at 1200 °C were dry pulverized through the use of high energy planetary ball milling process. The carbon nanoparticles produced after three hours of ball milling showed more uniform distribution of size. These particles were mixed with PLA to get conductive biopolymer films. The electrical conductivity not only depends on uniform distribution of particles but also to higher loading of particles. Not only electrical properties of PLA films increased by the addition of carbon nanoparticle but also thermal stability of films also increased due to more homogenous distribution and high temperature bearing characteristics of carbon particle. Furthermore the presence of activated carbon particles in the PLA films favors for improvement in thermal stability of PLA. Hence on the basis of above discussion it is clear that technique of dry milling of carbonized fibres from acrylic fibrous waste is simple, economical and environment friendly approach for the formation of conductive biopolymer films.

References

1. Evrim, B., Simona, J., & Fatma, K. (2015). A sustainable approach to collect post-consumer textile waste in developing countries. *Marmara Journal of Pure and Applied Sciences, 1,* 107–111.
2. Nahil, M., & Williams, P. (2010). Activated carbons from acrylic textile waste. *Journal of analytical and applied Pyrolysis, 89*(1), 51–59.
3. Williams, P., & Cunliffe, A. (1998). Properties of chars and activated carbons derived from the pyrolysis of used tyres. *Environmental Technology, 19*(1), 1177–1190.
4. Williams, P., & Buad, W. (2010). Activated carbons prepared from refuse derived fuel and their gold adsorption characteristics. *Environmental Technology, 31*(2), 125–137.
5. Nabais, J., Ribeiro Carrot, M., & Carrott, P. (2001). Preparation of activated carbon fibres from acrylic textile fibres. *Carbon NY, 39*(10), 1453–1555.
6. Rehman, M. S. A., Ismail, A., & Mustafa, A. (2007). A review of heat treatment on polyacrylonitrile fiber 2007. *Polymer Degradation and Stability, 92*(8), 1421–1432.
7. Donnet, J., & Wang, T. (1998). *Carbon fibers* (3rd ed.). London: CRC Press.
8. Charles, G., Gebelein, C., & Carraher, J. (1994). *Biotechnology and bioactive polymers*. New york: Plenum.
9. Scott, G. (1994). *Biodegradable plastics and polymers*. London, UK: Elsevier.
10. Laura, W. (2016). *Global bioplastics markets and technologies report*. Dublin, Ireland: Global Newswire.
11. Baheti, V., Militky, J., & Marsalkova, M. (2013). Mechanical properties of poly lactic acid composite films reinforced with wet milled jute nanofibers. *Polymer Composites, 34*(12), 2133–2141.
12. Baheti, V., Militky, J., & Marsalkova, M. (2013). Mechanical properties of poly lactic acid composite films reinforced with wet milled jute nanofibers. *Polymer Composites, 34*(12), 2133–2141.

13. Jonoobi, M., Harun, J., Mathew, A., & Oksman, K. (2010). Mechanical properties of cellulose nanofiber (CNF) reinforced polylactic acid (PLA) prepared by twin screw extrusion. *Composites Science and Technology, 70*(12), 1742–1747.
14. Petersen, K., Nielsen, P., & Olsen, M. (2001). Physical and mechanical properties of biobased materials starch polylactate and polyhydroxybutyrate. *Starch, 53*(8), 356–361.
15. Petersson, L., & Oksman, K. (2006). Biopolymer based nanocomposites: Comparing layered silicates and microcrystalline cellulose as nanoreinforcement. *Composites Science and Technology, 66*(13), 2187–2196.
16. Kumar, B., Castro, M., & Feller, J. (2012). Poly(lactic acid)–multi-wall carbon nanotube conductive biopolymer nanocomposite vapour sensors. *Sensors and Actuators B: Chemical, 161,* 621–628.
17. Philip, B., Abraham, J., Chandrasekhar, A., & Varadan, V. (2003). Carbon nanotube/PMMA composite thin films for gas-sensing applications. *Smart Materials and Structures, 12,* 935–939.
18. Feller, J. (2004). Conductive polymer composites: influence of extrusion conditions on positive temperature coefficient effect of poly(butylene terephthalate)/poly(olefin)-carbon black blends. *Journal of Applied Polymer Science, 91,* 2151–2157.
19. Knite, M., Tupureina, V., Fuith, A., Zavickis, J., & Teteris, V. (2007). Polyisoprene—Multi-wall carbon nanotube composites for sensing strain. *Materials Science and Engineering: C, 27*(1), 1125–1128.
20. Liu, Y., Choi, Y., Chae, H., Gulgunje, P., & Kumar, S. (2013). Temperature dependent tensile behavior of gel-spun polyacrylonitrile and polyacrylonitrile/carbon nanotube composite fibers. *Polymer, 54*(15), 4009–40003.
21. Siqueira, G., Abdillahi, H., Bras, J., & Dufresne, A. (2010). High reinforcing capability cellulose nanocrystals extracted from *Syngonanthus nitens*. *Cellulose, 17,* 289–298.
22. Cao, M., Song, W., Hou, Z., Wen, B., & Yuan, J. (2010). The effects of temperature and frequency on the dielectric properties, electromagnetic interference shielding and microwave-absorption of short carbon fiber/silica composites. *Carbon NY, 48,* 788–796.
23. Song, W., Cao, M., Hou, Z., Fang, X., Shi, X., & Yuan, J. (2009). High dielectric loss and its monotonic dependence of conducting-dominated multiwalled carbon nanotubes/silica nanocomposite on temperature ranging from 373 to 873 K in X-band. *Applied Physics Letters, 94,* 1–4.
24. Wen, B., Cao, M. S., Hou, Z. L., Song, W. L., Zhang, L.,& Lu, M. M. (2013). Temperature dependent microwave attenuation behavior for carbon-nanotube/silica composites. *Carbon NY, 65,* 124–139.
25. Ma, Y., Yin, X., & Li, Q. (2013). Effects of heat treatment temperature on microstructure and electromagnetic properties of ordered mesoporous carbon. *Transactions of the Nonferrous Metals Society of China, 23,* 1652–1660.
26. Tsuji, H., Kawashima, Y., Takikawa, H., & Tanaka, S. (2007). Poly(L-lactide)/nano-structured carbon composites: Conductivity, thermal properties, crystallization, and biodegradation. *Polymer, 48,* 4213–4225.
27. Ren, D., Zheng, S., Wu, F., Yang, W., Liu, Z., & Yang M. (2014). Formation and evolution of the carbon black network in polyethylene/carbon black composites: Rheology and conductivity properties. *Journal of Applied Polymer Science, 131*(7).
28. Liu, M., Zhang, Y., & Zhou, C. (2013). Nanocomposites of halloysite and polylactide. *Applied Clay Science, 75*(2), 52–59.
29. Kim, I. H., & Jeong, Y. G. (2010). "Polylactide/exfoliated graphite nanocomposites with enhanced thermal stability, mechanical modulus, and electrical conductivity," *Journal of Polymer Science Part B: Polymer Physics, 48,* 850–858.

Allocation in the Life Cycle Assessment (LCA) of Flax Fibres for the Reinforcement of Composites

John Summerscales and Nilmini P.J. Dissanayake

Abstract The ISO 14040 series of standards describe the principles and framework for the conduct of life cycle assessment (LCA). The system defines four phases: (i) definition of the goal and scope of the LCA, (ii) the life cycle inventory analysis (LCI), (iii) the life cycle impact assessment (LCIA), and (iv) the life cycle interpretation. The standards do not describe the LCA technique in detail, nor do they specify methodologies for the individual phases of the LCA. Dependent of the goal and scope, there can be very different outcomes from the analysis. This paper considers how the outcomes might change for the specific case of flax fibres for the reinforcement of composites. The study compares allocation of environmental burdens to two different primary products: (i) flax seed as a nutritional supplement with fibre generated from the waste stream, or (ii) flax fibre as the primary product.

Keywords Allocation · Composites · Flax fibre · Flax seed · Life cycle assessment · LCA · Reinforcement

Introduction

There is a considerable activity in the academic study and commercial exploitation of natural fibres as the reinforcement for polymer matrix composites (e.g. Dissanayake and Summerscales [1], Summerscales et al. [2–4], Summerscales and Grove [5], Summerscales and Gwinnett [6]). These NFRP materials are often referred to as "sustainable composites" although very few LCA are available to confirm or refute that description.

Ekvall and Finnveden [7] undertook a critical review of the adequacy and feasibility of methods recommended for allocation by the (then) current international standard on life cycle inventory analysis with a focus on multi-functional systems.

J. Summerscales (✉) · N.P.J. Dissanayake
ACMC, School of Engineering,
Plymouth University, Drake Circus, Plymouth PL4 8AA, UK
e-mail: jsummerscales@plymouth.ac.uk

R. Fangueiro and S. Rana (eds.), *Advances in Natural Fibre Composites*,
https://doi.org/10.1007/978-3-319-64641-1_19

They demonstrated that different approaches to the allocation problems result in different types of information. They recommended, "that all of the environmental burdens of the multifunction process be allocated to the product investigated". LCA results appear to be largely dependent on the chosen allocation methods used. ISO 14040 [8] and ISO 14044 [9] also recommend avoiding allocation whenever possible either through subdivision of certain processes or by expanding the system limits to include associated additional functions.

Dissanayake et al. undertook an investigation into energy use in the production [10], and Life Cycle Impact Assessment (LCIA) [11], of UK flax fibre for the reinforcement of composites. The Functional Unit (FU) was one tonne of fibre as either sliver (aligned mat) or yarn (twisted filament) based on an assumption of equivalent specific modulus. The respective moduli and densities for flax or glass were taken as 42 or 72 GPa and 1500 or 2500 kg/m^3. The analysis adopted the Ekvall and Finnveden allocation recommendation of all burdens assigned to the primary product. The harvested flax can produce seed, long fibre (used for composite reinforcement in this study), short fibre (used for paper-making or animal bedding) and dust (briquetted for solid bio-fuel).

Le Duigou et al. [12] conducted an environmental impact analysis on French flax fibres using a different set of underlying assumptions. They concluded, "without the allocation procedure the results from the two studies (France vs. UK) would be similar". The key differences were:

- UK plants desiccated at mid-point flowering but French plants allowed to set seed,
- UK yield only 6000 kg/ha, but French yield 7500 kg/ha at harvest,
- UK study excluded photosynthesis and CO_2 sequestration,
- Higher level of nuclear power in the French energy mix, and
- UK allocated all burdens to fibre, French allocated on mass of product and co-products.

Analysis

Table 1 shows the relative allocation of burdens for single products, then for multifunctional systems based on economic value allocation and on mass allocation. The chosen goal and scope can clearly influence the reported environmental credentials of the respective products. Mass allocation has lower variability whereas value allocation is susceptible to price fluctuations for the respective co-products with the time.

The environmental burdens identified by the International Organization for Standardization [13] have been translated to environmental impact classification factors (EICF) by Azapagic et al. [14, 15]:

- Global Warming Potential (GWP)
- Acidification Potential (AP)

Table 1 Yields, allocation and assigned relative burdens for flax products (highest value in bold)

Scenario			Allocation of burdens			
Single product	Mass/tonne yarn (kg)	Value/kg	Seed	Long fibre	Short fibre	Dust
Flax seed (FS)	500 (3%)	£3	**100%**	–	–	–
Flax long fibre (FLF)	1000 (6%)	£0.90	–	**100%**	–	–
Flax short fibre (FSF)	5000 (30%)	£0.10				
Shive/dust	10000 (61%)	£0.10				
Multiple products	Allocation					
FS and FLF	Value allocation		**62.5%**	37.5%	–	–
FS/FLF/FSF	Value allocation		**51.7%**	31.0%	17.2%	–
FS/FLF/FSF and shive/dust	Value allocation		**38.5%**	23.1%	12.8%	25.6%
FS and FLF	Mass allocation		33.3%	**66.7%**	–	–
FS/FLF/FSF	Mass allocation		7.7%	15.4%	**76.9%**	–
FS/FLF/FSF and shive/dust	Mass allocation		3.0%	6.1%	30.3%	**60.6%**

- Eutrophication Potential (EP)
- Human Toxicity Potential (HTP)
- Aquatic Toxicity Potential (ATP)
- Ozone Depletion Potential (ODP)
- Photochemical Oxidants Creation Potential (POCP)
- Non-Renewable/Abiotic Resource Depletion (NRADP).

The environmental burdens generated by flax fibres for the reinforcement of composites can be minimised when the plant is grown **primarily** for **seed** as a nutritional supplement, the flax is a co-product, and the Ekvall and Finnveden allocation recommendation is followed. Under those conditions, the high embodied energy in the agro-chemicals and the environmental burdens arising from the agricultural operations are allocated 100% to the seed. At the other extreme, when the plant is desiccated at mid-point flowering, there is no seed produced, so all environmental burdens must be allocated to the fibre.

The analysis in Table 1 assumes that seed-plus-fibre is a single co-product system with environmental burdens from all operations shared by the co-products. In the case where flax seed (health food supplement) is the primary product, the processing for fibre should not be included in the environmental burdens allocated to the seed. The flax fibre can be considered as a burden-free raw material resource as those burdens are already taken by the seed. However, burdens will arise from the post-harvest processing. Appendix Tables 3, 4, 5, 6, 7, 8 and 9 present the data compiled by Dissanayake et al. [10, 11] and then consider the proportionate burdens to be allocated to the fibre arising as a waste from seed production. Table 2 summarises the burdens arising from the fibre processing and their respective percentages.

Table 2 Proportionate environmental burdens arising from fibre processing when flax fibres are produced from the waste stream of flax seed production

Environmental burden	S1 sliver	S2 sliver	S3 sliver	S1 yarn	S2 yarn	S3 yarn
Global warming potential (GWP: kg CO_2)	1973 (16%)	2566 (14%)	14217 (63%)	6110 (37%)	6708 (29%)	18328 (68%)
Acidification potential (AP: in kg of SO_2)	ppm (0%)	ppm (0%)	0 (0%)	ppm (0%)	ppm (0%)	0 (0%)
Eutrophication potential (EP: kg of PO_4^3)	ppm (0%)	ppm (0%)	0 (0%)	ppm (0%)	ppm (0%)	0 (0%)
Human toxicity potential (HTP: kg)	ppm (0%)	ppm (0%)	0 (0%)	ppm (0%)	ppm (0%)	0 (0%)
Aquatic toxicity potential (ATP: in $m^3 \times 10^{12}$)	0 (0%)	0 (0%)	0 (0%)	0 (0%)	0 (0%)	0 (0%)
Ozone depletion potential (ODP: ng of CFC-11)	850 (10%)	6224 (26%)	0 (0%)	860 (10%)	6300 (26%)	0 (0%)
Photochemical oxidants creation potential (POCP: kg $\times 10^{-6}$ of ethylene)	1.2 (11%)	8.7 (27%)	0 (0%)	1.2 (11%)	8.7 (27%)	0 (0%)
Non-renewable/abiotic resource depletion potential (NRADP: parts per 10^{15})	Not available					

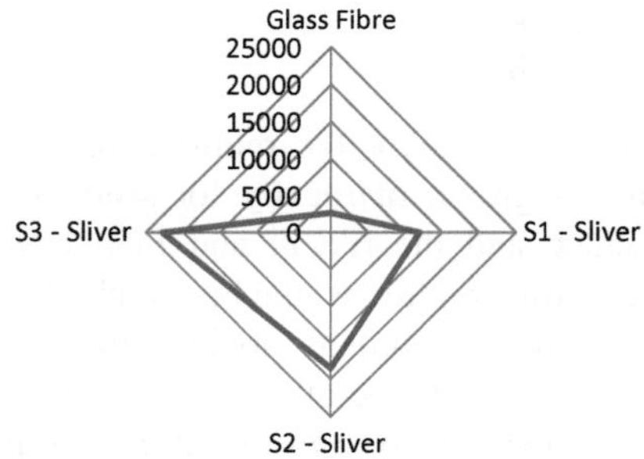

(a)

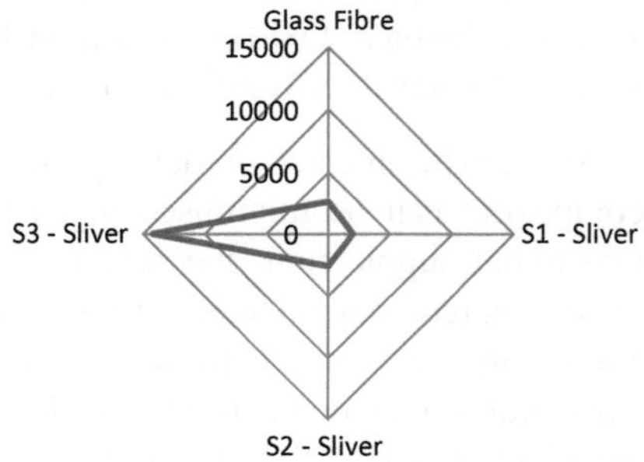

(b)

Fig. 1 Radar plot of the global warming potential (GWP) for flax fibre sliver referenced to glass fibres: **a** flax fibres as primary product (*left*), **b** flax seed as primary product (*right*)

Figures 1, 2, 3, 4, 5, 6 and 7 present "radar plots" for the EICF in Table 2. Appendix Table 11 presents LCIA results per tonne of glass fibre production derived from EcoInvent v2.0 [16].

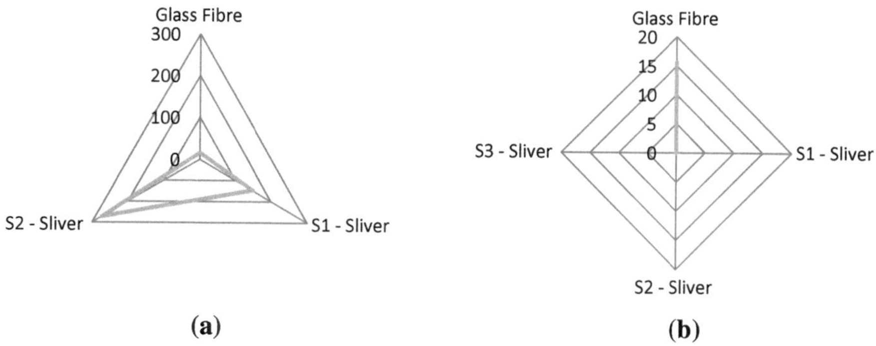

Fig. 2 Radar plot of the acidification potential (AP) for flax fibre sliver referenced to glass fibres: **a** flax fibres as primary product (*left*), **b** flax seed as primary product (*right*)

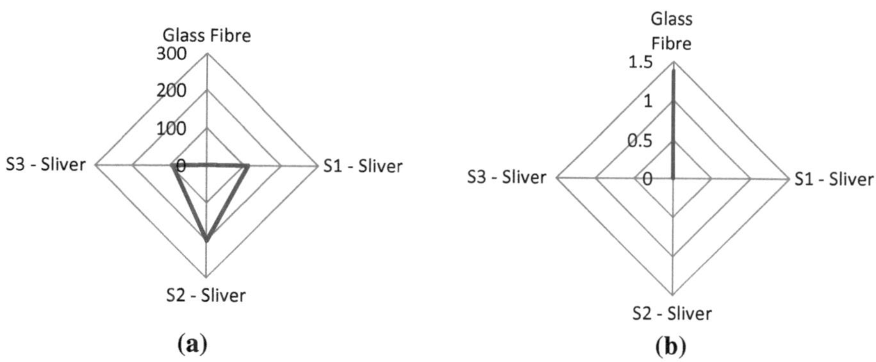

Fig. 3 Radar plot of the eutrophication potential (EP) for flax fibre sliver referenced to glass fibres: **a** flax fibres as primary product (*left*), **b** flax seed as primary product (*right*)

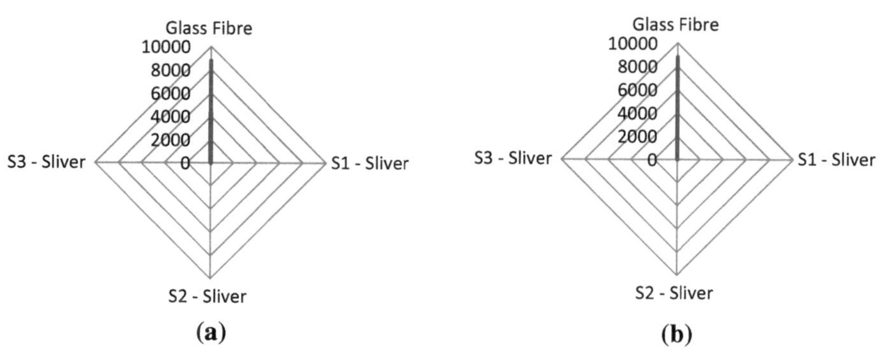

Fig. 4 Radar plot of the human toxicity potential (HTP) for flax fibre sliver referenced to glass fibres: **a** flax fibres as primary product (*left*), **b** flax seed as primary product (*right*)

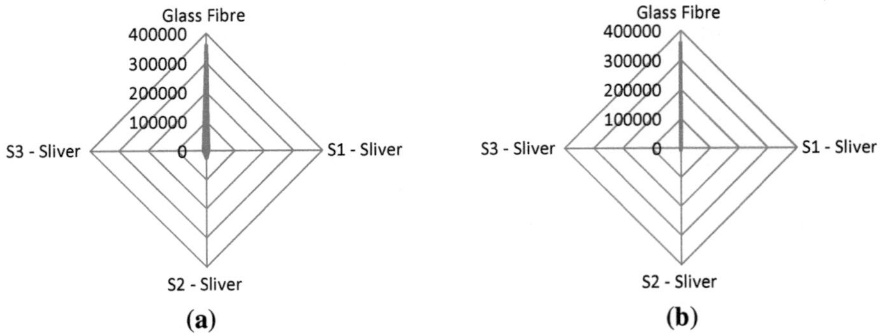

Fig. 5 Radar plot of the Ozone depletion potential (ODP) for flax fibre sliver referenced to glass fibres: **a** flax fibres as primary product (*left*), **b** flax seed as primary product (*right*)

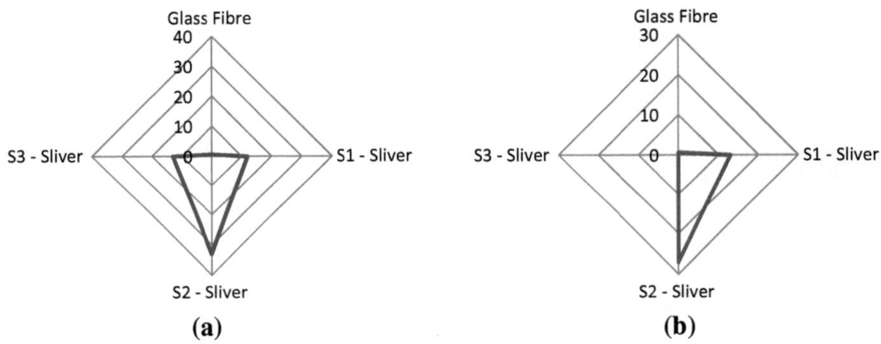

Fig. 6 Radar plot of the photochemical oxidants creation potential (POCP) for flax fibre sliver referenced to glass fibres: **a** flax fibres as primary product (*left*), **b** flax seed as primary product (*right*)

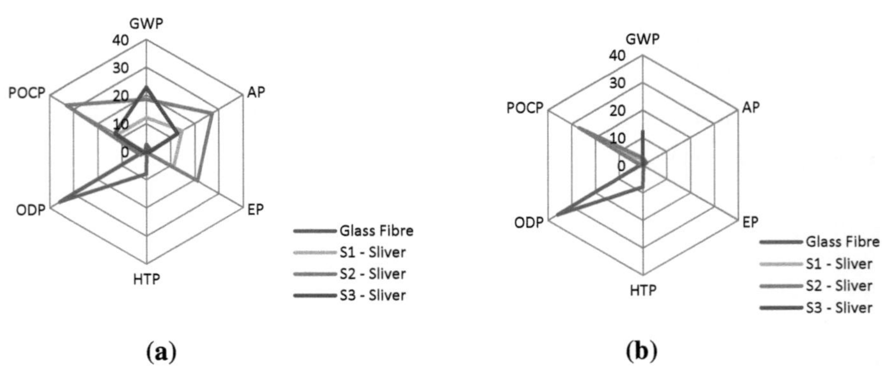

Fig. 7 Radar plot of the comparison of GWP, AP, EP, HTP, ODP and POCP for flax fibre sliver referenced to glass fibres: **a** flax fibres as primary product (*left*), **b** flax seed as primary product (*right*)

Discussion

When flax fibre is harvested at mid-point flowering (no seed produced), the global warming potentials (GWP) are similar to those for glass fibre [10, 11] for best agricultural practice (no-till then water-retting (S1: Scenario 1). However, the fibre is less "sustainable" for conservation-tillage then stand/dew-retting (S2: Scenario 2) and especially for conventional tillage then bio-retting (S3: Scenario 3). This study considers flax fibre derived as a waste product from the production of flax seed following the Ekvall and Finnveden [7] recommendation of all relevant environmental burdens assigned to the primary product. Figures 1, 2, 3, 4, 5 and 6 clearly show that assigning the environmental burdens to seed as the primary product enhances the "sustainable" credentials of flax fibre for the reinforcement of composites.

The dataset used for the analysis here is incomplete. There was no comparison data available for ATP for glass fibres and NRADP values were not available. The other 6 EICF were directly compared with glass fibres considering either flax fibres as the primary product or flax seeds as the primary product. Lignification of the flax plant during maturation of the seed after mid-point flowering will potentially increase the burdens from retting and decortication. Further, the fibre quality and properties may be compromised when sourced from older plants.

The dataset in the Appendices considers only oil, gas and coal under non-renewable/abiotic resource depletion. There is increasing concern that soil is becoming a more critical finite resource [17]. After the compilation of the dataset underlying the analysis in this paper, BS8905:2011 [18] on the sustainable use of materials identified land-use an additional factor to be analysed.

GWP for glass fibre production is lower than for flax sliver production when the flax fibres are considered to be the primary product and the impacts are allocated to the fibre. The values are comparable (slightly lower than glass fibre production, ~2635 to 1973 kg of CO_2) for S1 sliver production when the flax seeds are considered to be the primary product. The impacts arising from agrochemical intensive flax cultivation were allocated completely to the seed, not to the fibre. Nitrogen fertiliser used in crop production is the highest contributor in AP hence no values are recorded in all three secnaria for AP when the burdens are allocated for seed. The major contributors for EP are N and P fertiliser, and zero values are recorded in all three scenaria where flax seed is considered to be the primary product. The values recorded in this study for HTP and ODP for flax sliver production in both cases (flax fibre as primary product and flax seed as primary product) are negligible compared to the values obtained for glass fibre. The POCP values are dependent of the diesel consumption in crop production and retting, therefore higher values are recorded in flax sliver production than the glass fibre production.

The radar plots clearly show how the EICFs change with allocation procedures. The three scenaria considering flax seed as the primary products have improved sustainability credentials for flax fibre. S1 sliver production (no-till then water-retting) has the minimum environmental impact (GWP is lower than for glass fibre and POCP is higher than for glass fibre production). The fibre extraction processes might need improving (e.g. retting) as the mature stems are only available after seed extraction when seed is considered as the primary product. Glass fibre production clearly has ODP and HTP impacts, whereas flax fibre production has no impact on those categories. The flax sliver production has EP, AP and POCP impacts, whereas glass fibre has no impact on those categories. GWP seems to be the only impact category that could be compared with nominal values in this analysis.

The analysis need to be improved for other two impact categories (ATP and NRADP) for the comparison to be complete. The other impact categories that can be quantified and omitted from this analysis (e.g. land use, CO_2 sequestration etc.) also need to be addressed to fully understand the environmental implications in flax fibre and glass fibre production. The differences in fibre processing methods also need to be investigated when the mature stems (after seeding) are used to produce sliver.

A UN Human Rights Council [19] report states, "Reliance on hazardous pesticides is a short-term solution that undermines the rights to adequate food and health for present and future generations", and "Pesticides contaminate and degrade soil to varying degrees". In consequence, the analysis in this paper may underestimate the burdens arising from HTP, ATP and NRADP. The 8 EICF considered align with ISO/TR 14047:2003 [13], but the two toxicity burdens may not adequately address "loss of biodiversity", especially in the context of pollinators.

Conclusions

Variation is observed in the LCA study with the two different allocation methods in place. The analysis considering flax seed as the primary product has resulted in improved environmental credentials for the flax sliver production while assuming the fibre extraction and preparation methods are similar. The six EICFs were compared with glass fibre production and the GWP is the only category that could be directly compared. The values GWP for glass fibre production and S1 sliver production (no-till and water-retting) were very comparable. While glass fibre production is resulted in other environmental impacts such as HTP and ODP, flax fibre sliver production is resulted in AP, EP and POCP.

Acknowledgements JS is grateful to a series of research degree candidates (Nilmini Dissanayake, Amandeep Singh Virk, Aitor Hernandez Michelena and Hossein Mohammad Khanlou) who helped maintain his interest in natural fibre composites and to the delegates at various conferences for their insight into issues arising around LCA.

Appendix

See Tables 3, 4, 5, 6, 7, 8, 9, 10 and 11.

Table 3 Global warming potential (GWP) for the production of flax in the three scenaria

	Global warming potential (GWP) in kg of CO_2			Allocation
	Scenario 1	Scenario 2	Scenario 3	
Crop production	2.5	6.5	3.3	Seed
Agro-chemicals	10192	16077	8800	Seed
Retting	0.3	2.3	12228	Fibre
Scutching	1618	2098	1605	Fibre
Hackling	379	497	384	Fibre
Spinning	4113	4111	4111	Fibre
Sliver (pre spinning)	12045	18457	22744	
Yarn (post-spinning)	16305	22792	27131	
Sliver (post-harvest/pre spinning)[a]	1973 (16%)	2566 (14%)	14217 (63%)	
Yarn (post-harvest/post-spinning)	6110 (37%)	6708 (29%)	18328 (68%)	

[a]Sliver data decreased by 1.2% to correct for mass loss from the spinning operation

Table 4 Acidification potential (AP) for the production of flax in the three scenaria

	Acidification potential (GWP) in kg of SO_2			Allocation
	Scenario 1	Scenario 2	Scenario 3	
Crop production (diesel)	5.5×10^{-3}	14×10^{-3}	7×10^{-3}	Seed
Agro-chemical N fertiliser	142.3	268.6	125.9	Seed
Agrochemical P fertiliser	6.1	7.0	3.2	Seed
Agrochemical pesticides	24×10^{-6}	40×10^{-6}	22×10^{-6}	Seed
Retting (diesel)	640×10^{-6}	5100×10^{-6}	0	Fibre
Sliver (pre spinning)[a]	146.6	272.3	127.5	
Yarn (post-spinning)	148.4	275.6	129.1	
Sliver (post-harvest/pre spinning)[a]	0 (0%)	0 (0%)	0 (0%)	
Yarn (post-harvest/post-spinning)	0 (0%)	0 (0%)	0 (0%)	

[a]Sliver data decreased by 1.2% to correct for mass loss from the spinning operation

Table 5 Eutrophication potential (EP) for the production of flax in the three scenaria

	Eutrophication potential (EP) in kg of PO_4^{3-}			Allocation
	Scenario 1	Scenario 2	Scenario 3	
Crop production (diesel)	930×10^{-6}	2400×10^{-6}	1200×10^{-6}	Seed
Agro-chemical N fertiliser	62.4	119.8	55.2	Seed
Agrochemical P fertiliser	50.0	83.4	38.0	Seed
Retting (diesel)	110×10^{-6}	850×10^{-6}	0	Fibre
Sliver (pre spinning)[a]	111.1	200.7	92.1	
Yarn (post-spinning)	112.4	203.1	93.2	
Sliver (post-harvest/pre spinning)[a]	0 (0%)	0 (0%)	0 (0%)	
Yarn (post-harvest/post-spinning)	0 (0%)	0 (0%)	0 (0%)	

[a]Sliver data decreased by 1.2% to correct for mass loss from the spinning operation

Table 6 Human toxicity potential (HTP) for the production of flax in the three scenaria

	Human toxicity potential (HTP) in kg			Allocation
	Scenario 1	Scenario 2	Scenario 3	
Crop production (diesel)	6.2×10^{-3}	16×10^{-3}	8.3×10^{-3}	Seed
Agro-chemical N fertiliser	16.4	23.4	14.5	Seed
Agrochemical P fertiliser	10.0	11.5	5.2	Seed
Agrochemical pesticides	4900×10^{-6}	9500×10^{-6}	4400×10^{-6}	Seed
Retting (diesel)	720×10^{-6}	5700×10^{-6}	0	Fibre
Sliver (pre spinning)[a]	26.0	34.5	19.5	
Yarn (post-spinning)	26.4	34.9	19.7	
Sliver (post-harvest/pre spinning)[a]	0 (0%)	0 (0%)	0 (0%)	
Yarn (post-harvest/post-spinning)	0 (0%)	0 (0%)	0 (0%)	

[a]Sliver data decreased by 1.2% to correct for mass loss from the spinning operation

Table 7 Aquatic toxicity potential (ATP) for the production of flax in the three scenaria

	Aquatic toxicity potential (ATP) in $m^3 \times 10^{12}$			Allocation
	Scenario 1	Scenario 2	Scenario 3	
Agrochemical pesticides	1794	2067	942	Seed
Sliver (pre spinning)[a]	1772	2042	930	
Yarn (post-spinning)	0	0	0	
Sliver (post-harvest/pre spinning)[a]	0 (0%)	0 (0%)	0 (0%)	
Yarn (post-harvest/post-spinning)	0 (0%)	0 (0%)	0 (0%)	

[a]Sliver data decreased by 1.2% to correct for mass loss from the spinning operation

Table 8 Ozone depletion potential (ODP) for the production of flax in the three scenaria Units translated to nanograms for integer values

	Ozone depletion potential (ODP) in ng of CFC-11			Allocation
	Scenario 1	Scenario 2	Scenario 3	
Crop production (diesel)	7400	18000	9400	Seed
Retting (diesel)	860	6300	0	Fibre
Sliver (pre spinning)[a]	8161	24001	9287	
Yarn (post-spinning)	8260	24300	9400	
Sliver (post-harvest/pre spinning)[a]	850 (10%)	6224 (26%)	0 (0%)	
Yarn (post-harvest/post-spinning)	860 (10%)	6300 (26%)	0 (0%)	

[a]Sliver data decreased by 1.2% to correct for mass loss from the spinning operation

Table 9 Photochemical oxidants creation potential (POCP) for the production of flax in the three scenaria

	Photochemical oxidants creation potential (POCP) in kg $\times$ 10^{-6} of ethylene			Allocation
	Scenario 1	Scenario 2	Scenario 3	
Crop production (diesel)	10	24	13	Seed
Retting (diesel)	1.2	8.7	0	Fibre
Sliver (pre spinning)[a]	11.0	32.3	12.8	
Yarn (post-spinning)	11.2	32.7	13	
Sliver (post-harvest/pre spinning)[a]	13%	27%	0 (0%)	
Yarn (post-harvest/post-spinning)	10.7%	27%	0 (0%)	

[a]Sliver data decreased by 1.2% to correct for mass loss from the spinning operation

Table 10 Non-renewable/abiotic resource depletion potential (NRADP) for the production of flax in the three scenaria

	Non-renewable/abiotic resource depletion potential (NRADP: parts per 10^{15})		
	Scenario 1	Scenario 2	Scenario 3
Sliver			
Coal	1.0	1.3	6.8
Gas	600	800	4400
Oil	2000	4500	7400
Yarn			
Coal	3.1	3.4	9.2
Gas	2000	2200	5900
Oil	3900	6500	9600

Table 11 The LCIA results from EcoInvent v2.0 per tonne of glass fibre production

EICF	LCIA—Results from EcoInvent v2.0
GWP (kg CO_2-Eq)	2634.5
AP (kg SO_2-Eq)	15.66
EP (kg PO_4-Eq)	1.37
HTP (kg1, 4-DCB-Eq)	8763.7
ODP (kg-CFC11-Eq)	0.00036
POCP (kg ethylene-Eq)	0.59

References

1. Dissanayake, N. P. J., & Summerscales, J. (2014). Life cycle assessment for natural fibre composites, Chapter 8. In Thakur, V. K. (Ed.), *Green composites from natural resources* (pp. 157–186). USA: Taylor and Francis Group LLC. ISBN 978-1-4665-7069-6.
2. Summerscales, J., Dissanayake, N., Hall, W., & Virk, A. S. (2010). A review of bast fibres and their composites. Part 1: fibres as reinforcements. *Composites Part A Applied Science and Manufacturing, 41*(10), 1329–1335.
3. Summerscales, J., Dissanayake, N., Hall, W., & Virk, A. S. (2010). A review of bast fibres and their composites. Part 2: composites. *Composites Part A Applied Science and Manufacturing, 41*(10), 1336–1344.
4. Summerscales, J., Virk, A. S., & Hall, W. (2013). A review of bast fibres and their composites. Part 3: Modelling. *Composites Part A Applied Science and Manufacturing, 44*(1), 132–139.
5. Summerscales, J., & Grove, S. (2014) Manufacturing methods for natural fibre composites, Chapter 7. In Hodzic, A., & Shanks, R. (Eds.), *Natural fibre composites: Materials, processes and properties* (pp. 176–215). Cambridge: Woodhead Publishing. ISBN 978-0-85709-524-4 (print). ISBN 978-0-85709-922-8 (online).
6. Summerscales, J., & Gwinnett, C. (2017). *Forensic identification of bast fibres*. In Roy, D. (Eds.), *The industrial development of high-performance biocomposites* (pp. 125–164). Elsevier/Woodhead (in press). ISBN 978-0-08-100793-8.
7. Ekvall, T., & Finnveden, G. (2001). Allocation in ISO 14041—A critical review. *Journal of Cleaner Production, 9*(3), 197–208.
8. ISO 14040:2006. (2006). *Environmental management—Life cycle assessment—Principles and frameworks*. Geneva: International Organization for Standardization, July 31, 2006 (reviewed and confirmed in 2016). ISBN 0-580-48992-2.
9. ISO 14044:2006. (2006). *Environment management—Life cycle assessment—Requirements and guidelines*. Geneva: International Organization for Standardization, July 31, 2006 (reviewed and confirmed in 2016). ISBN 0-580-48677-X.
10. Dissanayake, N. P. J., Summerscales, J., Grove, S. M., & Singh, M. M. (2009). Energy use in the production of flax fiber for the reinforcement of composites. *Journal of Natural Fibers, 6* (4), 331–346.
11. Dissanayake, N. P. J., Summerscales, J., Grove, S. M., & Singh, M. M. (2009). Life cycle impact assessment of flax fibre for the reinforcement of composites. *Journal of Biobased Materials and Bioenergy, 3*(3), 245–248.
12. Le Duigou, A., Davies, P., & Baley, C. (2011). Environmental impact analysis of the production of flax fibres to be used as composite material reinforcement. *Journal of Biobased Materials and Bioenergy, 5,* 153–165.

13. ISO/TR 14047:2003(E). (2003). Environmental management—Life cycle impact assessment—Examples of application of ISO14042. Geneva: International Organization for Standardization, December 11, 2003 (revised by ISO/TR 14047:2012). ISBN 0-580-43112-6.
14. Azapagic, A., Emsley, A., & Hamerton, I. (2003). *Polymers, the environment and sustainable development*. Wiley. ISBN 0-471-87741-7.
15. Azapagic, A., Perdan, S., & Clift, R. (Eds.). (2004). *Sustainable development in practice—Case studies for engineers and scientists*. Wiley, May 2004. ISBN 0-470-85609-2. Second edition, 2011: ISBN 978-0-470-71872-8.
16. Ecoinvent Centre: Swiss Centre for Life Cycle Inventories. (2010). May, 24, 2010. Available from: http://www.ecoinvent.org/database/.
17. Montgomery, D. R. (2012). *Dirt; The erosion of civilizations*. Berkeley & Los Angeles CA: University of California Press April 2012. ISBN 978-0-520-27290-3.
18. BS 8905:2011. (2011). *Framework for the assessment of the sustainable use of materials*. London: Guidance, British Standards Institution.
19. United Nations (UN) Human Rights Council, Report of the Special Rapporteur on the right to food. (2017). UN General Assembly Report A/HRC/34/48, January 24, 2017.

Applications of Building Insulation Products Based on Natural Wool and Hemp Fibers

Lorenzo Savio, Daniela Bosia, Alessia Patrucco, Roberto Pennacchio, Gabriele Piccablotto and Francesca Thiebat

Abstract FITNESs, Fibre Tessili Naturali per l'Edilizia Sostenibile (Natural Textile Fibers for Sustainable Building), is a research project concerning an experimental hemp and sheep wool insulation panel. The new panel has two main innovative features: unlike the already existing hemp and wool insulation mats, it is a semi-rigid product and has low environmental impact, as shown by the Life Cycle Assessment. FITNESs panels are particularly suitable for eco-building sector, as they are 100% natural, recyclable and made with by-products from local production chains (Piemonte Region). The paper presents the production process of the panel from wool and hemp fibers and some experimental applications for sustainable architecture.

Keywords Hemp fibers · Sheep wool · Recycled materials · Natural building components · LCA

Introduction

FITNESs experimental research project realized in 2016 that lead to 50% recycled sheep wool and 50% technical hemp fibers, has outlined panels high thermal insulating and sound absorption performance [1].

Both fibers used for panels production have quite a low economic value on the market: used wool are recycled fibers coming from Piemonte regional sheep breeding not used in textile industry because of its dark color and poor quality [2], while Short Technical hemp fiber used is not meant for textile manufacture and is currently quite hard to place on the market (Assocanapa).

L. Savio (✉) · D. Bosia · R. Pennacchio · G. Piccablotto · F. Thiebat
DAD—Dipartimento di Architettura e Design, Politecnico di Torino, Viale Mattioli 39, 10125 Turin, Italy
e-mail: lorenzo.savio@polito.it

A. Patrucco
CNR ISMAC, Corso G. Pella 16, 13900 Biella, Italy

R. Fangueiro and S. Rana (eds.), *Advances in Natural Fibre Composites*,
https://doi.org/10.1007/978-3-319-64641-1_20

Interest in hemp as a natural component for panels production relies in its cultivation deep tradition in Italy, especially in Piemonte region, since medieval age, and from the recent hemp based building component development and sustainable construction materials market high interest.

During the research project an innovative and fully natural production method has been developed. The method allows to exploit keratin presence inside sheep wool fibers, in order to keep wool and hemp fibers linked together in a quite high density mixture.

LCA studies developed, also from research group previous works [3], in order to assess the related environmental impacts, underlined how washing wool process has great impact on the whole panels production process being not a completely industrialized process.

In the present work a scaling up scenario for FITNESs panels production was proposed, in order to evaluate the possibility to sensibly lower sheep wool fibers washing dependant costs. The possibility to merge panels production with raw materials washing and cleaning operations needed, was explored. The planned scenario also represents an opportunity to give back life to recently abandoned wool washing plants in Piemonte region, due to the recent economic crisis.

Moreover some application scenarios for the new wool-hemp panel, suitable to its experimentally measured characteristics, have been defined. The growing interest in dry building technologies market led to investigate possible installation solutions and fixing methods both in dry newly constructed building technologies both in energy retrofit solutions for traditional envelope technologies. This last scenario, on the other hand, remains a topic of great interest, particularly in our country, which, in the last fifty years has seen a quite important land consumption due to building industry development.

Assumed wall stratigraphy technology solutions thermal performances, in dynamic and steady state conditions, have been also investigated in order to verify panels effectiveness and thickness needed to accomplish with Italian legislation.

FITNESs Panels Production Process

Row Material Supply

The primary role of EU sheep farming is meat and milk production; sheep breeds for milk and meat generally produce low quality wool (coarse wool) not suitable for textile uses.

Therefore, the EU wool is generally a by-product of little value which is often illegally buried or burnt.

Similarly, hemp fibers became a by-product when hemp is cultivated for oil and seeds production, since it is a multi-purpose crop, producing fibers, shives, seeds and pharmaceuticals.

Fig. 1 Bio-composites specimen of wool and hemp board (10 × 10 cm)

The multi-purpose cultivation of hemp requires plants' ripening, which increases the lignin amount in the plant stem and negatively affects retting, scotching and the subsequent textile processing, such as spinning and weaving. Moreover, dew retting (on field) and water retting are not environmentally sustainable and chemical retting is too expensive for such a low quality, "technical" fibers. Nevertheless, hemp fibers show high tenacity and elasticity and low specific weight to be highly suitable as composites reinforcement material in place of artificial or man-made fibers.

Valorization of coarse wool wastes and technical hemp can be carried out in non-conventional textile application such as bio-building and eco-design where materials recycling and process sustainability are take into great consideration.

In this work, an innovative process was developed treating wool and hemp fibers with alkali with the aim to produce a totally bio-based composite for thermal and acoustic insulation.

Alkali treatments effectively swell hemp fibers, improving their hydrophilic behavior and mechanical properties, and partially hydrolyzing the wool protein material (keratin) which releases a sort of keratin-based resin able to stick the fibers each other (Fig. 1). In the resulting novel bio-composite, ligno-cellulosic fibers and the remaining wool, work as the reinforcement of the hydrolyzed keratin matrix.

In this study, coarse wool in the form of loose fibers (obtained from raw wool scouring) was supplied by Davifil Srl, while hemp technical fibers were supplied by Assocanapa Srl [4]. All chemicals were of analytical grade and purchased from Sigma-Aldrich, except otherwise specified.

Production of Specimens

Wool and hemp fibers were cut into snippets 2 cm length, blended and bathed in NaOH solutions ranging from 0.25 to 0.5 N with temperature from 50 to 80 °C and liquor ratio 1:20.

The proportion between hemp fibers ranged from 15 to 70% and the treatment time was from 15 to 120 min. The resulting material was neutralized with sulfuric acid until 7 pH, washed and then placed in a plastic container (10 × 10 cm) to be dried at 70 °C (Fig. 2).

The final density and performances of the resulting composite materials varied depending on the process conditions.

Production of Insulation Panels

Insulation panels (90 × 50 × 4.5 cm) were produced scaling-up the laboratory methodology with the aim of reducing the number of steps, water and energy consumption.

The best compromise between final density and the functional properties described below, was obtained with the following operative conditions.

Wool and hemp fibers (50:50) were bathed in 0.25 N NaOH, liquor ratio 1:20, at 45 °C for 3 h, then squeezed and dried without neutralization to reduce water consumption.

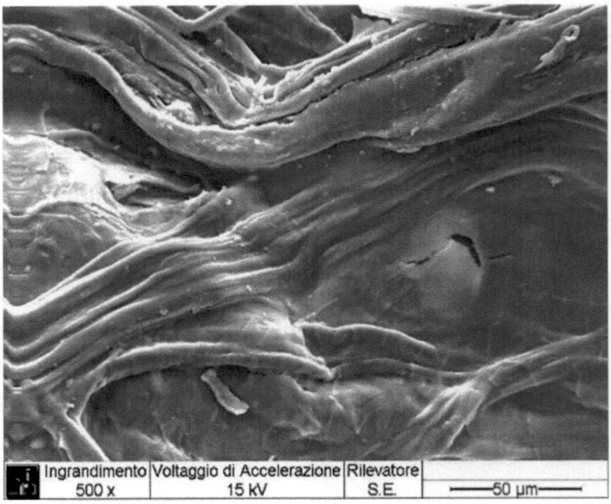

Fig. 2 SEM image of hemp fibers stuck by keratin-based matrix (500×)

Scale-Up Scenario

The valorization of agriculture waste represents an attempt to drive the building sector toward a circular economy goal. At the same time, industries that have lost their original function with the economic crisis, may regenerate by processing of waste materials.

In the area of Biella, in Piemonte, several wool scouring plants are mostly abandoned due to the wool washing process displacement in other countries, where environmental standards are more permissive and manufacturing is less expensive, and could fit to this purpose.

Thus, providing a scale-up scenario from laboratory to industrial production, the wool scouring might be realized along with the production of the wool- hemp panels in the same factory. Both the processes, in fact, make use the same machinery: opening, dusting and willowing machines for wool; cold and hot water tanks for washing and treatments; an oven to dry wool as well as the bio-composite panels.

Our research aims to join the life cycle approach with the circular economy using a "cradle to cradle" strategy. Figure 3 shows the energy/environmental flow-chart

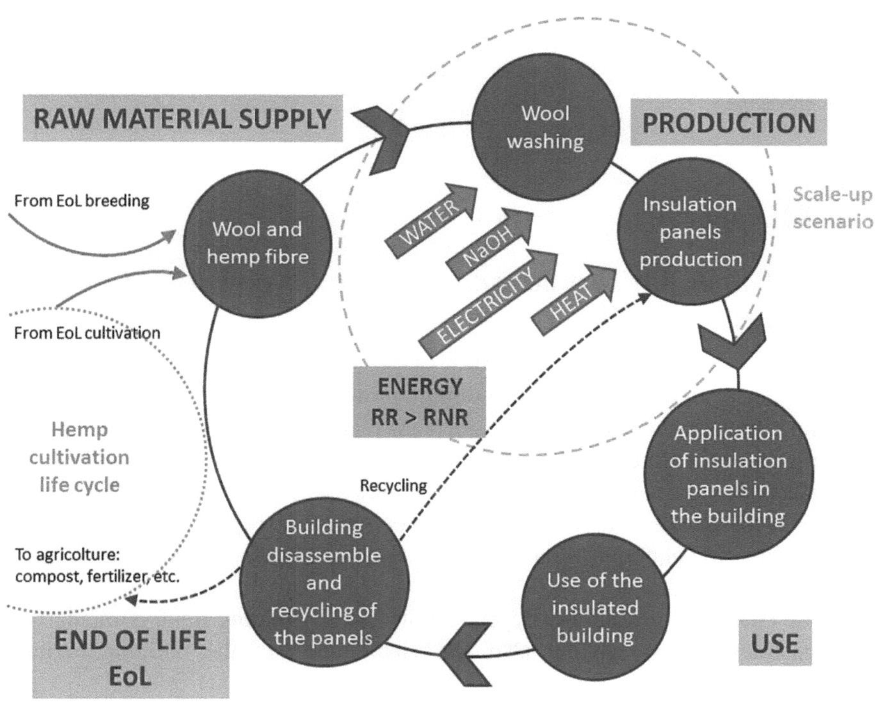

Fig. 3 A "cradle to cradle" life cycle of the bio-composite insulation panels with an hypothesis of an industrial scenario (*dotted circle*)

based on the hypothesis of an industrial production for building insulation panels, including both the construction/use phase and the end of life scenarios.

The study includes an environmental analysis based on the LCA following the standard ISO 14040/44 as reported in Pennacchio et al. Inventory data were collected directly from the industrial partners of the project and were elaborated with Ecoinvent 2.0 using the Cumulative Energy Demand as indicator for the assessment of the environmental impacts.

The functional unit is the mass necessary for 1 m^2 to reach the thermal Resistance of 2.5 m^2K/W.

Analyzing the results, in particularly, the non-renewable energy (RNR) demand values, they are lower if compared with other insulation materials with same characteristics, such as rockwool or glasswool. Furthermore, each manufacture process have been kept separate in order to define the more impactful stage. specifically, raw materials supply weight 13.95 MJeq, their transport at the production site 24.95 MJeq, the insulation panels manufacturing weight 147.35 MJeq and the packaging phase 16.77. Total non-renewable energy required to produce 1 m^2 of wool and hemp panel to reach the thermal Resistance of 2.5 m^2K/W is 202.64 MJeq.

Then, imaging the reuse of an old scouring plant for the panels manufacturing in an industrial scenario, the 147.35 MJeq associated to the production life cycle phase that represent almost 60% of the total energy demand, can be reduced through the use of renewable energy (RR). The flow-chart in Fig. 3 shows the whole life cycle of a building, where a dotted circle highlights the industrial scenario processes.

Panels' Application Scenarios

Fitness Panels for Thermal and Acoustic Insulation of an External Wall

During panels production process, keratin protein while slowly dissolving and spreading through wool and hemp fibers, tends to settle at the bottom of the bathtub, giving particular stiffness to panels bottom face.

Thanks to its stiffness and high density, unlike sheep wool panels at the moment present on the market, FITNESs is able to support its own weight, and has shown good workability; although certain attention in cutting operations is required due to hemp fibers strength.

These characteristics allows large use potential in building field, particularly in dry construction systems, but also for rehabilitation of the building heritage, through energy retrofit.

In Fig. 4, panels have been suggested as an infill in a common wooden frame wall structure, and as a further insulating coat, supported by a light steel profiles

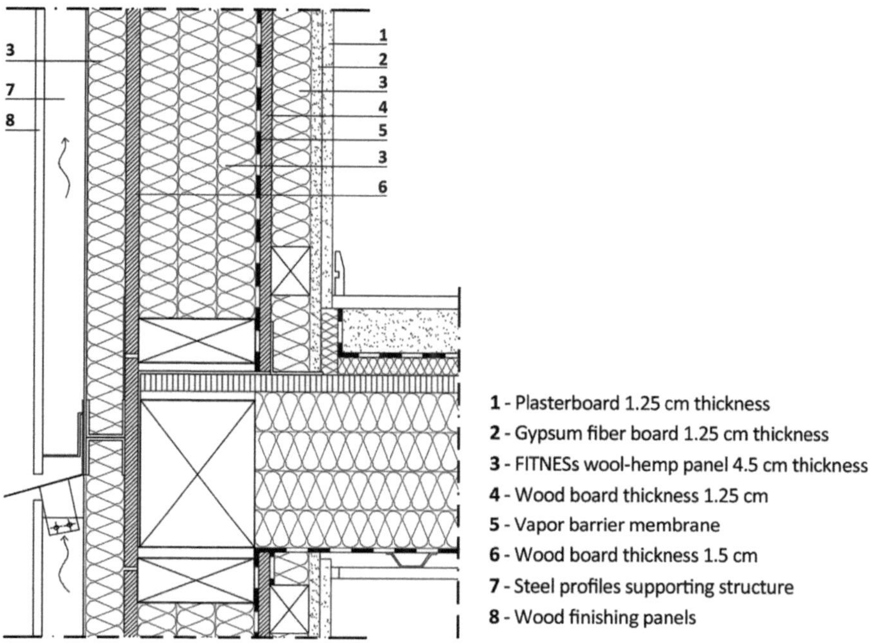

Fig. 4 Ventilated Wooden frame wall stratigraphy, dry assembled using FITNESs panels as insulating infill and coating

structure without any further dowelling need, protected by a ventilated wall. FITNESs suitability in dry building system has already been proved during a demonstration building wall assembly, at Environment park, Politecnico di Torino [1].

Moreover a dry technology system like the presented kind, fully disassemblable, permit to eventually recover and reuse panels, or easily dispose of it, as 100% natural and compostable.

FITNESs also it's a natural and eco-friendly alternative to synthetic insulating coating for building heritage energy retrofit. In Fig. 5 a hollow bricks wall FITNESs coating, in a ventilated facade system is shown; in this case insulating panels can be fixed to the existing hollow bricks wall, using proper dowels, as for common synthetic insulation coatings solutions, paying attention to protect air exposed panel face with a protective membrane film.

Assumed wall stratigraphies dynamics and stationary thermal performances have been calculated, in accordance with UNI EN ISO 13786 [5], both in the new wood-frame structure case and in the existing traditional wall retrofit scenario. FITNESs panels bulk density and thermal conductivity used in the have been obtained from experimental measurements in laboratory, while assumed specific heat was calculated by averaging sheep wool and hemp fibers value, very similar in any case.

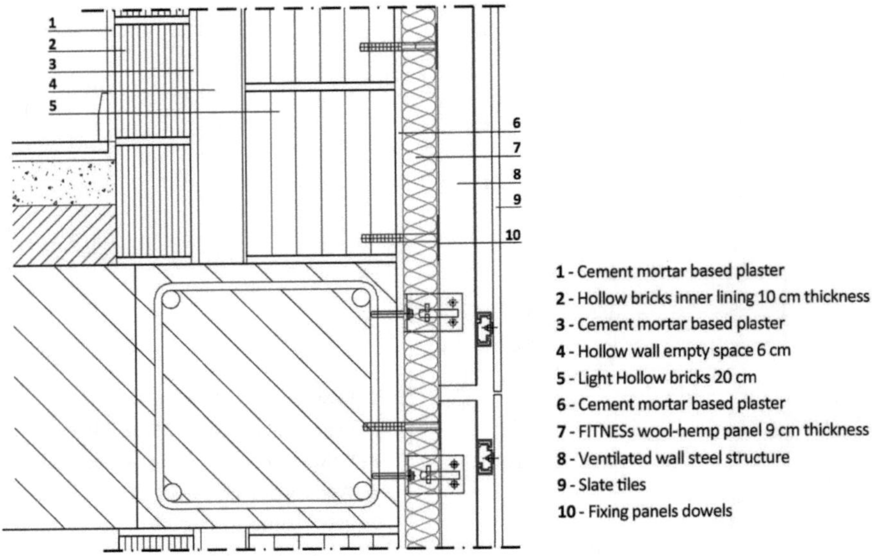

Fig. 5 Energy retrofit of a hollow wall proposing FITNESs panels as insulating coating in a ventilated wall system

Table 1 Wood-frame structure and hollow brick wall retrofit with FITNESs panels thermal performances

Wall samples	s (cm)	f_a (–)	φ (h)	Y_{ie} (W/m^2 K)	U (W/m^2 K)	C (W/m^2 K)
Wood-frame structure	34	0.08	16.22	0.013	0.16	0.17
Hollow brick wall Retrofit	57	0.08	15.53	0.025	0.30	0.30

Results are shown in Table 1 and outline wool-hemp panels effectiveness as an insulating material. Italian legislation requirements for thermal performance of newly constructed buildings are easily met, also with an interesting time lag performance, probably due to the relative insulating panels high density, that suggests a good thermal performance also in hot summer time.

On the other hand, results show how thermal transmittance values required by Italian legislation in Piemonte region, in case of building retrofit [6], can be reached with a not so important 9 cm overall insulating panels thickness.

New wool-hemp insulating panels allows to improve thermal and acoustic performance of traditional or ancient envelope technology, with the not negligible advantage to could be dry mounted, without the need of mortars or adhesives.

Fitness Panels for Internal Finishes: Walls and Ceiling

Nowadays, in Italy, existing building stock refurbishment is more important than new constructions, in terms of market and interventions number. Often, the building refurbishment requires not only energy retrofit interventions, with the external building envelope extra insulation, but also a deep renovation of the interiors, making them suitable for new functions and/or integrating new technological equipment. Considering the improvement of indoor environmental quality, a key point in existing building refurbishment, the research group designed a specific application for the Fitness panels as finishes for interior wall and ceilings.

A typical intervention was considered (Fig. 6): in an existing building with "traditional" interior partition walls, made with 8 cm thick bricks and plaster on both sides, a wallboard covering and a ceiling are realized in order to integrate new equipment and to improve the thermal and acoustic performances. In this application, Fitness panels surface is always protected, but it is in direct contact with the indoor air:

- for the wall, to a height of about one meter, panels are protected by a perforated sheet;
- for the wall, up to one meter height, and for the ceiling, panels are covered by an open-weave acoustically-transparent fabric.

Panels workability has been already tested by the research group during the construction of the test-wall for the thermal performance measurement in real conditions [1] and it makes the panels suitable for that specific application. Panels are stiff, but also flexible enough to be clamped between metal uprights, as shown in Fig. 6. Protection with the acoustic fabric or with the perforated sheet (made with wood or metal) is necessary to avoid material degradation and panel superficial scratching. The application of Fitness panels in interior finishes allows to improve the acoustic absorption thanks to the wall and ceiling absorbing surface.

Thanks to the additional insulation, the thermal transmittance of the interior partition decrease from 1.972 to 0.613 W/m^2K, under the limit defined by the national regulations for internal partitions between different thermal zones with different set-point temperatures or heating timing. The acoustic performance of the acoustic-fabric coated panels were tested by measuring the sound absorption coefficient in reverberation room [1]. From the acoustic absorption point of view, Fitness panels can be compared with other fibrous acoustic products, with a sound absorption coefficient $\alpha_w = 0.75$ (MH).

Furthermore, the designed application should improve the interior aesthetic quality, choosing between different kinds of perforated cladding and acoustic fabric. However, other two innovative features should be considered in this specific application:

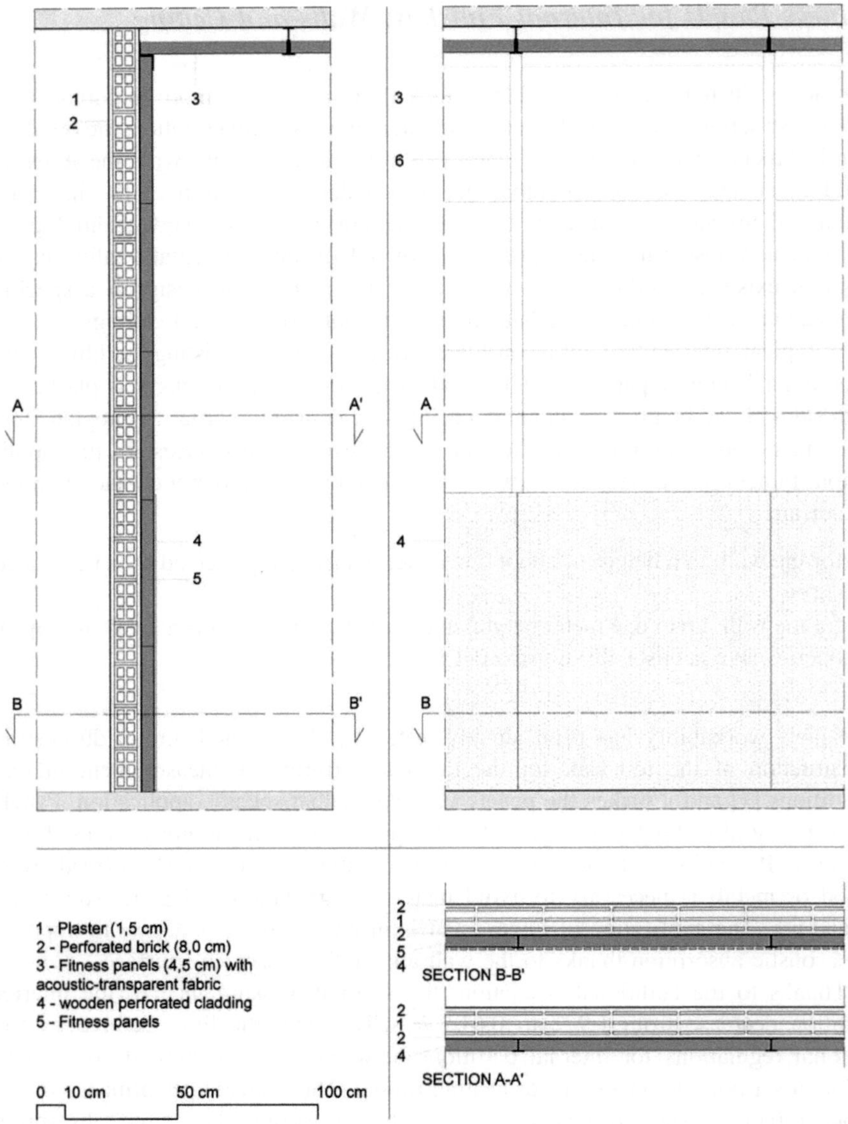

Fig. 6 Application of FITNESs panel for the insulation of an internal partition (between different thermal zones) and for the insulation of the ceiling

- wool fibers can absorb the formaldehyde, one of the most important indoor air pollutant. So, the application of Fitness panels for interior finishes can contribute to the indoor environmental quality improvement;
- wool fibers are strongly hygroscopic. Being directly in contact with the indoor air, Fitness panels may have a thermo-hygrometric regulation function, absorbing or giving back water vapor to ambient air, restoring more adequate air relative humidity levels.

Conclusions

The paper described how the production of FITNESs panels is made only with completely natural raw materials from agriculture waste and a low environmental impact production process.

Opportunity to integrate panels production in an existing industrial process, allowing a possible panels industrial production scaling up, have been investigated.

This allowed us to hypothesize a complete life cycle including panels usage and end life phases, where panels are recycled and employed again in the agricultural field as a fertilizer [7], in a cradle to cradle strategy (Fig. 3).

Furthermore, panel proved its suitability as an alternative to synthetic and higher environmental impact products, in most common technology dry and humid systems for new buildings construction as for existing building heritage retrofit.

In simulated applications has been demonstrated how, with limited thickness, FITNESs panels are able to meet Italian legislation envelope thermal insulation requirements. FITNESs panels thermal insulation performances, associated to its low environmental impact, make them particularly suitable also for zero energy building (nZEB) design. With respect to indoor applications the prospect to evaluate panels formaldehyde absorption and indoor air moisture regulation performance is of particular interest, as far as experimentation in larger and more complex demonstrative buildings.

References

1. Pennacchio, R., Savio, L., Bosia, D., Thiebat, F., Piccablotto, G., Patrucco, A., Fantucci, S. (2017). Fitness: Sheep-wool and hemp sustainable insulation panels. In: *Energy Procedia, 111*, 287–297. ISSN:1876-6102.
2. Tonin, C., Patrucco, A., Ramella Pollone, F., Bosia, D., Savio, L., & Giordano, R. (2011). *Processo di lavorazione della lana, materiali di lana prodotti con detto processo e articoli comprendenti detti materiali di lana*. Italian Patent 0001410156 September 5th 2014GE2012A000029.
3. Bosia, D., Savio, L., Thiebat, F., Patrucco, A., Fantucci, S., Piccablotto, G., et al. (2015). Sheep wool for sustainable architecture. *Energy Procedia, 2015*(78), 315–320. doi:10.1016/j.egypro.2015.11.650.
4. AssoCanapa; www.assocanapa.org.
5. UNI EN ISO 13786:2008. *Thermal performance of building components—Dynamic thermal characteristics—Calculation methods*.
6. DGR 46-11968. Deliberazione della Giunta Regionale 4 agosto 2009, n.46-11968. *Bollettino Ufficiale Regione Piemonte Parte I-II, 4° supplemento al numero 31-7 agosto 2009*.
7. Zoccola, M., Montarsolo, A., Mossotti, R., Patrucco, A., & Tonin, C. (2015, October). Green hydrolysis as an emerging technology to turn wool waste into organic nitrogen fertilizer. *Waste Biomass Valor Waste and Biomass Valorization, 6*(5), 891–897.

Hemp-Clay Concretes for Environmental Building—Features that Attribute to Drying, Stabilization with Lime, Water Uptake and Mechanical Strength

Monika Brümmer, Mª Paz Sáez-Pérez and Jorge Durán Suárez

Abstract There are numerous parameters which define the mechanical properties of hemp concretes. Several authors identified significant differences in mechanical properties due to the quality, particle size, form and distribution of the hemp aggregates, and also due to the energy of compaction, binder dosage, density, drying conditions and drying time. The other factors which influence the mechanical properties of hemp concretes are the type of binders and their influence on the setting process of mortars containing plant aggregates. This research work has been done to encourage the use of hemp in construction industries that can reduce the usage of industrial binders and shift the focus on natural binders. The experiments were carried out to observe the drying kinetics, capillary water uptake and mechanical strength of hemp-clay concretes made with de-fibered or non-de-fibered hemp materials of industrial and domestic origin. The influence of addition of the lime on the mechanical performance of hemp-clay composites was also studied.

Keywords Hemp · Concrete · Hurds · Fiber · Waste · Lime

M. Brümmer
Ph.D. Program in History and Arts (Territory, Heritage and Environment), University of Granada, Cannabric, Cañada Ojeda 8, 18500 Guadix, Spain
e-mail: mbrummer@correo.ugr.es; cannabric@cannabric.com

M.P. Sáez-Pérez (✉)
Department of Architectonic Constructions, Advanced Technical School of Building Engineering, University of Granada, Avenue Severo Ochoa, Fuentenueva Campus, 18071 Granada, Spain
e-mail: mpsaez@ugr.es

J.D. Suárez
Department of Sculpture, Faculty of Fine Arts, University of Granada, Granada, Spain
e-mail: giorgio@ugr.es

R. Fangueiro and S. Rana (eds.), *Advances in Natural Fibre Composites*,
https://doi.org/10.1007/978-3-319-64641-1_21

Introduction

Bio-based materials from rapidly renewable resources provide the opportunity to create a new generation of low energy and carbon storing building products [6] with improved insulating properties. In this manner they are reducing greenhouse gas emissions caused by human activities [20] and replacing natural resources made from mineral aggregates [27]. It is widely assumed that renewable materials are environmentally beneficial and should be preferentially used in building envelopes to reduce the energy needed for manufacturing processes and air conditioning in the building sector, which is responsible for about 50% of the energy consumption worldwide [7]. The economic production and processing of natural fibers [9] is an added motive for the increasing research in this field in the last decade [27]. Hemp (*Cannabis sativa* L.) stands for a fast growing [17, 24] and multi-beneficial [2, 11, 29, 33] fiber plant. Its world-wide spread as one of the earliest culture plants [2, 29] is also explained by its easy acclimation to different climates, altitudes and soil types and its important role in crop rotations [2, 33, 34]. The use of plant aggregates in concretes and mortars is a method that allows versatile applications in construction, and practiced since antiquity to improve or modify their qualities [4, 5]. To ensure their major appraisal and use, building-product and green building designers must understand hemp based building materials in terms of their thermal, acoustic, hygrothermal and mechanical performances, in particular the links between these features. Hemp concretes, based on hemp hurds and lime binders, are proving since the early years 90s to be a user friendly, high performance and fire-resistant construction material for light weight applications in new building and renovation works [8]. However, hemp concretes have an elastic-plastic behavior and are therefore, reported to function with a support structure [10]. A lower mechanical strength of plant based concretes as compared to the mineral ones is considered as the major drawback for their use in construction [27]. An interesting challenge is, therefore, the refinement of the mechanical performance of plant based concretes for structural purposes. By now, it is demonstrated that the compressive strength of hemp concretes, with unchanged hemp content, can be increased by compaction [28], through further reduction of void volume by adding small sized aggregate or by shortening the mean particle length [1, 32]. As reported by several authors, adding fibers could not improve significantly the mechanical strength of hemp concrete formulations with lime, cement or gypsum [15, 26] but could be an interesting option for regions without facilities to separate stem materials [14]. However, very few studies are carried out on this matter, and therefore, the present research investigates the mechanical properties of hemp concrete, comparing the performance of de-fibered and non de-fibered stem material in a lime stabilized hemp-clay concrete matrix. Improving mechanical strength of hemp concretes by augmenting the dosage of lime or other calcined binders is possible [26] but it is not favorable, as the highest negative contribution to their climate change indicator is caused by commercial binders [28]. These materials consume 49% primary energy and 68% water of the total consumption of a hemp concrete and cause 47% of corresponding air pollution, followed by the wooden structure, plasters and hemp hurds [20]. Furthermore, in plant based concretes, setting problems occur [16] through the release of polysaccharides [18, 28], that hinder the carbonation of aerial lime; while in mortars with mineral aggregates, up to 90% of the CO_2 emitted during its calcination is re-absorbed [21]. A design that takes into account the local raw clay binders with the lowest possible environmental impact [30] and involving minimum transport is decisive for the final carbon footprint of a hemp concrete. It also takes into account the dosage, constructive systems (e.g. in harmony with traditional and climate adapted building) and application technology [19]. The motive of this research work is the development of hemp concretes for alternative economic development, based on the local hemp waste material of an ancestral hemp variety found in the Moroccan Central Rif, and historically harvested for industrial as well as recreational purposes. The use of both *C. sativa* L. and *C. sativa* L. - *indica* species as fiber hemp is undertaken by numerous archeological [11, 22] and written [2] affirmations. Due to technological obstacles, plant's specific morphological reasons and chemical differences between hemp hurds and fibers [25], whole stem utilization has been taken into account and compared with de-fibered stem material. The present study also aims to reveal if the plant infusions of both species have influence on the drying process of clay based mortars and on the interactions between clay and lime.

Materials and Methods

Hemp Materials

There exist significant variations in wood anatomy, wood cell distribution, cell wall thickness and cell dimension between *Cannabis sativa* L. and *Cannabis sativa* L. - *indica* species [2, 3]. Also, the chemical composition of bark differs from the woody stem material and hemi-cellulose is more abundant in the hemp core (31%) than in the bark (13%) [34]. In addition, the absorption capacity is more in the core than in the bark [23] and water and alkali soluble components increase with the maturity of the plant [13]. The present research focuses on both, industrial *Cannabis sativa* L. hemp, collected in Cervera del Rio Alhama (Spain) and ancestral *Cannabis sativa* L. - *indica* hemp (Fig. 1), collected in Morocco's "historic hemp farming region" [12]. Absorption of 5 g sample from each species of hemp core and hemp fiber was observed in the intervals of 72 h, 7 days and 14 days, and for studying the drying kinetics, the total water loss after drying in T20 and HR80 conditions was recorded. The plant materials were prepared by manual chopping. In addition, plant infusions were produced by submersion for 14 days, for the elaboration of lime stabilized earth mortars. Furthermore, hemp-clay concretes were produced to understand the differences in the use of industrial hemp hurds from Cavac (France), with a mean particle length of

Fig. 1 Moroccan hemp stalks of "beldiya"

2–25 mm (Fig. 2) and whole stem material from Cervera del Rio Alhama (Spain), cut parallel to the stem section in pieces of similar length (Fig. 3), with a chopping machine used for agricultural waste.

Binder and Stabilizer

For the preparation of mortar samples, earth was collected from the same region where the ancestral *C. sativa* L. - *indica* hemp materials were collected. For the analysis of its clay minerals, the fractions smaller than 2 μm have been separated after decanting the suspension of 200 g earth in 500 ml of water for 8 h. Liquids, obtained from the upper 10 cm of the column of liquid with particles, were collected and centrifuged for 4 min at 6000 revolutions. Mineralogical X-ray diffraction (XRD) analysis was done with a PANalytical X'Pert Pro diffractometer, equipped with a X'Celerator solid state detector, and using Cu-Kα radiation at 45 kV and 40 mA. The diffraction patterns were obtained by continuous sweeping between 3 and 50° of 2θ, with a space of 0.01° of 2θ and 20 s of time in each step. Its clay minerals (fractions of 2 μm) contain 18.1% quartz, 44.1% illite, 22% chlorite and 15.8% amorphous particles. Illite is a clay mineral formed by alteration of feldspars. It is expensive and of medium plasticity and have unctuous touch and low permeability. Illite from Issaguen is sodic and it dissolves rapidly in water. Its red color is associated with iron oxides. Its structure is mica type, but it has less

Fig. 2 Industrial hemp hurds (France)

Fig. 3 Chopped hemp stem (Spain)

potassium than muscovite, more inter-laminar water, and less substitution of aluminum for silica in the tetrahedral layer. Chlorite is found in nature combined with illites. It is not very expensive, has less cut resistance, and is difficult to disperse in water. This clay has tetrahedral-octahedral-tetrahedral (TOT) packages. For the stabilization of the clay as reinforcement of the inter-laminar matrix to resist the weathering conditions, white natural hydraulic lime NHL3.5 "Calcia" from Socli (France) was used and showed superior results in a previous research with different hemp-lime mortars [4].

Experimental Program with Clay Mortars and Plant Infusions

Infusions from the 14-day submersion of two hemp varieties (Figs. 4 and 5) have been used to prepare earth mortars with the addition of NHL3.5 lime.

In addition, a control mold was also made, and kneaded with water. The hemp infusions were obtained from a quantity of plant material corresponding to this mold size, employing 1 m^3 of plant material per m^3. The infusions have been prepared using a water-plant material ratio in such a way that the suitable amount is obtained to knead the mortars, taking into account that part of the water is absorbed by hemp. Plant materials were previously cut into pieces to obtain a similar length that is found in commercial hemp hurds (maximum 25 mm) after separating fibers and hurds. After immersion in closed containers, hemp infusions got darkened and gave off a smell in hemp hurds distinct from the whole stem material. Before mixing, the red earth of Issaguen has been sieved at 3 mm maximum grain size. Four different specimens have been made, one of each class, in standardized molds of 4 × 4 × 16 (cm). The first specimen ("A") contained sieved earth, NHL3.5 lime and potable water. The second one ("FPK") was prepared with sieved earth, NHL3.5 lime and hemp infusion of the full stem section of the ancestral Moroccan hemp variety, called locally "beldiya" [12]. The third specimen ("PK") was made with sieved earth, NHL3.5 lime and hemp infusion from the de-fibered stem material of Issaguen. The fourth specimen ("FPE") was prepared like the second one but with the infusions obtained from industrial hemp of Cervera del Rio Alhama (Spain) (Fig. 6).

The kneading was performed with equal proportions of water or hemp infusion by weight. The percentage of NHL3.5 lime was high, 25% by volume in relation to

Fig. 4 Moroccan hemp core and fiber after 14 days of submersion in water and drying

Fig. 5 Spanish hemp core and fiber after 14 days of submersion in water and drying

Fig. 6 Earth mortar specimens kneaded with water or plant infusions

the earth in order to improve the results and to study if the substances released by the plant material also affect the drying kinetics, and to compare the results with the specimen mixed with water. After 28 days of curing at temperature of 20 °C and RH of 80%, the mortar samples were analyzed using mineralogical X-ray diffraction (XRD) in order to verify if the minerals of clay and lime were associated.

Experimental Program with Lime Stabilized Hemp-Clay Concretes

For both hemp-clay concrete (Fig. 7) formulations, with either hemp hurds or chopped hemp stem (including the fibers) plus sieved earth and water, two different percentages of NHL3.5 lime (8 and 16% in volume) were added. This stabilization aimed to reinforce the inter-laminar matrix of the clay through cationic exchange and transformation of peripheral links to provide good resistance to local weather conditions (Central Rif has the highest precipitation rate in Morocco). The use of lime is encouraged due to the fact that it was found in the formulation of all traditional earth concretes and mortars used in the region, although it is not produced locally. During a curing and drying period of 5 weeks at T20 and RH80 conditions, drying kinetics was studied. After 5 weeks, when all concrete samples were in equilibrium with the relative humidity of the ambient, capillary water uptake was measured, according to the Spanish norm UNE-EN1015-18:2003. Mechanical results were obtained at an age of 37 days instead of 28 days because drying progressed slower than in mineral concretes. Compressive and flexural strength of the concretes were tested at Applus laboratory (Barcelona), according to the Spanish norm UNE-EN1025-11:2000/A1:2007, with four standardized rectangular concrete samples (4 × 4 × 16 cm) of each type, obtaining four specimens for the test of the flexural strength and eight specimens for the test of the compressive strength.

Fig. 7 Lime stabilized hemp-clay concrete test specimens

Results and Discussion

Influence of Plant Infusions on Drying of Lime Stabilized Earth Mortars and the Interaction Clay-Lime

Among the four different lime stabilized earth-mortars, a higher viscosity has been observed in the ones prepared with infusions that contain whole stem material. Under conditions of T20 and RH80, the specimen mixed with water advanced faster during the drying process of the first week, but the total water loss after 4 weeks was lower than in the specimens with hemp infusions. In the specimen made with infusion of whole stem material, deviations from the mortar sample "A" were higher than in the specimen with infusion of hemp core material. Even so, the final water loss % (by weight), as compared to the initial weight of the fresh mortars, varied a little in the different specimens, being 23.74% in the mortar mixed with water, 24.06% in the mortar mixed with infusion of core of the ancestral hemp variety, 25.89% in the mortar with infusion of whole stem material of the same and 26.63% in the specimen made with whole stem material of industrial hemp. According to these numbers the mortars with higher initial viscosity showed higher final water loss. The XRD tests that were performed with 28 days cured specimens and with fractions of less than 200 μm (Fig. 8) showed that in mortars kneaded with hemp infusions, a delay occurred in the carbonation of the lime that is not combined with the clay minerals, as well as in the combination of the rest of the lime with the clay as compared to the mortar sample mixed with water.

In the mortar sample "A", which has been kneaded with water, all clay minerals have reacted with NHL3.5 lime (they no longer appeared as such). The lowest percentage of carbonated lime was observed in the sample "PK", made with hemp

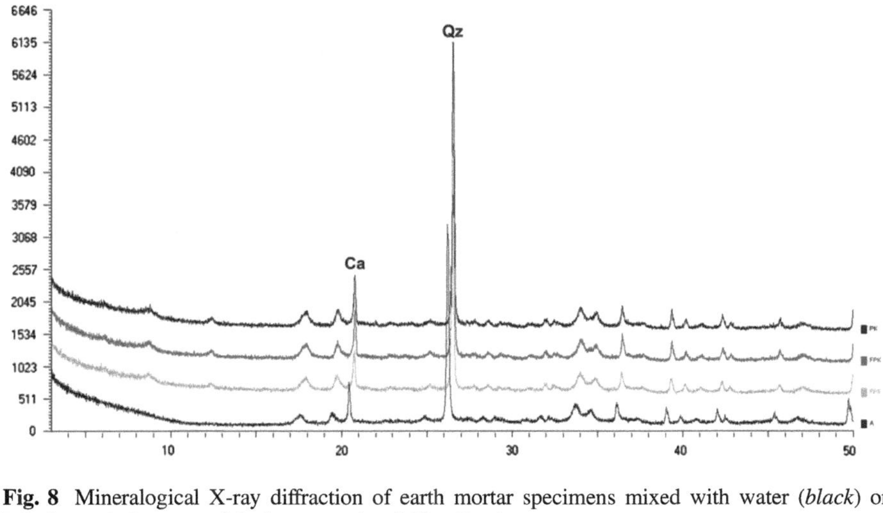

Fig. 8 Mineralogical X-ray diffraction of earth mortar specimens mixed with water (*black*) or infusions of plant materials (*green*, *red* and *blue lines*)

core infusion, i.e. 1.5 against 7.1% of calcite in the mortar mixed with water; while the samples with whole stem material showed 2 and 2.2% (Table 1).

This can be explained by the higher percentage of hemi-cellulose present in hemp core compared to the bark fraction [34] and the higher water absorption by the hemp stem material, compared to the bark [23], resulting in higher release of substances that hinder the hardening processes. The calcite content was slightly higher in "FPE" than in "FPK" specimen. This can be explained by variations in the wood anatomy, wood cell distribution, cell wall thickness and cell dimension between *C. sativa* L. - *indica* and *C. sativa* L. species [2, 3]. The submersion and drying tests of the plant materials confirmed a higher porosity of the ancestral *C. sativa* L. - *indica* hemp. A clearly higher and quicker water absorption has been observed in the ancestral hemp core, compared to *C. sativa* L. hemp core during the period of submersion, whereas no differences were observed in the fibers of both raw materials (Fig. 9). In all specimens the total water loss after drying was higher than the amount of water absorbed by submersion. While 5 g of *C. sativa* L. - *indica* hemp core absorbed 16.63 g of water in 14 days and lost 17.13 g in 6 days, *C. sativa* L. hemp core only absorbed 13.75 g and lost 14.25 g. These facts were probably attributed to the bigger pore size of the earlier matured *C. sativa* L. - *indica* hemp and an irreversible swelling of hemp core material during water absorption, as observed at micro-scale [5]. The quantitative analysis of lime-stabilized clay mortars (Table 1) verifies that specimens with infusions of Issaguen's hemp stem show the highest percentage of uncombined clay among all samples. In "FPE" specimen with industrial hemp stem it is mainly illite that got associated with lime, whereas in the specimen "FPK", with infusions of Issaguen's hemp stem, it is rather the chlorite. "FPE" specimen showed 27.7% free illite and 15.2% free chlorite and "FPK" specimen showed 42.2% free illite and 11.5% free chlorite, whereas in the sample mixed with water all the clay mineral matrixes got combined with lime.

Table 1 Quantitative analysis by mineralogical X-ray diffraction of earth mortar specimens

Quantitative analysis (%)	A	PK	FPK	FPE
	With water	With infusion of hemp core (Issaguen)	With infusion of whole stem material (Morocco)	With infusion of whole stem material (Spain)
quartz and silicates	73.2	36.3	30.1	42.2
illite	–	42.2	42.2	27.7
chlorite	–	10.4	11.5	15.2
calcite	7.1	1.5	2	2.2
amorphous	19.7	10.7	14.2	12.7

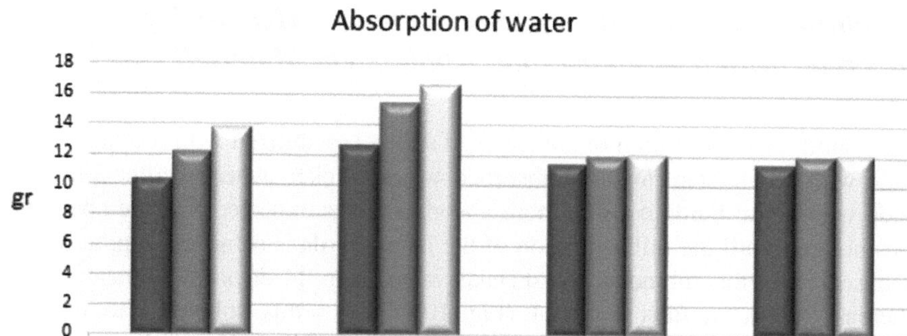

	hemp core ES	hemp core MA	hemp fiber ES	hemp fiber MA
72 h	10,38	12,63	11,50	11,50
7 days	12,25	15,50	12,00	12,00
14 days	13,75	16,63	12,00	12,00

Fig. 9 Absorption of water in grams by submersion in water of 5 g hemp material (origin: Spain and Morocco)

Drying Kinetics of the Different Hemp Concretes

In the first 72 h, the drying progressed faster in the molds of series 2, stabilized with 16% lime. Then the series 1, stabilized with 8% lime, advanced, until a slightly higher water loss was noticed in the series 2 after 5 weeks of drying. In the molds of series 1, those made with hemp hurds, dried faster in the beginning, while after 24 h the drying of specimens with chopped stem material was advanced. The total water loss of the last one was greater after 5 weeks. In the molds of series 2, those made with hemp hurds dried faster at first, while after 24 h the drying of specimens with chopped stem material was advanced. After the 3 weeks until 4 weeks of age, the specimens made with hemp hurds showed faster drying once more. Nevertheless, total water loss after 5 weeks was greater in the molds with chopped stem material of both series than in those with hemp hurds (in specimens with chopped stem material 347 g and 355.5 g in series 1 and 341.5 g or 334.5 g in series 2, in those with hemp hurds 317 g and 327 g in series 1 and 319 g and 305.5 g in series 2). Therefore, the final water loss was slightly higher in molds, stabilized with a lower amount of lime. According to these results it can be understood that in lime stabilized hemp clay concretes the hemp stem material releases more inner pore water through swelling than the hemp hurds material, possibly due to the higher contact area of the fibers.

Capillarity Water Uptake of Lime Stabilized Hemp-Clay Concretes with de-Fibered and Whole Stem Material

In the short-term absorption of concretes of series 1, stabilized with 8% lime, molds made with whole stem material absorbed water quicker, whereas after 3 h molds made with hemp hurds became more absorbent. It is expected that the absorption does not end until the full saturation of the hemp hurds is achieved, a process that, as observed in the submersion test, takes more than 14 days in the hemp hurds, whereas 7 days in fiber material (Fig. 10). The molds subjected to capillary absorption of 72 h (Figs. 10 and 11), take more than 3 weeks to dry under T20 RH80 conditions. The whole stem species dried earlier, as did the hemp fibers with respect to the hemp hurds. Comparing series 1 and 2, it can be observed that higher stabilization has a more favorable effect on the specimens with hemp hurds, because the absorption progresses slower from the beginning. However, a higher lime addition benefits practically nothing to the specimens with chopped stem. Till 24 h the absorption of series 2 progressed practically in the same rate as in the samples stabilized with half amount of lime, although after 24 h the absorption advanced somewhat slower than in the series 1. In series 2 the molds with hemp hurds showed more absorption than the molds with whole stem from 36 h and not after 3 h as in the series 1. Finally, the drying of the molds stabilized with twice the lime was slower than in series 1, although previously absorbed water was lower.

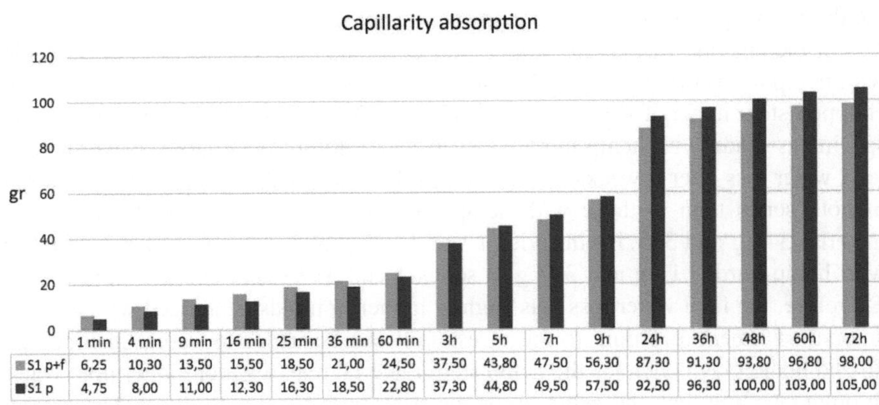

	1 min	4 min	9 min	16 min	25 min	36 min	60 min	3h	5h	7h	9h	24h	36h	48h	60h	72h
S1 p+f	6,25	10,30	13,50	15,50	18,50	21,00	24,50	37,50	43,80	47,50	56,30	87,30	91,30	93,80	96,80	98,00
S1 p	4,75	8,00	11,00	12,30	16,30	18,50	22,80	37,30	44,80	49,50	57,50	92,50	96,30	100,00	103,00	105,00

Fig. 10 Capillarity absorption in grams, in specimens of SERIES 1, stabilized with 8% lime, with hemp hurds (S1 p) or chopped stem (S1 p + f)

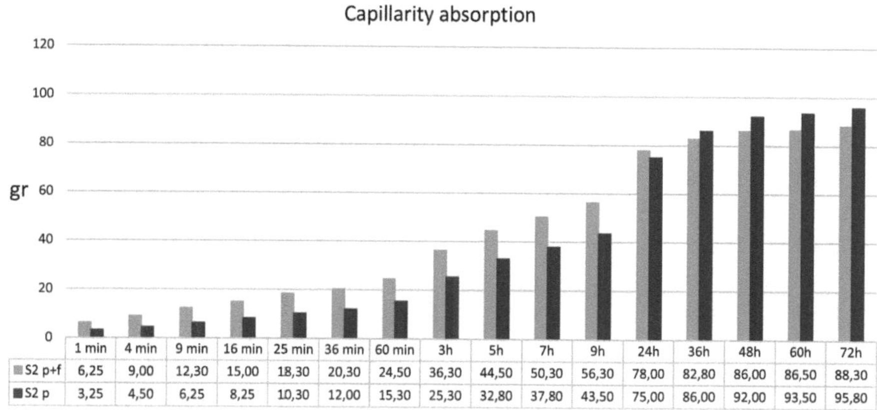

Fig. 11 Capillarity absorption in grams, in specimens of SERIES 2 stabilized with 16% lime, with hemp hurds (S2 p) or chopped stem (S2 p + f)

Mechanical Results of Lime Stabilized Hemp-Clay Concretes with de-Fibered and Whole Stem Material

Contrary to the results of another research carried out with lime, cement and gypsum concretes containing hemp hurds and fibers [14, 15], in the present work adding fibers successfully increased the compressive strength as well as the flexural strength of all tested lime stabilized hemp-clay concretes (Table 2 and 3).

Nevertheless, the samples of series 2, stabilized with 16% lime showed worse results in their compressive strength than samples of series 1 stabilized with 8% lime, which showed 1.3 MPa for the specimens with whole stem and 1.2 MPa for the specimens with hemp hurds. The specimens of series 2, stabilized with 16% lime showed 1.26 and 1.18 MPa, respectively. On the contrary, flexural strength was higher in series 2 stabilized with 16% lime than specimens of series 1, which exhibited 0.55 MPa for the specimens with whole stem and 0.43 MPa for the specimens with hemp hurds. The specimens of series 2 showed 0.58 and 0.45 MPa respectively. Although large amounts of plant material, equivalent to one m^3 per m^3 of concrete were used for the specimens of this research, their compressive strengths were within the upper range of adobe bricks that usually range from 0.32 to 2.46 MPa [31] and are frequently found in load-bearing applications. The mechanical performance of hemp concretes is limited due to meso-pores (from 0.1 to 1 mm) formed by trapped air within the hemp aggregates and the binder itself and micro-pores (lower than 0.01 lm) in the binder matrix. 40% in volume of lime binder would be needed to reach a compressive strength of 1.15 MPa by reduction of the macro-pores, that occupy spaces of up to 1 cm between particles of a hemp light concrete [10]. Earth is a cheaper and environmental friendly material to fill those macro-pores. Clay particles from 0.08 up to 2.5 μm, contrary to air and hydraulic lime binders [4], cover all the pores of the wall of hemp hurds but form a heterogeneous structure after drying and thus reduce the contact area with the same [7].

Table 2 Mechanical results series 1 (P + F = whole stem, P = hemp hurds)

Reference	Mass (g) specimens 4 × 4 × 16 (cm)	Flexural strength (37 days)		Compressive strength (37 days)	
		Load (N)	Tensile (MPa)	Load (N)	Tensile (MPa)
S1 P + F (a)	275.28	234	0.5	2.037	1.3
				2.026	1.3
S1 P + F (b)	274.59	230	0.5	2.163	1.4
				2.261	1.4
S1 P + F (c)	270.81	238	0.6	2.062	1.3
				2.137	1.3
S1 P + F (d)	280.02	253	0.6	2.192	1.4
				2.097	1.3
S1 P (a)	271.18	173	0.4	1.887	1.2
				2.070	1.3
S1 P (b)	269.91	176	0.4	1.846	1.2
				1.758	1.1
S1 P (c)	271.20	206	0.5	1.948	1.2
				2.103	1.3
S1 P (d)	270.75	175	0.4	1.880	1.2
				1.978	1.2

Table 3 Mechanical results series 2 (P + F = whole stem, P = hemp hurds)

Reference	Mass (g) specimens 4 × 4 × 16 (cm)	Flexural strength (37 days)		Compressive strength (37 days)	
		Load (N)	Tensile (MPa)	Load (N)	Tensile (MPa)
S2 P + F (a)	298.4	202	0.5	1.890	1.2
				1.923	1.2
S2 P + F (b)	288.14	206	0.5	1.991	1.2
				2.015	1.3
S2 P + F (c)	293.61	260	0.6	1.984	1.2.
				2.132	1.3
S2 P + F (d)	304.08	278	0.7	2.064	1.3
				2.257	1.4
S2 P (a)	272.00	181	0.4	1.957	1.2
				1.746	1.1
S2 P (b)	287.13	209	0.5	1.841	1.2
				1.987	1.2
S2 P (c)	281.24	196	0.5	1.896	1.2
				1.973	1.2
S2 P (d)	287.78	186	0.4	1.822	1.1
				1.859	1.2

Conclusions

This paper addresses the feasibility of the implementation of an ancestral hemp variety against industrial hemp in concrete. In this context, the drying kinetics and stabilization with lime of earth mortars, mixed with water or plant infusions were studied for both hemp hurds and whole stem material. The results confirmed different findings with de-fibered and non-de-fibered stem material and also with different hemp varieties. Those differences could be due to anatomic and chemical deviations. However, the differences between both species are not important enough to disregard the use of Moroccan ancestral *C. sativa* L. - *indica* hemp for construction purposes. The research rather demonstrates that in hemp-clay concretes, stabilized with lime, different varieties of hemp may need different formulations containing appropriate clays, in order to obtain maximum performance. Adapted formulations and treatments seem to be necessary to optimize results in all hemp varieties and those are more important in de-fibered than in whole stem material. Furthermore, clay based concretes made with de-fibered and non-de-fibered hemp stem, stabilized with two different percentages of lime were compared. Concerning the mechanical properties, the lower results of compressive strength with higher amount of lime addition could be explained by the alkaline medium created by higher lime content which is probably the reason for higher hemi-cellulose release, causing hardening problems and hindering the interaction between clay and lime. This resulted in weakening of clay-lime matrix. Why a higher amount of lime addition benefitted the flexural strength of both concretes can only be explained through additional macroscopic studies, or preparing control specimens without addition of lime. Higher additions of lime didn't reduce significantly the capillarity absorption and delayed the subsequent drying process of the specimens.

Acknowledgements The authors gratefully acknowledge the funding by REMINE Project, H2020-MSCA-RISE-2014 and Research groups HUM-629 and RNM 0179 (Regional Government of Andalusia). The authors thank to Antonio Sanchez Navas of University of Granada for technical assistance in analysis and the mineralogical X-ray diffraction of samples and also to Abdellatif Adebibe, whose request of collaboration in the alternative socioeconomic development of the Moroccan Central Rif region was an incentive for this research. Lime, hemp and clay materials were sponsored by members and collaborators of Cannabric (Spain) and Adrar Nouh (Morocco).

References

1. Ali, M., Liu, A., Sou, H., & Chouw, N. (2012). Mechanical and dynamic properties of coconut fibre reinforced concrete. *Construction and Building Materials, 30,* 814–825.
2. Amaducci, S., Scordia, D., Liu, F. H., Zhang, Q., Guo, H., Testa, G., et al. (2014). Key cultivation techniques for hemp in Europe and China. *Industrial Crops and Products.*

3. Anderson, L. C. (1974). A study of systematic wood anatomy in Cannabis. In *Botanical Museum Leaflets* (Vol. 24. pp. 29–36). Harvard University.
4. Arizzi, A., Cultrone, G., Brümmer, M., & Viles, H. (2015). A chemical, morphological and mineralogical study on the interaction between hemp hurds and aerial and natural hydraulic lime particles: Implications for mortar manufacturing. *Construction and Building Materials, 75,* 375–384.
5. Arizzi, A., Brümmer, M., Martín-Sanchez, I., Cultrone, G., & Viles, H. (2015b). The Influence of the type of lime on the hygric behaviour and bio-receptivity of hemp lime composites used for rendering applications in sustainable new construction and repair works. *PlOS ONE.*
6. Arnaud, L., & Boyeux, B. (2011). La chènevotte comme nouveau Granulat écologique et performant pour une utilisation dans le bâtiment. Académie d'Agriculture de France – Séance du 12 Janvier.
7. Balčiūnasa, G., Vėjelisb, S., Vaitkusc, S., & Kairytėd, A. (2013). Physical properties and structure of composite made by using hemp hurds and different binding materials. *Procedia Engineering, 57,* 159–166.
8. Bevan, R., & Wooley, T. (2008). *Hemp lime construction: A guide to building with hemp-lime*. Bre-Press.
9. Biagiotti, J., Puglia, D., & Kenny, J. M. (2004). A review on natural fiber-based composites-part I: Structure, processing and properties of vegetable fibers. *Journal of Natural Fibers.*
10. Cérézo, V. (2005). Propriétés mécaniques, thermiques et acoustiques d'un matériau à base de particules végétales: approche expérimentale et modélisation théorique. L'Institut National des Sciences Appliquées de Lyon. L'Ecole Nationale des Travaux Publics de l'Etat, Thèse doctorale: Génie Civil.
11. Chen, T., Yao, S., Merlin, M., Mai, H., Qiu, Z., Hu, Y., et al. (2014). Identification of *Cannabis* fiber from the Astana Cemeteries, Xinjiang, China. *Economic Botany, 68*(1), 59–66.
12. Chouvy, P.-A., & Afsahi, K. (2014). Hashish revival in Morocco. *International Journal of Drug Policy, 25,* 416–423.
13. Crônier, D., Monties, B., & Chabbert, B. (2005). Structure and chemical composition of bast fibers isolated from developing hemp stem. *Journal of Agriculture and Food Chemistry, 53* (21), 89–8279.
14. De Bruijn, P. B. (2008). Hemp concretes: Mechanical properties using both shives and fibres. Faculty of Landscape Planning, Horticulture and Agricultural Sciences Alnarp.
15. De Bruijn, P. B., Jeppsson, K.-H., Sandin, K., & Nilsson, C. (2009). Mechanical properties of lime-hemp concrete containing shives and fibres. *Biosystems Engineering, 103*(4), 474–479.
16. Diquélou, Y., Gourlay, E., Arnaud, L., & Kurek, B. (2015). Impact of hemp shiv on cement setting and hardening: Influence of the extracted components from the aggregates and study of the interfaces with the inorganic matrix. *Cement & Concrete Composites, 55,* 112–121.
17. Elfordy, S., Lucas, F., Tancret, F., Scudeller, Y., & Goudet, L. (2008). Synthèse des connaissances sur les bétons et mortiers de chanvre. Construire en Chanvre.
18. Gourlay, E., & Arnaud, L. (2011). Des matières premières au béton de chanvre: optimisation des propriétés thermiques et mécaniques. 20ième Congrès Français de Mécanique Besançon.
19. Hustache Y., & Arnaud L. (2008). Synthèse des connaissances sur les bétons et mortiers de chanvre.
20. Ingrao, C., Lo Giudice, A., Bacenetti, J., Tricase, C., Dotelli, G., Fiala, M., et al. (2015). Energy and environmental assessment of industrial hemp for building applications: A review. *Renewable and Sustainable Energy Reviews, 51,* 29–42.
21. Ip, K., & Miller A. (2012). Life cycle greenhouse gas emissions of hemp–lime wall constructions in the UK University of Brighton, School of Environment and Technology, Cockcroft Building, Lewes Road, Brighton BN2 4GJ, United Kingdom.
22. Jiang, H., Wang, L., Merlin, M. D., Clarke, R. C., Pan, Y., Zhang, Y., et al. (2016). Ancient Cannabis Burial Shroud in a Central Eurasian Cemetery. *Economic Botany, 70*(3), 213–221.

23. Kymäläinen, H.-R., Hautala, M., Kuisma, R., & Pasila, A. (2001). Capillarity of flax/linseed (*Linum usitatissimum* L.) and fibre hemp (*Cannabis sativa* L.) straw fractions. *Industrial Crops and Products, 14*(1), 41–50.
24. Kymäläinen, H.-R., Koivula, M., Kuisma, R., Sjöberg, A.-M., & Pehkonen, A. (2004). Technologically indicative properties of straw fractions of flax, linseed (*Linum usitatissimum L.*) and fiber hemp (*Cannabis sativa L.*). *Bioresource Technology, 94*(1), 57–63.
25. Kymäläinen, H.-R., & Sjöberg, A. (2007). M. Flax and hemp fibers as raw materials for thermal insulations. *Building and Environment, 43*(2008), 1261–1269.
26. Nguyen, T., Picandet, V., Sofiane, A., & Baley, C. (2011). Influence of compactness and hemp hurd characteristics on the mechanical properties of lime and hemp concrete. *European Journal of Environmental and Civil Engineering,* 1039–1050.
27. Niyigena, C., Amziane, S., Chateauneuf, A., Arnaud, L., Bessette, L., & Collet, F. (2016). Variability of the mechanical properties of hemp concrete. *Materials Today Communications, 7,* 122–133.
28. Nozahic, V., Amziane, S., Torrent, G., Saïdi, K., & De Baynast, H. (2012). Design of green concrete made of plant-derived aggregates and a pumice–lime binder. *Cement & Concrete Composites, 34,* 231–241.
29. Sawler, J., Stout, J. M., Gardner, K. M., Hudson, D., Vidmar, J., & Butler L. (2015). The genetic structure of marijuana and hemp. *PlOS ONE, 10.*
30. Senga Kiessé, T., Ventura A., Van der Werf H. M. G., Cazacliu B., Idir R., & Andraina, A. (2016). Introducing economic actors and their possibilities for action in LCA using sensitivity analysis: Application to hemp-based insulation products for building applications. *Journal of Cleaner Production,* 1–12.
31. Silveira, D., Varum, H., & Costa, A. (2007). Análise do comportamentos sísmico de construções existentes em adobe. -7 Congresso de sismologia e engenharia sísmica. SÍSMICA.
32. Stevulova, N., Kidalova, L., Cigasova, J., Junak, J., Sicakova, A., & Terpakova, E. (2013). Lightweight composites containing hemp hurds. *Procedia Engineering, 65,* 69–74.
33. Struik, P. C., Amaducci, S., Bullard, M. J., Stutterheim, N. C., Venturi, G., & Cromack, H. T. H. (2000). Agronomy of fibre hemp (*Cannabis sativa* L.) in Europe. *Industrial Crops and Products, 11*(2–3), 107–118.
34. Van der Werf, H. G. M. (1994). *Crop physiology of fibre hemp (Cannabis sativa L.)* (Ph.D. thesis). Agricultural University, Wageningen, The Netherlands, p. 153.

The Use of Hemp in Building Components for the Development of a Modular House in a Rural Area of Cauca—Colombia

Elena Piera Montacchini, Mónica Alexandra Muñoz Veloza, Roberto Pennacchio, Lorenzo Savio and Silvia Tedesco

Abstract Nowadays, people are looking for sustainable materials for construction industries due to many social, economic and environmental problems related to usage of non-renewable materials. The Colombian government and the main guerrilla group (Revolutionary Armed Forces of Colombia—FARC) has kept the development of rural areas as one of the principal task in their agenda. This work has been carried out to explore and analyze the different ways in which industrial hemp can be used in rural construction. The main focus is to develop a modular house prototype in the rural area of Toribío Cauca, by using industrial hemp as the main material. This will be used as a strategy for sustainable development of rural communities in the post-conflict period in Colombia.

Keywords Hemp · Building components · Sustainability · Rural development

Introduction

It is well known that the world population has grown rapidly in the past century, and social and environmental problems have become widespread alongside this growth. Extreme poverty, for example, has become one of the greatest humanitarian crises of our generation: in 2016 1.6 billion of the world's population were living in extreme poverty and 85% of the 1.6 billion lived in rural areas [1].

Most of the time, rural communities face transportation difficulties and precarious housing situations that affect the quality of life. Due to the isolated character of rural settlements and the current need to transport many of the construction technologies and long-established materials—such as masonry, concrete and steel—from the city, many of the conventional solutions to rural housing problems tend to have greater financial and environmental costs. Since adaptive capacity and resi-

E.P. Montacchini · M.A. Muñoz Veloza · R. Pennacchio · L. Savio (✉) · S. Tedesco
DAD Department of Architecture and Design, Politecnico di Torino, viale Mattioli 39, Turin 10125, Italy
e-mail: lorenzo.savio@polito.it

R. Fangueiro and S. Rana (eds.), *Advances in Natural Fibre Composites*,
https://doi.org/10.1007/978-3-319-64641-1_22

lience depend mostly on access to financial, material and social resources [2], rural population groups will be far more vulnerable than others against the impacts of global changes, especially the climate change. To be able to tackle these problems, innovation is required.

The increasing demand for developing biodegradable, sustainable, and recyclable solutions as a response to the environmental and social concerns [3] has led to a renewed interest in hemp materials in building and construction components. These bio-materials have many advantages that, compared to the production of conventional ones, make them more suitable for production in developing rural communities. The versatility of hemp can be seen in a wide range of building materials made from its seed, fiber, shives and leaf stock. The goal of this research is to investigate the possibility of building a modular house prototype for rural communities in Toribio—department of Cauca, Colombia—using hemp as main material, to improve living conditions, income and employment in this post-conflict rural area and to increase community empowerment and entrepreneurial capacities to engage with competitive markets.

Colombia: Background

Colombia possesses a prodigious natural wealth. After Brazil, it is the second country in the world with the highest biodiversity and is one of the eleven countries that still preserve vast areas of their original forests [4]. Unfortunately, despite having one of the richest bio-diversities on the planet, poverty continues to be a big problem. In Colombia, 27.8% of the population live in poverty, and this rate increases to 40.3% in rural areas, making the South American country one of the most unequal countries in the world [5].

In many ways, this problem is linked to the 50-year internal conflict between the Colombian State and guerrilla groups, which is estimated to have displaced more than 5.7 million people from their homes and lands, [5, 6] affecting mostly the rural population. This long internal conflict has been aggravated in recent decades by the expansion of drug trafficking—opium, *Cannabis*, and Cocaine—which has financed and created new forms of violence. Currently, the internal conflict and drug trafficking, combined with the impoverishment, unemployment and high informality, sanitation problems, the lack of state presence in rural areas, and the arrival of multinational companies have created a risky situation for the economy of rural populations that are not prepared to resist and combat these challenges.

Methodology

In order to propose a modular house prototype that satisfies the housing needs of a family in a rural community of Toribio, Cauca and, at the same time, provides a source of work to the inhabitants, the following analyses were carried out:

- identification of the main properties of hemp and the benefits it brings as a cultivation and building material;
- identification of the different types of construction products and materials that use industrial hemp as the main material and are currently in the market or being studied;
- checking the current availability of hemp in the study area and how it is being used by the population, and relate this with the information found about current hemp construction materials, to verify the real possibility of developing sustainable rural housing based on hemp in Toribio, Cauca;
- characterization of rural residents based on socio-cultural and environmental aspects.

These steps have led us to the definition of a matrix, which correlates socio-cultural and environmental aspects with design strategies, and is useful for the development of the prototype.

Properties and Benefits of Industrial Hemp

Hemp is a fast-growing grass-like plant containing fibers of great strength. The fibers have been used since Roman times for rope and sails, clothing and construction [7]. The choice of working with hemp as main material in the construction of a modular house prototype is based on the recognition of its physical characteristics, versatility and potential. Being a natural material that is easy to grow and is cheaper than other established materials such as concrete or steel, industrial hemp is an excellent solution for rural housing where the economic and environmental costs of conventional construction solutions often increase due to the difficulties in transporting building materials—which mostly arrive from the big cities—and the need to employ skilled labor. The benefits of hemp could be summarized as follows:

- hemp is a renewable and environment friendly material, breathing in 4 times the carbon dioxide (CO_2) of trees during its quick growing cycle. It has a great potential to act as sustainable 'sink' for atmospheric carbon dioxide and at the same time, save non-renewable resources;
- its manufacturing processes (materials, transport, used energy) do not contribute to global warming;
- hemp-based building materials reduce energy consumption due to its excellent insulation and airtightness [8].

Hemp Construction Materials

Hemp Fiber Panels for Insulation

This kind of panels are obtained by the means of processing and pressure of hemp fibers, with which it is possible to get semi rigid panels or rolls of hemp. Due to their excellent thermal capacity, hemp fiber panels can be used in perimeter insulation cavities of flats and buildings with outstanding results, filling the void and creating an effective barrier. They can also be used for other types of intervention as internal or external thermal envelope, ventilated roofs, ceilings and floors. Hemp is also not targeted by parasites, therefore doesn't need insecticide treatments.

Hemp and Lime Bio-Composite

This bio-composite can sequester 0.3 kg of carbon per kg and up to 20 tons of carbon in a typical house [7]. It is obtained by combining the wood-pulp core of the hemp's stem -also known as shives—and a hydraulic lime based binder with the addition of water. The shives are usually a by-product of hemp fiber processing and being naturally rich in silica, they help with the hardening of the lime. Once cured, the bio-composite is transformed into a rigid and lightweight durable material with excellent insulation properties, [9] that can store up to 135 kg of CO_2/m^3 [10]. This characteristic makes it one of the most carbon friendly building materials currently available on the market. This bio-composite can be used as a filler in wooden beams and pillar structures, and in screed floors and roofs. It can also be used as a natural mortar, an insulating plaster or an insulating board. Currently there are four different methods for the application of hemp-lime bio-composite [12] (Fig. 1):

- Mechanical spraying: it is used on larger projects because of the specialist equipment and operatives cost. Using a temporary waterproof plywood where it is being applied, this technique can be used to fill walls, floors or roof cavities. Although it has its advantages, this technique creates higher levels of waste.
- Casting: this is the fastest and most common method in most countries for small and medium size projects. In this process, the components—hemp, lime and water—are mixed in batches and inserted by hand into the wall cavity using shuttering and then manually tamped.
- Blocks: in this method, production is carried out with a cold process. It has become widespread in the construction industry in the developed countries because of the familiarity of its construction technique. Rigidity, lightness, strength and breathability combined with recyclability and resistance to freezing, insects and rodents. It has flame retardant properties and therefore does not need additional flame retardant substances. Although these blocks are not strong enough for structural use, they can be combined with a brick, timber, or steel frame structure. Timber frame is the most common choice since it avoids condensation issues and cold joints.

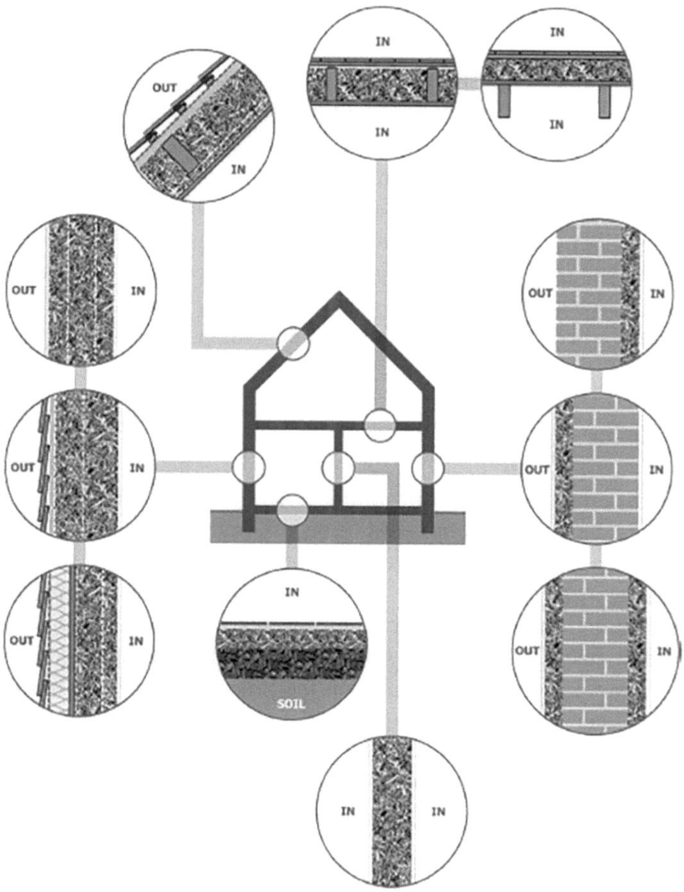

Fig. 1 Numerous uses of hemp and lime bio-composite in buildings [11]

Currently there are products on the market that allow to use a dry construction system. The Italian company Banca della Calce, for example, has launched the hemp-lime block with a particular geometry that allows a perfect joint between continuous elements and does not require the use of bedding mortar. The supporting function is attributed to a wooden structure.

- Prefabrication: This method has several advantages. It has greater consistency, higher accuracy, and can improve building efficiency. An example of this method is the Modcell® technology developed in the UK. Unfortunately, this technology also has some disadvantages. For example, it can be more expensive than other methods and its process can require more time.

Reinforcement in Composites

The use of hemp fibers as reinforcement in composite materials has increased in recent years. The mechanical properties of hemp fibers are comparable to those of glass fibers. However, their biggest disadvantage is the variability in their properties [13]. The blending materials range from thermoplastics to thermoset. Plant based resins made from soy, canola or corn are also used so that a 100% bio composite is feasible. Hemp fiber is also used to produce mineral based composites, in the same way polypropylene or glass fiber is used to reinforce concrete. The use of injection molding technology is also starting to gain ground using hemp fiber as a reinforcing material, allowing it to be used in more complex shapes and structures [14].

Structural Hemp

Although hemp-lime bio-composite offers a low embodied energy and negative carbon footprint, the timber frame could be be less environment friendly. That is why the English studio Hemp Architecture is exploring the potential of hemp in its most natural state, as a stalk, using it in construction as structural element. This new construction method attempts to achieve this by creating a space-frame of hemp stalk within a lime-hemp wall. The hemp stalk frame could be constructed for the whole building, creating a skeletal form and then the lime-hemp cast around it [15]. The results show higher potential in compression and tension (up until the frame fails) but less potential in bending. One of the advantages of the construction technique proposed by Hemp Architecture is the design freedom by not constraining the blocks to a frame structure in typical orthogonal hemp-lime construction.

Cannabis in Colombia

Industrial hemp is a genetic variant of the *Cannabis* with lower levels of the psychoactive compound tetrahydrocannabinol-THC (< 0.3% on a dry weight basis) [16]. In 2015, the *Cannabis* varieties cultivated in the most important production centers in Colombia were updated. 52 samples were collected by The National Police—Anti-Narcotics Division, in the departments of Cauca (10), Magdalena (26) and Norte de Santander (16). These samples were taxonomically classified and it was determined that the analyzed samples belong to the same species *Cannabis sativa L* [17] generally used to produce fiber and oil.

There are two modalities of *Cannabis* plantations in Colombia: full exposure cultivation and greenhouse. Each modality contemplates different challenges for detection. Unfortunately, to date, there is no official information about the number of hectares of *Cannabis* cultivation in Colombia that allow us to establish levels of production as well as identify the characteristics of the transformation processes and agricultural practices in each of the producing regions. It should be mentioned that the understanding of the dynamics of production, processing, marketing and trafficking of *Cannabis* presents great challenges because it can be grown easily both outdoors and indoors with optimum results in terms of quality [17]. However, as reported by the Cauca Police, in 2015 there were 215 ha of *Cannabis* plantations in the northern Cauca indigenous reserves [18].

In 2016, with the Congress approval for medicinal and scientific use, Colombia became the fourth Latin American country—after Chile, Puerto Rico and Uruguay—to have a legislation for the therapeutic and experimental use of *Cannabis* for scientific purposes [19]. With the current Colombian political framework, the legislation and regulation of *Cannabis* would allow the use of hemp as a solution for the post-conflict agreement key points that promote the substitution of illicit crops and the implementation of comprehensive rural reforms [20]. The local production of industrial hemp for construction materials would also contribute to the development of new rural Colombian enterprises that seek to mitigate the environmental impact and create employment in rural areas.

Study Area

The study area is in a rural zone of the Colombian Pacific region, specifically in the province of Toribio at north of the Cauca department. This area was chosen because of the environmental characteristics that make it particularly suitable for the cultivation of hemp. It was also chosen because, being one of the poorest and most vulnerable areas of the country, the realization of such a project could contribute to building a sustainable development model for the rural communities that live there.

Cauca department is one of the poorest in Colombia. More than 60% of its population lives in rural areas [22], such as Toribio, where poverty is particularly severe due to remote location and internal conflicts. The inhabitants of Toribio—in majority Nasa indigenous, Afro-indigenous and mestizo communities—are mainly engaged in agriculture, commerce and artisanal mining. In fact, since 2002 the National government has declared several mining areas in north Cauca exclusive to indigenous people, prohibiting multinationals from mining in those areas and increasing community entrepreneurial capacities (Fig. 2).

Since ancient times, these communities have cultivated cannabis for recreational purposes. Unfortunately, illicit crops have become an important part of their fictitious economy. With drug trafficking, cannabis, that is considered an ancestral plant along with the coca, have become a tragedy for Cauca's inhabitants because of illicit uses. Now, an alliance between the public sector, scientific community, local authorities and the rural community itself seeks to leave behind this stigma and expand its use for medical and industrial purposes.

In recent years, they have started several projects that aim to develop programs for legal use of this plant. Food, cosmetics and natural medicine are some of the explored industries so far. In 2016 in Corinto, a northern Cauca town near Toribio, was presented Caucannabis, the first hemp cooperative for medicinal purposes. This is a part of the "Integral management and alternatives for Cannabis cultivation in

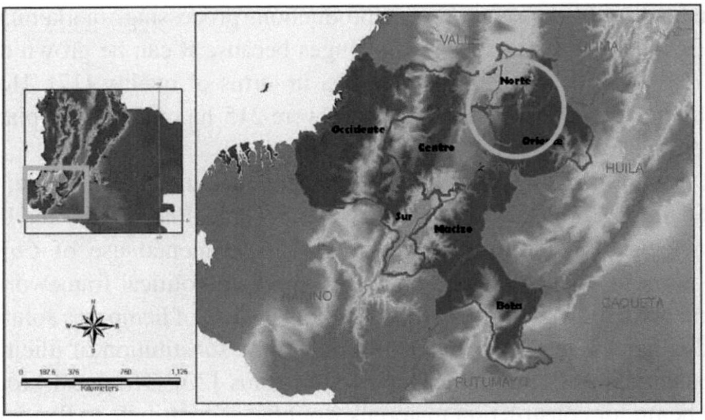

Fig. 2 Location of Cauca department [21]

five provinces of north Cauca" project. In addition to Corinto, the other four provinces involved are: Miranda, Caloto, Jambalo and Toribio. This project, which has the support of the National government, also aims to create a research center to identify new legal uses for this plant as well as reducing the levels of violence in this region of the country by replacing illicit crops [23].

That is why the possible use in construction of the hemp grown here and the artisanal but regulated extraction of lime, another important component of hemp bio-composites, should therefore integrate with what until now has started, contributing to the improvement of rural housing as well as representing a source of local economic and social development.

Housing Needs

For the analysis of the rural housing's identity and design, three basic criteria have been taken into consideration [24]:

- Relationship between home and environment. It indicates the adaptation of the construction to external conditions such as climate, topography and landscape. This criterion is especially important when considering the characteristic high slope of the areas where coffee is grown.
- Architectural features of rural housing: the predominant forms, the traditional construction techniques and materials, the use of space and the housing adaptation to the cultivation of coffee.
- The overlapping of identity elements and the architecture of the house: these identity elements are not part of the house structure, but they can change its appearance: color, plants, furniture, personal objects, etc. This is an important part of the rural house character because it represents the family or individual values.

Results

To design the prototype, we realized a matrix. We cross information about existing and experimental hemp materials, environmental, cultural and lifestyle needs of rural areas in Colombia, and environmental and technical requirements for industrial hemp cultivation and production. The following is an extract of the matrix (Table 1):

Conclusions

The rural areas of Colombia possess enormous wealth and potential. Therefore, sustainable hemp plantations and hemp composites production for rural construction can improve the living conditions of the population (better home comfort), and provide direct employment for many rural, unskilled people, (both in plantation and in processing facilities) contributing to build a sustainable economy model in rural communities in Colombia, where economic development opportunities are scarce. [27].

A dry construction system based on hemp is more environment friendly (especially in rural areas) as the elements, thanks to the ease of assembly-disassembly, can be recycled, replaced (for degraded elements) or reused to decrease the impact of any future disposal.

Although it is possible to build a modular house by using hemp composites in most of the building components, further tests should be performed to assess its thermal and acoustic performances. For example, due to the moisture absorption of hemp fiber composites [28], some components such as the roof cover need to be studied further.

The main outlooks of the research are

- Use of *guadua* instead of timber: *Guadua* is a Neotropical genus and it belongs to the same family as bamboo. Due to its quality, *guadua* has been widely used for the construction of houses along the inter-Andean rivers of Colombia. Removing the timber structural frame of the current hemp-lime construction and using a guadua structural frame would reduce both the building's embodied energy and the strain on the tree stock. It could also provide a more economical method of construction since *guadua* is a local agricultural crop and using a local material for the construction in this area would be far cheaper than timber.
- Roofing panels: Natural fibers and resin material can be compressed into laminates that can act as skins for sandwiched structures [29]. In rural societies, due to use of local plant fibres these panels can be used as a more durable and economical roofing option.

Table 1 Correlation between socio-cultural and environmental aspects and design strategies

Toribio, Cauca (Colombia)				
Tropical rainforest climate: typically hot and wet throughout the year				
Users	Needs		Requirements	Design strategies
Socio-cultural aspects	Individual		• Identification of individual characteristics such as age, gender and preferences	• Survey methodology to be able to describe the rural community accurately
	Collective: Family—Community		• Identification of family composition	• Adaptable house approach facilitates flexible design - both on the inside and out
			• Identification of productive activities carried out by the families, and their needs	• The house must be easily accessible and properly located according to agricultural activities (hemp cultivation) • Community participation in hemp cultivation • Relate building and rural context • Relate home and workplace
			• Use of space	
			• Recognition of ethnic and racial diversity	• Application of traditional design principles and architectural elements • Community participation in decision-making processes and building design and construction • Family participation in housing design and construction
			• Use of typical architectural features	
			• Identification of symbolic and religious elements	
			• Identification of rural house as an expression of identity	
Environmental aspects [25, 26]	Environmental protection		• Reduce air pollutant emissions (greenhouse gases—GHG)	• Use of hemp: it breathes in 4 times the carbon dioxide (CO_2) of trees during its life cycle
			• Protection of native plants	• Hemp farming: *Cannabis* is a native plant in this area. The inhabitants of Toribio have cultivated it since ancient times
	Rational use of resources	Water	• Reduce water consumption	• Use hemp fiber panels for insulation: hemp production requires less water than cotton (about half the amount)
			• Reuse rainwater	• Hip roof with a pitch of 35° so that rain water would not be accumulated on the roof and can be collected
			• Reuse gray water	• Separating gray water from black water collection system

(continued)

Table 1 (continued)

Toribio, Cauca (Colombia)				
Tropical rainforest climate: typically hot and wet throughout the year				
Users	Needs		Requirements	Design strategies
			• Reuse black water	• Separating gray water from black water collection system
		Soil	• Smart internal space distribution	• Rectangular shape with perimeter circulation (square shapes in small solutions)
			• Relationship between building and topography	• The house is designed as a horizontal platform. It can be in any place, extending the structural supports to the ground. The space below the platform can be used as storage, animal shelter or work place per the land slope
		Materials	• Use of local materials and techniques	• Use of hemp and lime, timber and guadua
			• Use of low-environmental impact materials	• Use of hemp: hemp manufacturing processes do not contribute to global warming
			• Use of recyclable materials	• Use of hemp lime blocks: they are recyclable and biodegradable at the end their life cycle
			• Design for Disassembly (DfD)	• Modular design
		Energy	• Passive solar building design (passive cooling)	• Use hemp for insulation: hemp-based building materials reduce energy consumption because of their excellent thermal properties
			• Thermal insulation	
			• Thermal inertia for natural air conditioning	
			• Passive ventilation (natural ventilation)	• Use of passive stack ventilation (PSV) system
	User's welfare, hygiene and health		• Reduction of thermal radiation between users and surrounding surfaces in overheating periods	• Protection against sun formed by wide porch and wide overhangs • House oriented with the long axis north-south and with most windows facing north

(continued)

Table 1 (continued)

Toribio, Cauca (Colombia)			
Tropical rainforest climate: typically hot and wet throughout the year			
Users	Needs	Requirements	Design strategies
		• Control of air flow across the building (moisture control, energy savings, comfort and health)	• Linear configuration of internal spaces allowing ventilation from both sides
		• Adaptive control of thermal comfort	• Use of hemp lime blocks: they offer thermal, hygrometric and acoustic comfort
		• Acoustic comfort (noise control)	
		• Visual comfort	• Use of wide porch and wide overhangs as sunscreens decreasing direct sunlight (sun dazzle)
		• Natural light	• Window sizes should be proportional to the space area, the frequency of use and the number of people that use it
		• Use of low emission materials (reducing toxic indoor emissions)	• Use of hemp and lime bio-composite in internal finishes. This bio-composite doesn't need additional toxic flame retardant substances
		• Reduce radon levels	• Stilt house (the typical wooden palafitte are often used in Cauca's rural settlements)

References

1. Global Multidimensional Poverty Index Databank. (2016). Oxford Poverty and Human Development Initiative (OPHI). University of Oxford. http://www.dataforall.org/dashboard/ophi/index.php/. Accessed 17 Jan 2017.
2. NCCARF National Climate Change Adaptation Research Facility. (2015). National Adaptation Network. Vulnerable Communities. https://www.nccarf.edu.au/vulnerable-communities/content/ vulnerable-communities. Accessed 02 Dec 2016.
3. United Nations. (1997). Convention on climate change. Kyoto protocol to the united nations framework convention on climate change, Kyoto, December 11, 1997.
4. SIB Sistema de Información sobre la biodiversidad de Colombia. (2016). Biodiversidad en cifras. http://www.sibcolombia.net/biodiversidad-en-cifras/. Accessed January 10, 2017.
5. IFAD International Fund for Agricultural Development. (2016). Investing in rural people in Colombia. https://www.ifad.org/documents/10180/6577c077-cac7-4c59-8854-4b9289b9a01a . Accessed November 13, 2016.
6. CNMH Centro Nacional de Memoria Historica. (2013). ¡Basta ya! Colombia: memorias de Guerra y dignidad. http://www.centrodememoriahistorica.gov.co/descargas/informes2013/bastaYa/basta-ya-colombia-memorias-de-guerra-y-dignidad-2016.pdf. Accessed October 26, 2016.
7. Ashby, M. F. (2009). Materials and the environment: eco-informed material choice, (pp. 231–245). Oxford.
8. Johnston, S. (2016). The Environmental Benefits of Industrial Hemp. Virginia Industrial Hemp Coalition. http://vahemp.org/resources/lobby-packet/routedownload/lobby-packet/the-environmental-benefits-of-hemp. Accessed October 20, 2016.
9. Ronchetti, P. (2014). Canapa e calce in bio-edilizia: storia, applicazioni e proprietà. http://www.canapaindustriale.it/2014/07/11/il-cemento-di-canapa-e-calce-storia-applicazioni-e-proprieta/. Accessed September 06, 2016.
10. Bevan, R., & Woolley, T. (2008). *Hemp lime construction* (pp. 77–84). Bracknell: A guide to building with hemp lime composites.
11. Evrard, A. (2008). Transient hydrothermal behavior of Lime-Hemp Materials. Universite Catholique de Louvain, Louvain. https://dial.uclouvain.be/pr/boreal/object/boreal%3A19675/datastream/PDF_00/view. Accessed November 02, 2016.
12. Hirst, E. A. J. (2013). Characterization of hemp-lime as a composite building material. University of Bath. http://opus.bath.ac.uk/43277/1/UnivBath_PhD_2013_E_Hirst.pdf. Accessed October 05, 2016.
13. Shahzad, A. (2011). Hemp fiber and its composites—a review. *Journal of Composite Materials*. vol. 46–8.
14. Real Hemp. https://www.realhemp.com/hemp-industries/hemp-biocomposites/. Accessed December 13, 2016.
15. HA Hemp Architecture Innovation Research Design. http://www.hemparchitecture.com/#/hemp-lime-with-hemp-stalk-frame/. Accessed October 14, 2016.
16. Johnson, R. (2015). Hemp as an Agricultural Commodity. Congressional Research Service. http://www.fas.org/sgp/crs/misc/RL32725.pdf. Accessed October 21, 2016.
17. UNODC United Nations Office on Drugs and Crime. (2016). Colombia: monitoreo de territorios afectados por cultivos ilícitos 2015. http://www.unodc.org/documents/crop-monitoring/Colombia/Monitoreo_Cultivos_ilicitos_2015.pdf. Accessed December 02, 2016.
18. Noticias, R. C. N. (2015). Policía: marihuana en Cauca se encuentra en los resguardos indígenas. http://www.noticiasrcn.com/nacional-regiones-pacifico/policia-marihuana-cauca-se-encuentra-los-resguardos-indigenas. Accessed October 10, 2016. (in press).
19. Law Nº 1787, July 6th 2016. Colombian Republic. National Government.
20. OACP High Commissioner for Peace. (2016). Summary of Colombia's Agreement to End Conflict and Build Peace. http://www.altocomisionadoparalapaz.gov.co/herramientas/Documents/summary-of-colombias-peace-agreement.pdf. Accessed February 07, 2017.

21. IGAC Agustín Codazzi Geographic institute. (2005).
22. Vergara, J. R. G. (2007). *La economía del departamento del Cauca: concentración de tierras y pobreza, Centro de estudios económicos regionales (CEER)* (pp. 7–23). Cartagena: Banco della República.
23. Popayán. (2016). Caucannabis y 12 productos a base de marihuana con fines medicinales. http://www.eltiempo.com/colombia/cali/cooperativa-de-marihuana-de-uso-medicinal/16638449. Accessed October 10, 2016. (in press).
24. Martinez, L. F., & Roa, A. S. (1984). La arquitectura de la vivienda rural en Colombia. Vol. 2. *Minifundio cafetero en Antioquia, Caldas, Quindio y Risaralda, Bogotá, 1984,* 180–181.
25. Colombia Ministerio de Ambiente y Desarrollo Sostenible. (2012). Criterios ambientales para el diseño y construcción de vivienda urbana/ Unión Temporal Construcción Sostenible S.A y Fundación FIDHAP (Consultor), Bogotá, D.C, pp. 66–140. http://www.minambiente.gov.co/images/AsuntosambientalesySectorialyUrbana/pdf/Sello_ambiental_colombiano/cartilla_criterios_amb_diseno_construc.pdf. Accessed October 26, 2016.
26. UNI 11277 Building sustainability. Ecocompatibility requirements and needs of new and renovated residential and office buildings design. February 2008.
27. Ashby, M. F., Balas, D. F., & Coral, J. S. (2016). Materials and sustainable development, (pp. 197–210), Oxford.
28. Panthapulakkal, S.,& Sain, M. (2007). Studies on the water absorption properties of short hemp–glass fiber hybrid polypropylene composites. *Journal of Composite Materials, 41*(15).
29. Van Rijswijk, K., Brouwer, W. D., & Beukers, A. (2011). Application of natural fiber composites in the development of rural societies. Structures and Materials Laboratory, Faculty of Aerospace Engineering, Delft University of Technology, Delf. http://www.fao.org/docrep/007/ad416e/ad416e00.htm. Accessed September 14, 2016.

MIX
Papier aus verantwortungsvollen Quellen
Paper from responsible sources
FSC® C105338

If you have any concerns about our products,
you can contact us on
ProductSafety@springernature.com

In case Publisher is established outside the EU,
the EU authorized representative is:
Springer Nature Customer Service Center GmbH
Europaplatz 3, 69115 Heidelberg, Germany

Printed by Libri Plureos GmbH
in Hamburg, Germany